# Basic Cell Biology

**Charlotte J. Avers**
RUTGERS UNIVERSITY

**D. Van Nostrand Company**
NEW YORK   CINCINNATI   TORONTO   LONDON   MELBOURNE

D. Van Nostrand Company Regional Offices:
New York     Cincinnati

D. Van Nostrand Company International Offices:
London     Toronto     Melbourne

Library of Congress Catalog Card Number: 77-86700
ISBN: 0-422-20383-7

Published by D. Van Nostrand Company
450 West 33rd Street, New York, N. Y. 10001
10  9  8  7  6  5  4  3  2  1

TO MY MOTHER

# Preface

This book has been written for introductory courses in Cell Biology in which students have a minimal background in biology and chemistry. I have made every effort to present each theme in its perspective of a dynamic, challenging thrust at understanding the amazing complexity and versatility of the cell and its activities, while still viewing the biology of the cell in the common patterns and economical ways that maintain life in all its variety.

Following an introduction to cellular life, there are chapters which provide adequate background to appreciate how organic molecules, cellular energetics, enzymes, and membranes contribute to cellular structure, function, and regulation. These chapters lay the foundations for the remainder of the book, which deals with systems for energy transformation, packaging of cellular protein products, cellular movements, and molecular and subcellular control over reproduction and other properties of life. The last two chapters bring together some of the ways in which modern cell biology has provided new approaches to problems in certain fields of medicine and in the search for cellular origins in evolution.

This book is an outgrowth of my earlier *Cell*

*Biology*. The new book has been considerably rewritten and reorganized to provide a shorter, simpler, but still current view of major themes in cell biology for a different audience of students.

It is my pleasure to acknowledge the kindness of many friends and colleagues who provided photographs for this book and the help of Faye Schwelitz and Jerry Brand, who read the entire manuscript and made many substantive suggestions and comments. Once again, I found the project to be both enjoyable and educational. I hope it will be a similar experience for the student.

Charlotte J. Avers

# Contents

# Basic
# Cell Biology

# Chapter 1

# Cellular Organization: Prokaryotes and Eukaryotes

Cell biologists concentrate on the microscopic features of the living world. Cell parts are made up of highly organized groups of molecules, so that cell biologists are as concerned with molecules as with cells or parts of cells. Cell biology is the modern outcome of a blend of studies using the microscope as a basic tool along with tools from genetics, biochemistry, and physiology. Using all the methods, ideas, and perspective we can get from a variety of approaches, cell biology has developed as a way of examining cells from the most comprehensive and dynamic points of view.

In the past ten to twenty years a picture has emerged showing the cell to be a dynamic living unit whose structure and functions are based on incredibly complex but interacting features. The cell biologist attempts to observe, understand, and explain how structure and function are correlated and how functions are regulated within the maze of activities going on at every moment in a living system. All together, we want to know what makes the cell tick, and how it manages to accomplish its functions within a structural framework in space and a regulatory framework in time.

## HISTORICAL BACKGROUND

The historical and modern basis for cell studies is microscopy. We cannot see most cells with the naked eye because they are too small to see as individual units and because we cannot distinguish the individual cells within a group even though the whole group may be large enough to see without magnification. The human eye cannot detect that two objects are separate entities if they are closer together than about 200 micrometers ($\mu$m). We would still see a fuzzy single object rather than two separate objects even when magnified, unless the magnifying lenses were good enough to provide sharp focus, or **resolution,** as well as enlarging the image observed. We must therefore go back to the invention of useful lenses and microscopes in tracing cell studies to their beginnings.

### Microscopy and the Cell Theory

Most cells are about 1 to 100 micrometers ($\mu$m) in size (Table 1.1). The first useful microscope was invented in 1590 by Z. and H. Janssen. They developed a **compound microscope,** that is, a magnifying system with two lenses providing the total enlargement. Their microscope could magnify an object 30 times its actual size with the object in sharp focus, so there was magnification with good resolution. With two lenses in a microscope there is less distortion of the magnified image, since there is less aberration of the optical system than in a **simple microscope** (having one magnifying lens).

Robert Hooke is credited with reporting the first important information using a compound microscope (Fig. 1.1). Hooke described "cells" or "pores" in cork and other plant tissues, and he also noted that some kinds of cells were filled with "juices". His observations in 1665 drew particular attention to the highly visible, thick cell wall rather than to the "juices", and cell walls continued to be the focus of cell studies by biologists for almost 200 years afterward.

The foundations for the idea that the cell is the underlying unit of structure were laid by Nehemiah Grew in 1672–1682 and by Antonie van Leeuwenhoek between 1673 and the early 1700s. Grew published extensive and detailed descriptions of microscopic plant anatomy, and van Leeuwenhoek wrote frequently about the many objects and moving forms he saw using a simple microscope which he put together himself.

**TABLE 1.1  Units of measurement used in biology and biochemistry**

| A. LENGTH | | | | |
|---|---|---|---|---|
| Meter (m) | Millimeter (mm) | Micrometer (Micron) ($\mu$m) | Nanometer (Millimicron) (nm) | Ångstrom (Å) |
| 1 | 1,000 ($1 \times 10^3$) | 1,000,000 ($1 \times 10^6$) | 1,000,000,000 ($1 \times 10^9$) | $1 \times 10^{10}$ |
| 0.001 | 1 | 1,000 | 1,000,000 | $1 \times 10^7$ |
| 0.000001 | 0.001 | 1 | 1,000 | $1 \times 10^4$ |
| $1 \times 10^{-9}$ | $1 \times 10^{-6}$ | 0.001 | 1 | 10 |
| $1 \times 10^{-10}$ | $1 \times 10^{-7}$ | $1 \times 10^{-4}$ | 0.1 | 1 |

| B. WEIGHT | | | | |
|---|---|---|---|---|
| Gram (g) | Milligram (mg) | Microgram ($\mu$g) | Nanogram (ng) | Picogram (pg) |
| 1 | 1,000 | 1,000,000 | $1 \times 10^9$ | $1 \times 10^{12}$ |
| 0.001 | 1 | 1,000 | $1 \times 10^6$ | $1 \times 10^9$ |
| $1 \times 10^{-6}$ | 0.001 | 1 | $1 \times 10^3$ | $1 \times 10^6$ |
| $1 \times 10^{-9}$ | $1 \times 10^{-6}$ | 0.001 | 1 | $1 \times 10^3$ |
| $1 \times 10^{-12}$ | $1 \times 10^{-9}$ | $1 \times 10^{-6}$ | 0.001 | 1 |

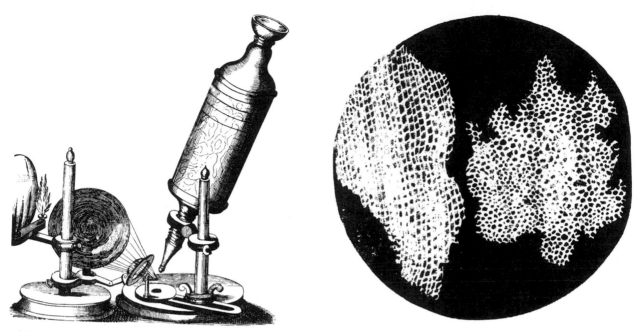

**Figure 1.1**
The crude microscope used by Robert Hooke to view materials such as cork tissue, in which he observed boxlike compartments he called ''cells.''

It is amazing that van Leeuwenhoek could see so clearly and describe in such detail using only a simple microscope for his studies.

Little new information about cellular structure became available until the 1830s when improvements in microscope lenses permitted higher magnification and greater resolution. At this time separate objects could be seen in sharp focus even if they were only 1 $\mu$m apart. Improvements continued steadily during the nineteenth century, eventually leading to lenses that provided 0.2 $\mu$m resolution. This is the resolution we achieve to the present day using ordinary white light as the source of illumination for a good compound microscope. Much higher magnification and resolution, 0.001 $\mu$m or less, can be achieved using the electron microscope (Fig. 1.2).

In addition to advances in microscope optics there also were new or improved methods and equipment which were important aids in studying cells at high magnifications. With the invention of the microtome in 1870, tissues could be sliced into thinner sections than had been possible earlier, and more light could pass through these sections. Cell structures were more clearly visible, but there was very little contrast between parts of the cell. This problem was solved when better stains became available in the latter part of the nineteenth century, mainly through the efforts of German industrial chemists. Cells and their parts are made much more distinct after staining, which increases contrast between different structures.

Plant and animal cells looked very different when viewed through a microscope, which led to the belief that these kinds of organisms were constructed differently. Plant cell walls provide conspicuous boundaries whereas individual animal cells are hard to see distinctly in a tissue. This impasse was overcome when T. Schwann showed that there were well-defined cell outlines in animal cartilage tissue, similar to the

Light microscope

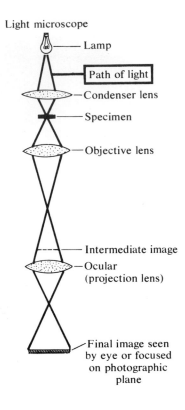

- Lamp
- Path of light
- Condenser lens
- Specimen
- Objective lens
- Intermediate image
- Ocular (projection lens)
- Final image seen by eye or focused on photographic plane

Electron microscope

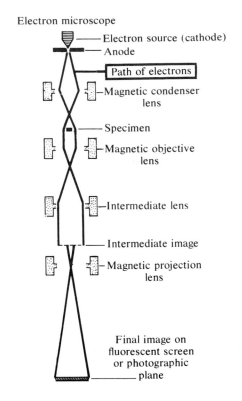

- Electron source (cathode)
- Anode
- Path of electrons
- Magnetic condenser lens
- Specimen
- Magnetic objective lens
- Intermediate lens
- Intermediate image
- Magnetic projection lens
- Final image on fluorescent screen or photographic plane

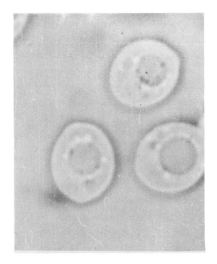

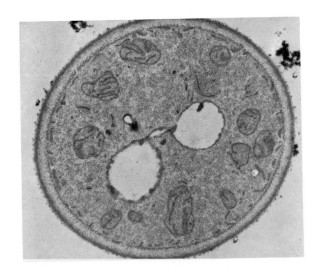

cell wall boundaries in plant tissues. There are thick fibrous deposits marking individual cell boundaries in cartilage. Based on their studies of plant and animal tissues, M. J. Schleiden and T. Schwann proposed a simple and unifying **cell theory** for all organisms. In 1838–1839, Schleiden and Schwann postulated that the cell is the basic unit of structure in all life, and that all organisms were constructed from one or more cells. The cell theory was refined further in 1855 when R. Virchow presented the very important idea that *all cells arise from pre-existing cells.* This idea helped to direct attention to the cell as an important factor in transmission of inherited traits from one generation to the next.

In the early 1830s Robert Brown had suggested that the **nucleus** was an important part of the cell. His suggestion was not generally accepted until the 1870s, by which time it had clearly been shown that nuclei maintained physical continuity from one generation to another and were not formed anew in each generation, as had been believed earlier. Later, the process of nuclear division by **mitosis** was described by W. Flemming, and it was also shown that each species had a particular number of chromosomes which remained constant for the species. Both these observations provided further support for the preservation of the nucleus from generation to generation, and helped to focus attention on the contribution made by the nucleus to heredity. The significance of the nucleus was demonstrated dramatically in 1875–1876 when Oskar Hertwig showed that two nuclei must fuse if an egg is to develop into an embryo. He also showed that one nucleus was present in the egg cell of the animal while the second nucleus was introduced by the sperm at fertilization of the egg. Similar observations were reported for plants by E. Strasburger in 1877.

To complete the modern version of the cell theory,

chromosome behavior had to be related to the nucleus as the physical link between generations. Hertwig had shown that two nuclei fused at fertilization, which initiates the next generation, but it was also known that chromosome number remained the same for a species in every generation. How could chromosome numbers remain constant if fertilization led to twice as many chromosomes after the fusion of two nuclei? There had to be a compensating event in which the chromosome number was reduced by one-half in sperm and eggs, so the original number would be restored at fertilization. There were several reports in the 1880s showing that such a *reduction division* of the nucleus did indeed take place during sperm and egg formation events. The details of this nuclear division, called **meiosis,** were described in 1905. It was therefore established before the turn of the century that continuity between generations depended on the cell nucleus, and the chromosome set of a species remained constant in every generation of a species. Cells arise only from other cells like themselves because of nuclear constancy based on transmission of a set of chromosomes from generation to generation.

Because of these discoveries and others during the nineteenth century there was enough information to propose a unifying cell theory, which embraced the entire living world known at the time. The cell theory states that:

1. All living organisms are made up of one or more living units or cells.
2. Each cell can maintain its living properties independently of the rest, but the properties of life of any organism are based on the properties of life of its individual cells.
3. Cells arise only from other cells.
4. The smallest clearly defined unit of life is the cell.

**Figure 1.2**
Schematic drawings showing the basic optical systems of the light microscope and the electron microscope. Resolution of structure is much lower in the yeast cells photographed through the light microscope (*left*, × 3,000) than in a comparable cell that has been sectioned and photographed using the electron microscope (*right*, × 25,000).

### Genetic and Biochemical Aspects of the Cell

Various proposals were made in the later 1800s to relate the control of heredity to particular factors in the cell nucleus. The studies of inheritance in garden peas that were reported by Gregor Mendel in 1866 went largely unnoticed in his lifetime. Biologists of the time were poorly prepared to appreciate an abstract approach to heredity, and few biologists in the 1860s believed that the nucleus was a vital part of the cell. By 1900 the state of biology had advanced enough for Mendel's concepts to be accepted readily when they were quoted in independently conducted studies reported by Hugo de Vries, Carl Correns, and Ernst von Tschermak. The study of **genetics** as a biological discipline is dated as beginning in 1900. As early as 1902 it was formally proposed by W. Sutton in the Chromosome Theory of Heredity that the hereditary factors, or **genes,** *were located within chromosomes in the nucleus.* Supporting evidence was collected by various geneticists in the years that followed, and by 1916 the Chromosome Theory had been widely accepted.

Although the chemical basis of heredity could be dated as beginning in 1871 when Friedrich Miescher described "nuclein" (now known to be **deoxyribonucleic acid** or **DNA**) he had isolated from while blood cell nuclei, DNA was not acknowledged to be the genetic material until the early 1950s. Before 1950 many scientists held to the belief that genes were proteins because only proteins were then known to be of great enough variety to correspond with the great diversity of genes in organisms. Since chromosomes were made of proteins as well as DNA there was little reason to hesitate in considering proteins as the genetic material. DNA was then believed to be an unlikely candidate because DNA molecules seemed to be identical to one another rather than a highly diversified compound.

By the early 1950s the scientific community was more disposed to consider DNA than proteins as the genetic material. Several important experimental approaches combined to show clearly that DNA was a highly variable type of molecule and that DNA rather than protein was transmitted from one generation to the next in a genetic succession. In 1953 when James D. Watson and Francis H. C. Crick proposed a model for DNA as a double helix molecule, it was accepted overwhelmingly. The incredibly rapid advances in molecular genetics afterward were due in large measure to the usefulness of the Watson-Crick model of DNA in experimental studies in many laboratories.

Advances in chemistry were of crucial importance in development of the cell theory and in laying the foundation for understanding cell function in relation to structure. When Friedrich Wöhler synthesized the organic compound urea from inorganic ammonium cyanate in 1828, he revolutionized biology as well as chemistry. Before his experiments it was generally believed that the living and nonliving worlds were entirely different and that the laws of physics and chemistry did not apply to living systems. The similarities between living and nonliving chemistry were made clear when Wöhler showed that the same substance known to be made in living systems could also be made from inorganic raw materials in the laboratory, following the same laws of physics and chemistry. From this and later studies it was gradually accepted that organic compounds were formed in ordinary chemical reactions in cells, rather than by vague and mysterious "vital" processes.

In 1871 Louis Pasteur demonstrated that fermentation of sugar to form alcohol would take place under natural conditions only if living yeasts were present and active. In this way he made clear that living organisms were associated with chemical activities involving organic compounds. On the other hand, his work also strengthened the belief that such reactions represented vital processes to be found only in living cells. In 1897, however, H. and E. Buchner accidentally discovered that sugars could be fermented by extracts made from yeast cells; that is, the yeasts did not have to be alive to carry out the process. The factors in the yeast extracts which were responsible for the chemical activities were **enzymes** (earlier known as "ferments"), which are unique proteins that influence chemical reaction rates in cells under natural conditions. Enzymes can retain their activity outside cells if carefully extracted and placed in suitable conditions.

These and other chemical studies helped to establish the foundations for chemical analysis of the major groups of organic compounds, and to provide methods for studying chemical processes responsible for cellular functions. The search for relationships between cell structure and function continued to provide a focus for experimental studies in the twentieth century. As more became known about these relationships and the dependence of form and function on one another, biologists and chemists gradually came to examine both aspects of cells, thereby blurring the distinction between study of cell chemistry only by biochemists and physiologists and study of cell structure only by cytologists (microscopists). Cell biology, molecular biology, and other comprehensive disciplines are outcomes of these unified approaches to cell study. This blurring of boundaries between biological disciplines is a positive indication of advances in our understanding of interrelationships in the living world. Just as the study of biology allows us to see fundamental similarities among plants, animals, and microorganisms which we might miss if we studied botany, zoology, or microbiology separately, the union of biological studies has allowed us to seek the fundamental qualities common to all living systems, rather than simply cataloguing their superficial differences.

## CELL STRUCTURAL ORGANIZATIONS

In traditional systems of classification, all organisms have been assigned either to the plant or the animal kingdom. While many species can easily be accommodated within these major categories, others pose difficulties because they have a blend of plantlike and animallike features. For example, the unicellular *Euglena* has chloroplasts like plants but lacks the rigid plant cell wall, and it also moves about by means of whiplike flagella as some animals do. Bacteria and fungi have no chloroplasts, but these organisms have a rigid cell wall and have therefore been classified as plants in traditional systems. The difficulties of sorting organisms in evolutionarily related groups have prompted various attempts to modify kingdom classifications so that evolutionary relationships are emphasized rather than selected structural traits (Table 1.2).

Some recent revisions in kingdom classifications are based on the radically different cell plans found in **prokaryotes** and **eukaryotes.** Prokaryotes lack a membrane-bound nucleus while eukaryotic cells have a nucleus which is separated by a membrane system from the surrounding cytoplasm of the living cell. The

difference in nuclear organization is the fundamental distinction, but other features also identify an organism as prokaryotic or eukaryotic (Table 1.3). The fundamental evolutionary divergence of prokaryotes and eukaryotes is an important aspect of modern cell studies. The terms prokaryote (*pro:* before; *karyon:* nucleus) and eukaryote (*eu:* true) were suggested by Hans Ris in the early 1960s and have been widely accepted since then.

### Prokaryotic Cell Organization

Blue-green algae and bacteria constitute the major prokaryotic groups. These organisms have been grouped into the kingdom Monera by several authorities. There is widespread acceptance of this classification, since the prokaryotic cell organization distinguishes monerans from all other cellular life forms.

The average size of a prokaryotic cell is 1–10 $\mu$m, but some bacteria are as small as 0.2–0.3 $\mu$m and some blue-green algae cells are as large as 60 $\mu$m in diameter. In some species, individual cells remain associated in chains, filaments, or other groupings, but many prokaryotes exist as single, ungrouped cells. In either situation we still consider prokaryotes to be **unicellular** rather than multicellular organisms, since each cell can exist independently of all others and there is no difference between cells associated together in a group. Each prokaryotic cell can produce new cells like itself, usually by a division process called **binary fission,** in which an enlarged parent cell divides in half to produce two equivalent daughter cells in which all the living material has been partitioned about equally.

The living material of a cell is called **protoplasm.** The protoplasmic contents of a prokaryotic cell include an enveloping **plasma membrane,** or **plasmalemma,** within which are two recognizable regions called **cytoplasm** and **nucleoid.** The living protoplasm in turn is surrounded by a rigid or semirigid **cell wall,** which provides support and shape but which itself is not considered to be protoplasmic in nature; the cell wall usually is considered to be a secretion of the living material within the cell (Fig. 1.3).

Every living cell must have and maintain an intact plasma membrane if it is to survive. The plasma

**TABLE 1.2 Systems of kingdom classifications***

| *"Traditional"* | *Dodson, 1971* | *Stanier et al., 1970* | *Copeland, 1956* | *Whittaker, 1969* |
|---|---|---|---|---|
| **Plantae** | **Monera** | **Protista** | **Monera** | **Monera** |
| Bacteria | Bacteria | Bacteria | Bacteria | Bacteria |
| Blue-green | Blue-green | Blue-green | Blue-green | Blue-green |
| algae | algae | algae | algae | algae |
| Chrysophytes | | Protozoa | | |
| Green algae | **Plantae** | Chrysophytes | **Protoctista** | **Protista** |
| Red algae | Chrysophytes | Green algae | Protozoa | Protozoa |
| Brown algae | Green algae | Red algae | Chrysophytes | Chrysophytes |
| Slime molds | Red algae | Brown algae | Green algae | |
| True fungi | Brown algae | Slime molds | Red algae | **Fungi** |
| Bryophytes | Slime molds | True fungi | Brown algae | Slime molds |
| Tracheophytes | True fungi | | Slime molds | True fungi |
| | Bryophytes | **Plantae** | True fungi | |
| **Animalia** | Tracheophytes | Bryophytes | | **Plantae** |
| Protozoa | | Tracheophytes | **Plantae** | Green algae |
| Metazoa | **Animalia** | | Bryophytes | Red algae |
| | Protozoa | **Animalia** | Tracheophytes | Brown algae |
| | Metazoa | Metazoa | | Bryophytes |
| | | | **Animalia** | Tracheophytes |
| | | | Metazoa | |
| | | | | **Animalia** |
| | | | | Metazoa |

*References:
Copeland, H. F. 1956. *Classification of the Lower Organisms.* Palo Alto: Pacific Books.
Dodson, E. O. 1971. The kingdoms of organisms. *Systematic Zoology* 20:265–281.
Stanier, R., Adelberg, E., and Doudoroff, M. 1970. *The Microbial World.* Englewood Cliffs, New Jersey: Prentice-Hall.
Whittaker, R. H. 1969. New concepts of the kingdoms of organisms. *Science* 163:150–160.

**TABLE 1.3 Some major differences between prokaryotes and eukaryotes**

| *Characteristic* | *Prokaryotes* | *Eukaryotes* |
|---|---|---|
| Cell size | Mostly small (1–10 $\mu$m) | Mostly larger (10–100 $\mu$m) |
| Genetic system | DNA not associated with proteins in chromosomes | DNA complexed with proteins in chromosomes |
| | Nucleoid not membrane-bounded | Membrane-bounded nucleus |
| | One linkage group | Two or more linkage groups |
| | Little or no repetitious DNA | Repetitious DNA |
| Internal membranes (organelles) | Transient, if present | Numerous types and differentiations, e.g., mitochondrion, chloroplast, lysosome, Golgi, etc. |
| Tissue formation | Absent | Present in many groups |
| Cell division | Binary fission, budding, or other means; no mitosis | Various means, associated with mitosis |
| Sexual system | Unidirectional transfer of genes from donor to recipient, if present | Complete nuclear fusion of equal gamete genomes; associated with meiosis |
| Motility organelle | Simple flagella in bacteria, if present | Complex cilia or flagella, if present |
| Nutrition | Principally absorption, some photosynthesizers | Absorption, ingestion, photosynthesis |

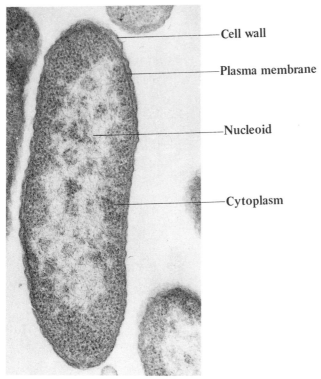

Cell wall

Plasma membrane

Nucleoid

Cytoplasm

**Figure 1.3**
Electron micrograph of a thin section of the rod-shaped
bacterium *Pseudomonas aeruginosa*. The central nucleoid
region is surrounded by a denser cytoplasmic area. A
relatively thin cell wall encloses the living protoplast.
× 60,000. (Photograph by H.-P. Hoffmann)

membrane of a cell is about 10 nanometers (nm) thick,
and when seen in cross section (view of the cut edge) it
displays a characteristic three-layered staining pattern
in photographs taken with the electron microscope
(Fig. 1.4). The plasma membrane may be infolded at
some sites, with these infoldings extending into the cytoplasm. In photosynthetic prokaryotes such as the
blue-green algae and purple bacteria, the infolded
membranes contain pigments and enzymes involved in
the light-capturing processes of photosynthesis (Fig.
1.5). While plasma membrane infoldings generally are
believed to be attached at one or more sites to the
plasma membrane from which they are derived, it is
possible that the folded photosynthetic membranes are

entirely separate from the plasma membrane. If this is
true, then the photosynthetic membranes would be an
exception to the general rule of *noncompartmentation*
in prokaryotes. By this we mean that prokaryotic cells
have no permanent membrane-bounded compartments; the only *permanent* membrane is the plasma
membrane (with the possible exception of photosynthetic membranes in some species).

The cytoplasmic region contains many kinds of
chemicals, including enzymes active in metabolism,
and numerous **ribosomes.** Ribosomes are tiny particles
constituted of RNA and proteins, and they are
essential for protein synthesis in cells. Since
ribosomes measure only 15–20 nm in diameter, they
can only be seen at high magnifications achieved with
the electron microscope. The nucleoid region within
prokaryotic cells appears as an irregularly-shaped
lighter area within the surrounding cytoplasm. Each
nucleoid region contains one molecule of DNA, which
represents the entire set of genes for the species. The
DNA molecule is not complexed with proteins in
prokaryotes, but exists instead as a naked double-helix
molecule only 2 nm wide but many thousands of nanometers long.

All prokaryotic cells therefore contain a ribosome-
rich cytoplasm, DNA located within the nucleoid
region, and an enclosing plasma membrane. There is a
cell wall boundary in all prokaryotes except for a
group of very small bacteria called mycoplasmas.
Various other features are also found in some but not
all prokaryotes. Some have a surrounding sheath or
capsule outside the cell wall; many do not. Some
prokaryotes produce spores; many do not. Some
prokaryotes move about by means of whiplike flagella;
many do not. Some prokaryotes produce storage
granules or gas vacuoles, but many do not. Prokaryotes are relatively simple in construction and
cellular organization, but there is considerable diversity within the group with regard to some structural components and to their metabolic activities.

**Eukaryotic Cell Organization**

The major trademark of the eukaryotic cell is its
membrane-bounded compartments which are

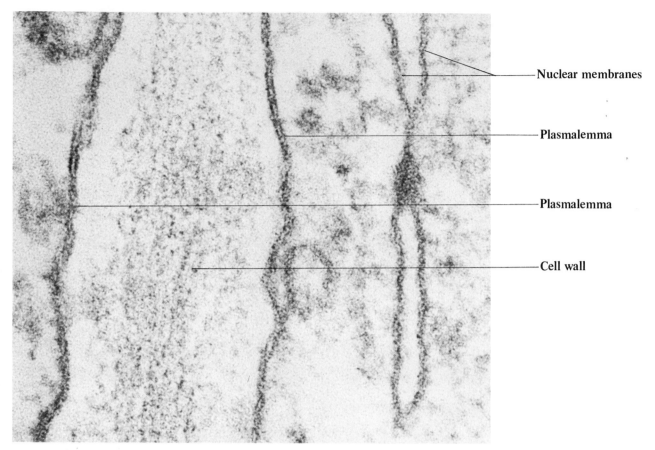

Nuclear membranes

Plasmalemma

Plasmalemma

Cell wall

**Figure 1.4**
Electron micrograph of a thin section through parts of two adjacent cells of *Arabidopsis thaliana* root. The three-layered staining pattern is evident in the plasmalemma of each cell and in the nearby membranes of the nuclear envelope at the right. × 274,000. (Courtesy of M.C. Ledbetter)

physically separate from the plasma membrane (Fig. 1.6). The most conspicuous compartment is the **nucleus** in which DNA is organized into complex nucleoprotein bodies called **chromosomes.** Each chromosome is made of a molecule of DNA which is complexed with proteins, and there are never fewer than two chromosomes in eukaryotic species. Unlike prokaryotes whose set of genes occurs within a single molecule of naked DNA, genes in eukaryotes are distributed among two or more DNA molecules within two or more chromosomes in the nucleus.

The nuclear membrane system, called the **nuclear envelope,** is a two-membrane complex which is pockmarked with openings called **nuclear pores.** The outer surface of the nuclear envelope, facing the cytoplasm, is studded with ribosomes (Fig. 1.7). Within the confines of the nuclear envelope there are other recognizable components in addition to chromosomes. A nucleus always includes at least one **nucleolus,** but there may be two or more nucleoli present. These are globular structures which are essential for the continued existence of the cell because they are the centers for synthesis of ribosomes. Unless the cell can maintain its ribosomal machinery, replacing used

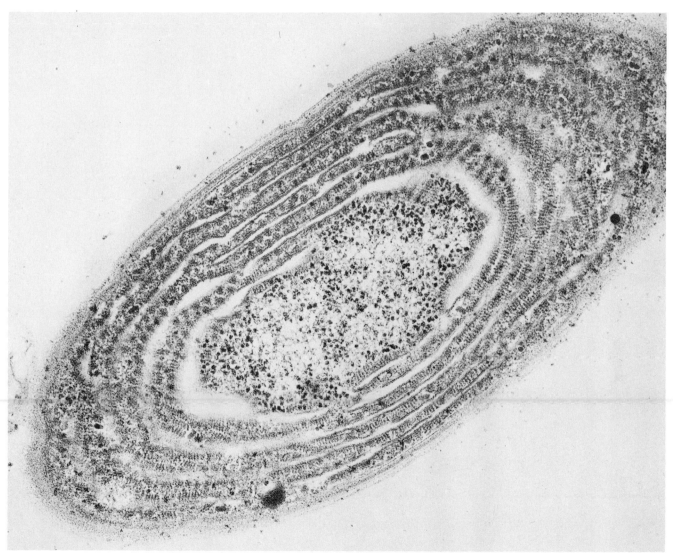

**Figure 1.5**
Electron micrograph of a thin section through the blue-green alga *Synechococcus lividus*.
The concentrically folded photosynthetic membranes are infolded differentiations of the
plasma membrane. × 62,400. (Courtesy of E. Gantt, from Edwards, M. E., and E. Gantt,
1971, *J. Cell Biol.* **50**:896–900, Fig. 1.)

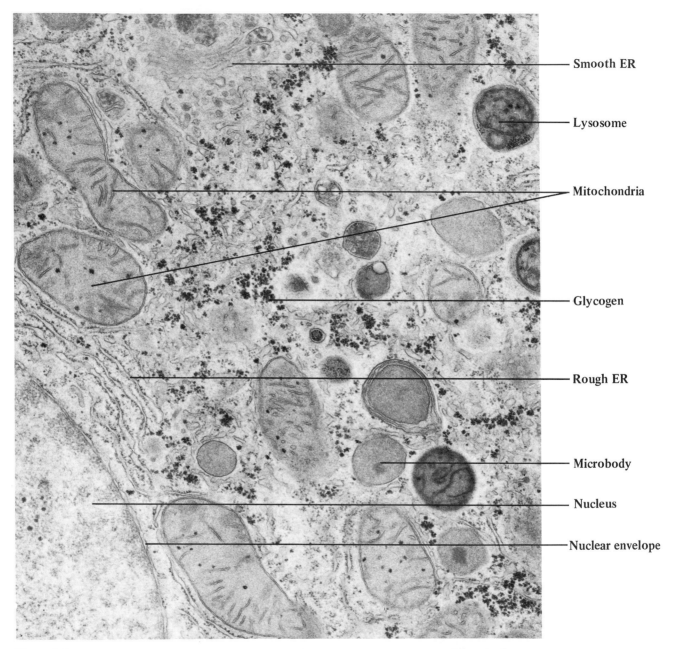

Smooth ER

Lysosome

Mitochondria

Glycogen

Rough ER

Microbody

Nucleus

Nuclear envelope

**Figure 1.6**
Thin section of rat liver cell. A small portion of the nucleus is visible along with various membranous and fibrous cytoplasmic structures. The opaque granules of glycogen are stored carbohydrate food material in the cytoplasm of these cells. × 24,300. (Courtesy of K. R. Porter)

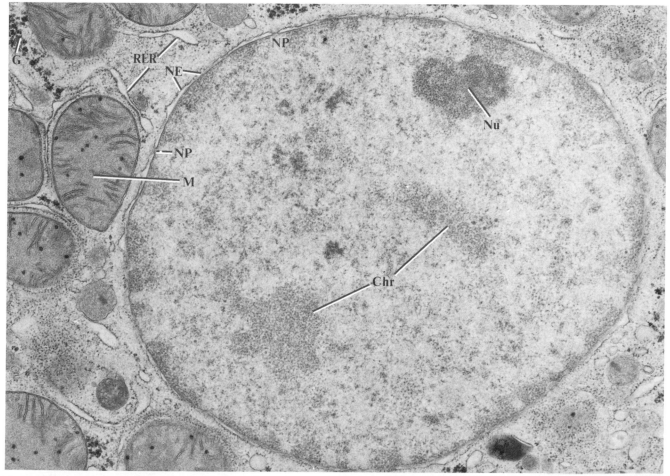

**Figure 1.7**
Electron micrograph of the eukaryotic nucleus as seen in a cell from bat liver, fixed in osmium tetroxide, which preserves fibrous components. The nucleolus (Nu) and two chromosome (Chr) areas are within the nucleoplasm of the nucleus. Ribosomes are attached only to the outer membrane of the nuclear envelope (NE). Note the resemblance between the rough endoplasmic reticulum (RER) elements and the nuclear envelope. Two nuclear pores (NP) are present in the plane of this section through the nuclear envelope. Glycogen (G) and mitochondrial profiles (M) are in the cytoplasm surrounding the nucleus. × 23,660. (Courtesy of K. R. Porter)

ribosomes and making new ones for new cells, there is no way to replenish proteins lost by wear and tear or to synthesize new proteins. Apart from the structurally defined chromosomes and nucleoli the remainder of the nuclear contents consists of a chemically complex but shapeless material called **nucleoplasm.** Of all the parts of the nucleus, only the chromosomes and their

gene contents are transmitted regularly from one generation to another, fully accounting for the continuity of life.

Between the plasmalemma boundary and the nuclear envelope is the **cytoplasm** with its profusion of membranes and particles, all bathed in a granular **cytosol.** None of the cytoplasmic membranes is connected to the plasma membrane in eukaryotic cells. The many kinds of components in the cytoplasm carry out unique functions, but their activities are coordinated and regulated such that a harmony emerges despite the diversity.

The **endoplasmic reticulum,** or **ER,** is a system of membranous channels and passageways which extends and branches throughout the cytoplasmic volume. When the ER is seen in cross section in an electron micrograph, it usually appears to be a multi-layered membrane system or to have tubular regions here and there. This view is somewhat deceptive because we actually see only the cut ends of a sheet of membrane folded back on itself many times over (Fig. 1.8). Regions of the ER may have ribosomes attached on one surface of the folded membrane sheet, in which case we refer to these regions as **rough ER.** Where there are no attached ribosomes and therefore no jagged appearance to the membrane region, the area is called **smooth ER.** The endoplasmic reticulum, including its attached ribosomes, forms the structural basis for synthesis and distribution of proteins in living cells. Movement of newly made proteins is much faster and more directed an activity when these molecules travel

through ER passageways than when these same molecules move from place to place by diffusion alone. Some proteins do move from one place to another without the help of the ER system, and cells with a minimum ER development may distribute molecules primarily by diffusion or similar processes.

The **Golgi apparatus** is a compartment constituted of smooth membranes, whose function is to process and package certain proteins made at the rough ER in cells. Some of these proteins are **secretions,** such as mucus, which are chemically modified during their passage through the Golgi apparatus. In addition to processing these molecules, the Golgi apparatus packages the modified proteins in **vesicles** (membrane "bubbles"). These vesicles may remain within the cell or they may move toward the cell surface, where their contents are discharged into the surrounding space (Fig. 1.9).

Another type of vesicle containing proteins processed and packaged in or near the Golgi apparatus is the **lysosome,** which is a container for powerful digestive enzymes involved in intracellular reactions. These enzymes can break down almost every known organic substance of biological importance, but diges-

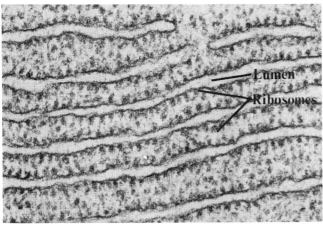

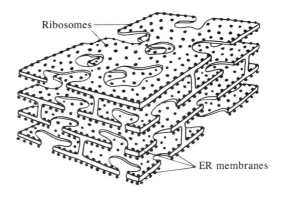

**Figure 1.8**
The two-dimensional perspective of an electron micrograph is contrasted with a three-dimensional representation of the system of membranes and channels that form the rough endoplasmic reticulum of eukaryotic cells. × 85,000. (Courtesy of K. R. Porter)

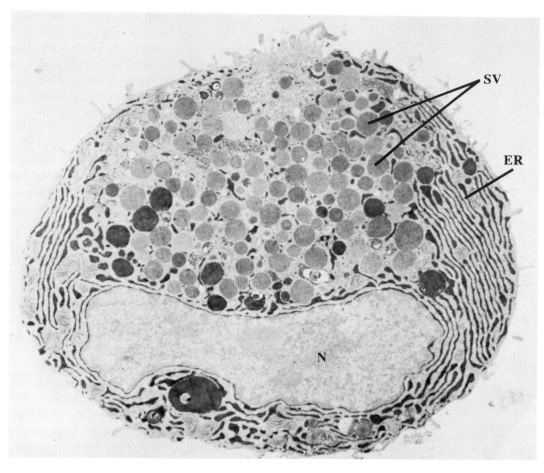

**Figure 1.9**
Isolated stained cell of rat lacrimal gland containing numerous secretion vesicles (SV), a prominent nucleus (N), and displays of endoplasmic reticulum (ER). The secretions are released at the cell apex (top of photo). × 7,300. (Reproduced with permission from Herzog, V., H. Sies, and F. Miller, 1976, *J. Cell Biol.* **70**:692, Fig. 10.)

tion is controlled because the enzymes are separated by the lysosome membrane from the rest of the cell. The enzymes do not attack the lysosome membrane itself under ordinary conditions. Digestion usually takes place within the lysosome after it has fused with some cellular or foreign material. Intracellular digestion is an important feature of metabolism in cells, but lysosomes also provide one line of defense against invasion by bacteria and other harmful agents. When such foreign materials do enter a cell they generally encounter and interact with lysosomes, fuse with the organelles, and proceed to be digested inside the lysosome space. If the lysosome does release its enzymes directly into the cytoplasm, the cell usually dies shortly afterward. Lysosome breakdown and

enzyme release is a normal and recurring event in certain tissues which regularly disintegrate during cyclic events, such as tissues involved in some phases of the menstrual cycle. Unlike secretion vesicles, lysosomes remain within the cells in which they were formed.

Eukaryotic cells generally contain one or more **mitochondria** (sing., mitochondrion), which are centers for synthesis of high-energy **ATP** (adenosine triphosphate) molecules during food breakdown reactions of **respiration**. The principal identifying structural feature of the mitochondrion is its inner membrane infolded into numerous tubular projections called **cristae** (sing., crista) (Fig. 1.10). Many of the enzymes of ATP synthesis and respiration are situated

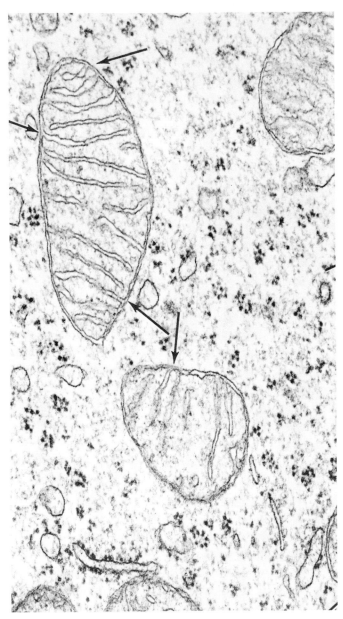

**Figure 1.10**
High magnification electron micrograph showing
mitochondrial profiles in a spermatogonial cell of rat testis.
The cristae can be seen to be infolded regions of the inner
membrane in favorable planes of the section (at arrows). ×
65,000. (Courtesy of H. H. Mollenhauer)

within the inner mitochondrial membrane, and other
enzymes are present in the internal space bounded
by this membrane. The energy released from food
molecules during respiration is trapped and stored as
chemical energy in ATP molecules, which can then
subsidize many kinds of biosynthesis reactions in cells
by providing the necessary energy for these reactions
to go forward. Respiration reactions within mitochondria act on smaller chemical units produced after the
larger food molecules have been partially processed in
the cytosol or lysosomes. These smaller molecules are
then dismantled completely within the mitochondrion
if oxygen is present. Basically, all oxygen-using
(**aerobic**) life derives the bulk of its energy for all
cellular activities from the processes of mitochondrial
respiration in eukaryotes and from respiratory
activities of the plasmalemma enzymes in prokaryotes.

Certain cell features are found uniquely in some
eukaryote groups and not in others. Plants, eukaryotic
algae, and some kinds of **protists** (unicellular
organisms that are not conveniently classified as
plants, animals, or fungi and are grouped into the
kingdom Protista) possess one or more kinds of **plastids.** The most familiar type of plastid is the **chloroplast**
of green cells. Chloroplasts have two encircling
membranes and a third set of internal membranes
bathed within a granular stroma or matrix material
(Fig. 1.11). The processes of photosynthesis, by which
foods are manufactured from carbon dioxide and
water using light energy, take place in chloroplasts of
eukaryotic species. Photosynthetic species are essential to the existence of almost all life on the planet
for two reasons: (1) oxygen in our atmosphere is replenished continuously by photosynthetic cells which
release this gas as a byproduct of their light-capturing
activities, and (2) organic foods required by organisms
are provided by green cells which alone are able to
convert light energy to chemical energy and use this
chemical energy to manufacture food molecules;
photosynthetic species are the ultimate component in
the food chain for life on Earth.

Algae, fungi, and land plants characteristically
have a thick **cell wall** surrounding the living materials
of their cells. These walls may remain as a structural

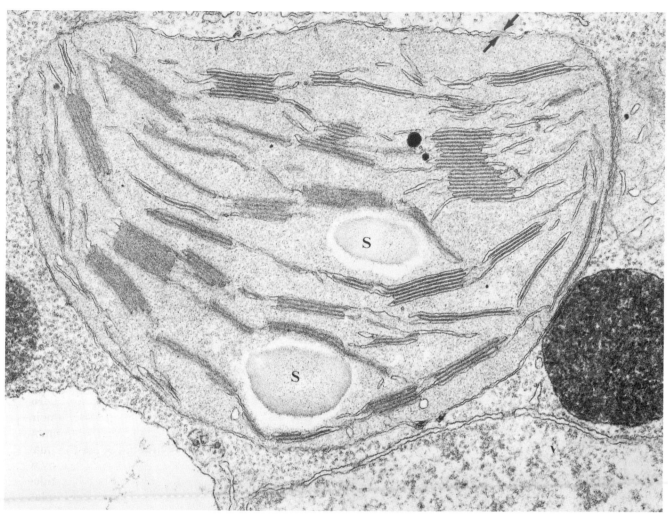

**Figure 1.11**
Thin section through a chloroplast from tobacco leaf mesophyll cell. The two closely
appressed enveloping membranes (at arrows) enclose a third system of stacked internal
chloroplast membranes. Photosynthetic reactions take place within these stacked
membranes and in the amorphous matrix in which they are suspended. Two sites of starch
(S) deposit can also be seen within the chloroplast matrix. × 46,000. (Courtesy of E. H.
Newcomb, from Frederick, S. E., and E. H. Newcomb, 1969, *J. Cell Biol.* **43**:343.)

framework after cells have died, as we see in woody
plants. The woody parts are composed of dead cells
for the most part, but their cell walls provide a system
of structural support as well as a system for water
transport between the roots and the rest of the plant.
Another typical feature in algae, fungi, and land plants
is the presence of one or more **vacuoles,** which are
membrane-bordered regions containing various kinds
of substances in dilute solution or in suspension (Fig.
1.12). Almost the entire volume of a mature plant cell
may be filled by a vacuole which is surrounded by a
narrow band of protoplasm. Animal cells often contain

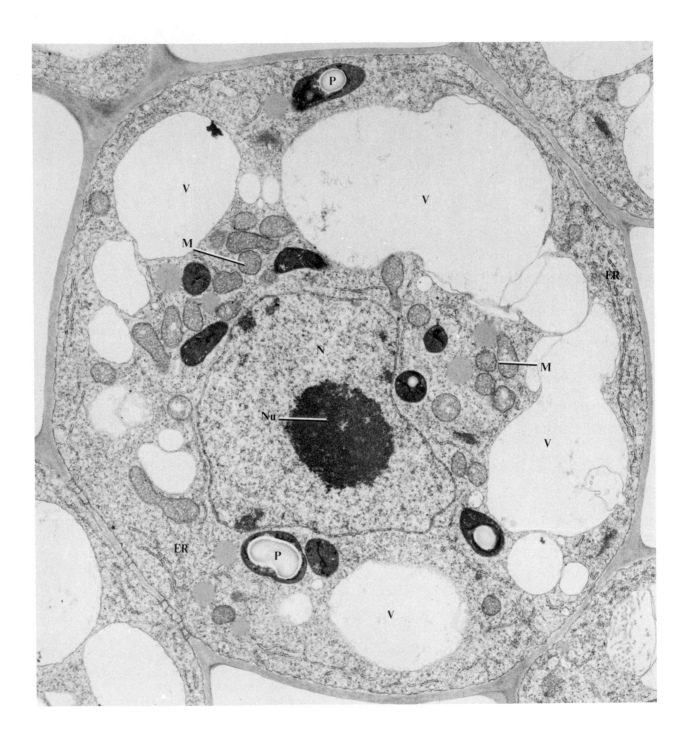

small vacuoles, but more typically the cytoplasm is densely packed with cellular components.

The animal cell surface appears to be a remarkable system which contributes to many kinds of interactions between cells. Since there is no wall covering the plasma membrane, the cell surface can establish close contact and communication with neighboring cells in tissues. One kind of specialized surface feature is the **desmosome,** which provides a site involved in adhesion between adjacent cells. Desmosomes are especially prominent between cells in tissues that are subjected to stresses, such as skin. Presumably there is less problem resulting from such stresses when cells are held together more firmly by common sites of adhesion as represented by desmosomes. In addition to various surface specializations, such as desmosomes, the animal cell plasmalemma may be variously folded, sometimes into fingerlike projections called **microvilli** (Fig. 1.13). These folds lead to a substantial increase in cell surface area and undoubtedly contribute to more efficient transfer of molecules into and out of cells bordering some intercellular space.

### Evolutionary Relationships

As we have divided all cellular life into prokaryotes and eukaryotes, it seems appropriate here to ask how these two groups of life forms are related. Prokaryotes are generally accepted as the most ancient kinds of organisms, which gave rise to eukaryote descendant forms during evolution. We might infer that prokaryotes are the ancestral types, since they are much simpler in structure and organization than the eukaryotes. The most convincing evidence, however, comes from the fossil record. Prokaryotes have been found in deposits well over 3 billion years old, while the earliest record we have for eukaryotes is about 1.3

**Figure 1.12**
Large vacuoles (V) are a distinctive feature of mature plant cells, as this one from *Potamogeton natans* root. The cytoplasm also contains profiles of mitochondria (M), starch-filled plastids, (P), endoplasmic reticulum (ER), and ribosomes. There is a prominent nucleolus (Nu) in the centrally situated nucleus (N). × 13,000. (Courtesy of M. C. Ledbetter)

billion years old. Since the fossil record provides a reasonable sampling of former life, we judge that prokaryotes are ancestral and eukaryotes are descendants of ancient prokaryotic lineages. Each group has continued to evolve through all these eons of time, and modern species are, of course, different in many details from their ancient ancestors, but sufficient distinction remains even now to clearly recognize the two subdivisions of cellular life.

There is considerable similarity between particular molecules and particular functions in prokaryotes and eukaryotes, which implies that they are related rather than being unique forms of life that appeared entirely independently of one another. All cellular life is characterized by having DNA as genetic material, a ribosomal machinery for protein synthesis, enzymes that influence chemical reaction rates in metabolism, various metabolic pathways which are similar or even identical, and other features that point to a common ancestry for modern cellular life. The particular lineages, however, are the subjects of lively controversies.

R. H. Whittaker has proposed a detailed lineage for the five kingdoms of organisms which he recognizes (Fig. 1.14). Although this is only one proposal out of several that are under consideration at present, it is a scheme that takes many observations into account. The prokaryotic Monera include species that are heterotrophic and others that are autotrophic. **Heterotrophs** derive energy and carbon-based building blocks for growth from their breakdown of complex organic foods; **autotrophs** are able to get energy either from light or from inorganic chemicals and to derive carbon from carbon dioxide in the air, using these to make organic compounds in metabolism. The most familiar prokaryotic autotrophs are the photosynthetic bacteria and the blue-green algae. The heterotrophic bacteria all *absorb* organic nutrients in solution from their environment. The simplest and probably the most ancient types of eukaryotes are in the kingdom Protista, which includes heterotrophs and autotrophs. The autotrophs are photosynthetic, but among the heterotrophs some only absorb organic nutrients and others have the ability to *ingest* solid foods rather than

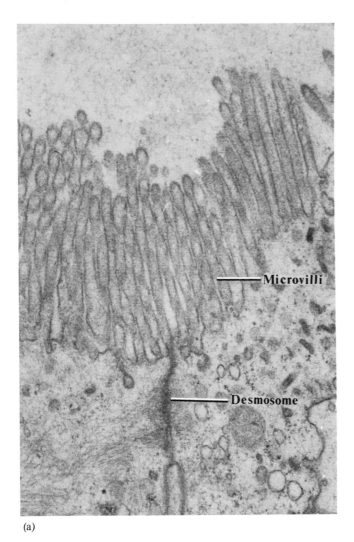

(a)

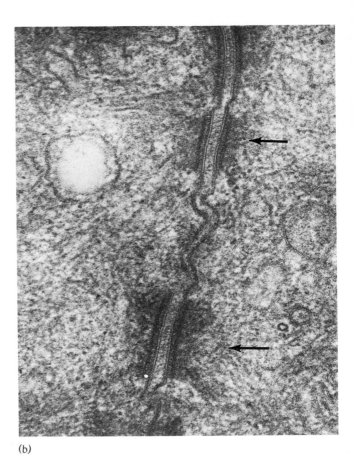

(b)

**Figure 1.13**
Cell-surface differentiations: (a) Microvilli provide increased absorptive surface area in these cells from proximal tubule of frog kidney. A desmosome is visible below the microvilli, probably serving as an adhesion region between the two contiguous cell membranes. × 30,000. (b) Higher magnification view showing two desmosomes (at arrows) in rat intestinal epithelium. The desmosomes occur as symmetrical plaques between two adjacent cells. × 156,000. (Courtesy of N. B. Gilula, from Gilula, N. E., 1974, *Cell Communications* [ed. R. P. Cox], John Wiley & Sons, pp. 1–29, Fig. 12.)

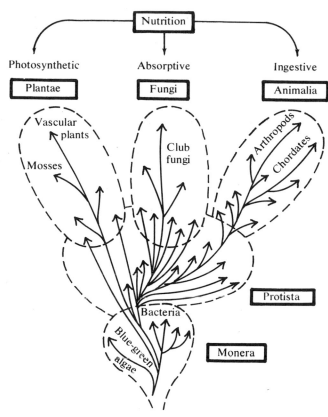

**Figure 1.14**
The classification scheme suggested by R. H. Whittaker. This scheme organizes all cellular life into five separate kingdoms. All prokaryotes are assigned to the Monera, while different types of eukaryotes are members of the remaining four kingdoms. (Adapted with permission from Whittaker, R. H., *Science* **163**:150–160, Fig. 3, 10 January 1969. Copyright© 1969 by the American Association for the Advancement of Science.)

obtaining their needs only from materials that move across the cell boundary in dissolved form.

Whittaker and others have proposed that different kinds of protists gave rise to new eukaryotic lineages which are represented today by species belonging to the three remaining kingdoms of eukaryotes: plants, animals, and fungi. Except for a few species, plants are photosynthetic. Fungi are heterotrophic and absorb nutrients in solution from their environment. Animals are heterotrophic, and almost all ingest solid

foods. As we will discuss in Chapter 12, the invention of the mobile cell surface which is typical of animal cells may well have been the key factor that contributed to the ability to obtain solid foods by ingestion. Larger quantities of food can be taken in per unit time by ingestion than by absorption, which may have been advantageous during eukaryote evolution which ultimately led to animals as a novel life form.

**VIRUSES**

Since viruses are not cellular they are neither prokaryotic nor eukaryotic. It is still debated, in fact, whether or not they are living systems. Viruses are inactive outside a living host cell, and some viruses may even be crystallized like many chemicals. After entering a host cell, many kinds of viruses will become active and begin to direct the production of new virus particles. To do this they must rely partly on their own programmed genetic information which is brought into the host cell, and partly on exploiting the biosynthetic machinery of their host.

Viruses are quite a varied group. They range in size from 30 to 300 nm, or from about the size of a ribosome to about the limit of visibility in the light microscope. The simpler viruses are made only of nucleic acid and protein, but more complex viruses have additional kinds of organic molecules in their construction. Viruses sometimes are enclosed within a membrane, but it is membrane derived from the host cell and not membrane which is made by or specified by the virus. Virus shapes are specific for each species, some being spherical, others rodlike, others like a lollipop, and so forth (Fig. 1.15). Viruses tend to be named in a random fashion according to the disease caused (poliomyelitis virus), the tissue affected (adenoviruses infect adenoid tissue), the host organism (bacterial virus, also called bacteriophage or **phage**), or some coded system (T1, T2, P1 phages of the colon bacillus *Escherichia coli*).

Virus multiplication is very different from cell replication mechanisms. Cells produce their own chromosomes, proteins, membranes, and other constituents, and these materials are partitioned into

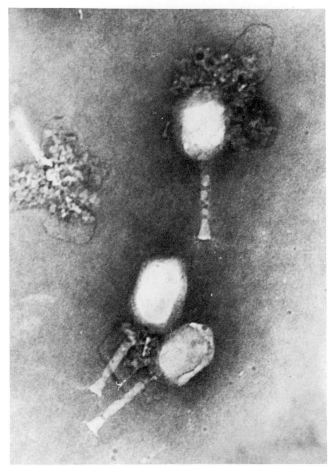

**Figure 1.15**
Negatively stained T4 bacteriophages, partially disrupted by the method of preparation. × 139,000.

progeny cells after a division process in the parent cell. As stated in the cell theory, cells arise only from other cells. Viruses, on the other hand, only provide the genetic information or blueprints for new viruses, and the progeny particles are assembled from molecules made in the host cell by host cell machinery, as specified by virus genes. The assembled viruses are then released from their host, each virus being able to initiate another cycle like the one in which it was produced (Fig. 1.16).

Living cells share three basic characteristics which permit them to live independently of one another, or at least to possess the *potential* for an independent existence.

1. A cell has **a set of genes** which constitutes the blueprints for making new cells with all their components and parts.
2. There is **a cell membrane** around each cell that acts as a physical boundary between the cell and everything around it, but which permits controlled exchange of matter and energy with the outside world.
3. Each cell has **a metabolic machinery** by which energy can be obtained and by which this energy can be used to subsidize living processes such as growth, reproduction, and repair or replacement of parts.

Viruses have only one of these characteristics and must therefore rely on host cells to furnish the remaining requirements for virus multiplication. Viruses certainly are not cellular, but are they living entities? There is no clear answer to this question because there is no single definition of life which will satisfy everyone. If life is defined as being cellular, then viruses are not alive. If life is defined as being capable of making new life directly through its own metabolic efforts, then viruses are not living. If life is defined as being able to specify each new generation according to its own genetic instructions, then viruses are living systems. Each new crop of viruses is made specifically according to a set of genetic instructions which is unique for each kind of virus. Each cellular life form also has a unique set of genes. A system may borrow a membrane or borrow metabolism, but it cannot borrow a set of genes from another species. Since the one indispensible and unique feature of every life form is its genes, most biologists would agree that viruses are living organisms of an unusual type, but with the full potential for continuity of the species and for evolutionary change.

### SUMMARY

1. Cell biology is a blend of microscopy, biochemistry, genetics, and physiology. Cell biologists are concerned with the ways in which cells accomplish their functions and

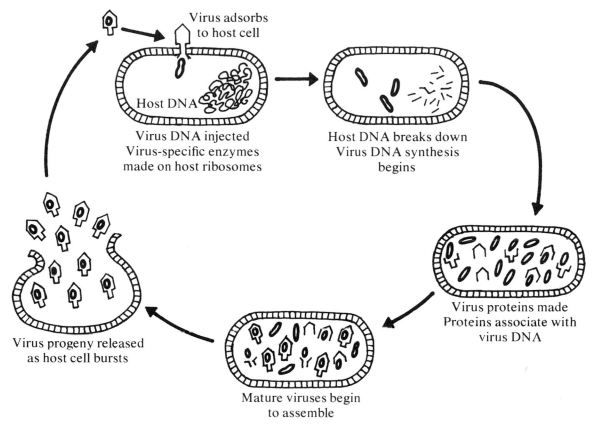

**Figure 1.16**
Virus multiplication in a host cell. Virus DNA directs synthesis of new virus-specific DNA and proteins, made by host biosynthetic machinery. The newly-assembled virus particles are released when the host cell bursts at the end of the infection cycle. These virus progeny particles can then establish new infection cycles.

activities within the structural framework of space and the regulatory framework of time.

2. Cell study became possible when microscope lenses were invented in 1590. The first important studies during the seventeenth century by Robert Hooke and Nehemiah Grew drew attention to the cell wall in plant tissues, and attention was not shifted to cell contents until the mid-nineteenth century. In 1838–1839 Matthew Schleiden and Theodor Schwann postulated the cell theory, which stated that: (a) All living organisms are made up of one or more living units or cells; (b) each cell can maintain its living properties independent of other cells, but the properties of life of any organism are based on the properties of life of its individual

cells; (c) the smallest unit of life is the cell. In the 1850s, it was further stated by Rudolf Virchow that (d) cells arise only from pre-existing cells. By the end of the nineteenth century there was a solid appreciation of the role of the nucleus in the continuity of life, and the foundations were laid for incorporating Mendelian genetic principles into the cell theory.

3. By the last quarter of the nineteenth century, the optical systems of the light microscope were equal to modern instruments in producing magnification and resolution. Along with biological advances there were parallel advances in understanding the nature of cellular organic chemistry. It was shown that organic molecules were the same in the living and the nonliving worlds, and that organic

reactions followed the same rules of physics and chemistry in both worlds. Nucleic acids were first analyzed in the 1870s, but were not finally accepted as the genetic material until 1953 when James Watson and Francis Crick proposed a molecular model for DNA.

4. The introduction of electron microscopy in the 1950s opened a new era of cell study. High magnification and resolution of the electron microscope permitted new levels of observation, including the basic distinction in cellular organization between prokaryotes and eukaryotes. Emphasis shifted from viewing life in terms of plants and animals to viewing evolutionary lineages in terms of cellular construction plans, and according to their chemical and genetic organization.

5. Prokaryotes are the bacteria and blue-green algae, organized into the kingdom Monera. Eukaryotes can be grouped into four kingdoms: Protista, Fungi, Plantae, and Animalia. Prokaryotes are ancestral to eukaryotes. Prokaryotic cells are not compartmented. The only permanent membrane system in prokaryotes is the plasmalemma, which may be differentiated into functionally distinct regions. Prokaryote DNA is not separated from the cytoplasm by a membrane, but does exist in a distinct region called the nucleoid. Eukaryotes are all characterized by a membrane-bounded nucleus, along with a number of membranous compartments in the cytoplasm surrounding the nucleus. An intact plasmalemma and ribosomes in the cytoplasm are characteristics of all cellular life, as is a set of genes situated in one or more DNA molecules.

6. Viruses are not cellular. They are usually too small to be seen with the light microscope, but can be observed in some detail by electron microscopy. Viruses differ from cellular organisms in various ways, including their mode of reproduction. Viruses do not give rise directly to new viruses. Instead, they must subvert the biosynthetic machinery of their host so that virus-specific proteins and nucleic acids are made, according to viral genetic information. Viruses then assemble from newly-made molecules in host cells and are released when the host cells burst. They may then initiate new cycles of infection in other host cells. There is some dispute about whether viruses are living organisms, since they lack metabolism. Viruses borrow metabolism and a sheltering membrane from their host, but they provide the genetic instructions that ensure continuity of their species from generation to generation. Viruses can evolve by the same mechanisms that are responsible for evolution of cellular life. Viruses are living, according to most biologists.

## STUDY QUESTIONS

1. What were some of the technical and conceptual problems that delayed the formulation of the cell theory for more than 200 years after the invention of the first useful microscope? What did the following scientists contribute to the cell theory: Robert Hooke, Nehemiah Grew, Robert Brown, Theodor Schwann, Matthew Schleiden, Rudolf Virchow, Oskar Hertwig?

2. What were the scientific observations that showed that (a) organic compounds were not made only by living organisms; (b) cell chemistry followed the same rules for reactions as chemistry outside the living cell?

3. What were the important discoveries leading to a better understanding of the nucleus? (a) before 1900 and (b) after 1900? What is the role of the nucleus in the perpetuation and continuity of life?

4. What are the major statements of the cell theory?

5. What is the basic difference in cellular organization of prokaryotic and eukaryotic cells? If you examined a section through a cell which happened to be cut in such a way as to exclude the nucleus from the section, how could you determine whether the source was indeed a eukaryotic cell?

6. What structural features are shared by prokaryotic and eukaryotic cells? What are the basic components required for the cell to live and reproduce, whether prokaryotic or eukaryotic?

7. What is the identifying structure and function of mitochondria? chloroplasts? lysosomes? What kinds of cells and species contain each kind of organelle? What are some of the differences between animal and plant cell-surfaces?

8. What is the probable evolutionary history of eukaryotes? What are the sources of energy and building blocks for growth among organisms? How do we sort out and identify different life forms?

9. Are viruses living or nonliving? What are the arguments for each view? How do viruses differ basically from cellular organisms? How do viruses multiply?

## SUGGESTED READINGS

Campbell, A. M. How viruses insert their DNA into the DNA of the host cell. *Scientific American* **235**:102 (Dec. 1976).

Claude, A. The coming of age of the cell. *Science* **189**: 433 (1975).

DuPraw, E. J. *The Biosciences: Cell and Molecular Biology*. Stanford: Cell and Molecular Biology Council (1972).

Fawcett, D. W. *The Cell. An Atlas of Fine Structure*. Philadelphia: Saunders (1966).

Fernández-Morán, H. Cell fine structure and function—past and present. *Experimental Cell Research* **62:**90 (1970).

Ledbetter, M. C., and Porter, K. R. *Introduction to the Fine Structure of Plant Cells*. New York: Springer-Verlag (1970).

Lenhoff, E. S. *Tools of Biology*. New York: Macmillan (1966).

Mirsky, A. E. The discovery of DNA. *Scientific American* **218:**78 (June 1968).

Porter, K. R., and Bonneville, M. A. *Fine Structure of Cells and Tissues*. Philadelphia: Lea & Febiger (1968).

Porter, K. R., and Novikoff, A. B. The 1974 Nobel prize for physiology or medicine. *Science* **186:**516 (1974).

Roller, A. *Discovering the Basis of Life: An Introduction to Molecular Biology*. New York: McGraw-Hill (1974).

Rosenberg, E. *Cell and Molecular Biology: An Appreciation*. New York: Holt, Rinehart & Winston (1971).

# Chapter 2

# Cellular Chemistry: Organic Molecules

Chemistry is a basic feature of the living as well as the nonliving world. To understand the interactions between atoms and molecules that make life possible we must look at some basic properties of atoms and aggregates of atoms, or molecules. One of the important properties to examine is *bonding* between atoms or molecules, so that we can better understand the special features of cellular chemistry. In addition, we will briefly examine the energy relations which underlie bonding and the capacities of bonds in chemical interactions. Afterward we can discuss those kinds of organic molecules that are basic to cell structure, function, and regulation.

## CHEMICAL BONDS

Two or more atoms are held together in an ag-

gregate by chemical bonds. Any chemical bond has a certain chance of being broken at any time in any system, but some bonds are more stable than others in a given system. A *stable* bond is one which leads to more energy being released when the bond is made and more energy being needed to break such a bond, in comparison with other chemical bonds in the system. It is convenient to refer to the more stable bonds as strong chemical bonds, and to less stable ones as weak chemical bonds.

In biological systems we refer to **covalent bonds** as strong chemical bonds, and to other chemical bonds as being weaker than, or secondary to, covalent bonds. The main, biologically important weak bonds are **hydrogen bonds, ionic bonds,** and **van der Waals forces.** A fourth type of association, often called a hydrophobic bond or **hydrophobic interaction,** is really not a bond at all, but is actually a recognized tendency of certain chemical groups to associate together and exclude water when water is present.

## Covalent Bonds

The important stable bonds of organic chemistry are covalent bonds, in which *two atoms share one or more pairs of electrons*. Every atom has a certain stable number of electrons which can be achieved if the atom loses or gains one or a few electrons. Some atoms, such as helium and neon, exist in the most stable state and do not readily undergo change. When two hydrogen atoms share their electrons, each atom assumes the more stable helium configuration, that is, each atom in the H—H unit has the more stable configuration of two electrons even though only two electrons are present in the unit (Fig. 2.1). The two atoms are bonded together by their shared electron pair and have thus achieved stable configurations. It is conventional to indicate a single covalent bond, or pair of shared electrons, by a dash between the bonded atoms (e.g., H—H, H—O—H). A covalent double bond is formed when two atoms share two pairs of electrons (e.g., O=C=O), and a triple covalent bond involves three pairs of shared electrons between two atoms (e.g., H—C≡N).

Since more energy is needed to break stable, covalent bonds than is needed to break the weaker chemical bonds, covalent bonds do not readily come apart at normal cell temperatures. Covalent bonds generally are responsible for keeping molecules intact or relatively intact, so that organic molecules do not fly apart or easily dissociate into their constituent atoms. Weaker bonds, on the other hand, are more easily made and broken since less energy is involved. Weaker chemical bonds allow substantial flexibility in chemical interactions, which is important in modifying molecules during metabolism. The combination of covalent and weaker chemical bonds in an organic molecule provides both a basis for relative stability and relative flexibility in rearranging atoms and groups of atoms during chemical reactions.

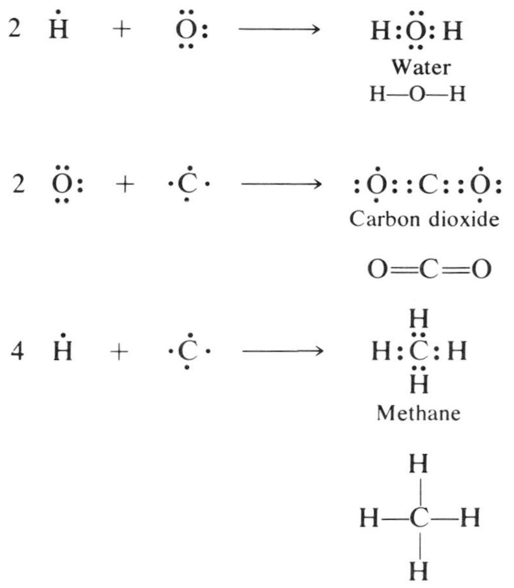

**Figure 2.1**
Covalent bonds form when electrons of different atoms share a common electron shell. Carbon compounds are very stable when the outer shell of the carbon atom is completely filled with eight electrons. Covalent interactions are very strong when compared with hydrogen bonds or other atomic interactions over greater atomic distances.

## Weak Chemical Bonds

When chemical bonds are made or broken there is an exchange of energy between the atoms and their surroundings. The most commonly used unit of measuring energy is the **calorie,** which is defined as the amount of energy needed to raise the temperature of one gram of water from 14.5°C to 15.5°C. The amount of reaction material which is involved is usually expressed in **moles** (gram molecular weight). Energy changes in chemical reactions are expressed in kilocalories per mole **(kcal/mole),** since thousands of calories (1000 cal = 1 kcal) usually are involved in breaking a mole of chemical bonds between atoms. The amount of energy that must be added to break a chemical bond is equal to the amount which is released upon its formation.

The most useful way to express change in energy during bond-making and bond-breaking is through the concept of **free energy,** signified as *G*. Free energy is the energy that can perform work. A *change* in free energy, signified by Δ*G,* occurs in all situations except those at equilibrium, when the number of bonds forming equals the number of bonds breaking, per unit time. During formation of covalent bonds from free atoms the free-energy change is negative (indicating free-energy release) and relatively large; that is, Δ*G* = −50 to −110 kcal/mole. Free-energy change during formation of weaker bonds is in the range of Δ*G* = −1 to −7 kcal/mole. These Δ*G* values provide a quantitative basis for the designation of bonds as being strong or weak in comparison with each other.

A **hydrogen bond** develops from sharing a hydrogen nucleus (proton). The shared proton is usually found between two nitrogens or two oxygens or between one atom of each. The hydrogen nucleus itself is covalently bonded to an atom of nitrogen or oxygen, so that the shared proton or hydrogen bond is a chemical bond that holds two covalently bonded units together (Fig. 2.2).

An **ionic bond** develops from electrostatic forces operating between atoms or groups of atoms having opposite charges, as between a negatively charged carboxyl group (COO⁻) and a positively charged

amino group (NH₃⁺) or between a Na⁺ ion and a Cl⁻ ion. Ionic groups of opposite charge attract each other, but these bonds generally develop between strong electron donors and strong electron acceptors whose outermost electron orbitals become stabilized by gaining or losing from 1 to 3 electrons.

**Van der Waals bonding** arises from attractive forces due to two atoms that come close to each other. Fluctuating charges develop because of the nearness of atoms to each other. Since size and shape of interacting units are more important than the particular atoms involved, van der Waals forces are relatively nonspecific and can therefore operate for many kinds of molecules. A repulsive van der Waals force develops when atoms in the molecules are close enough for their outer electron orbitals to overlap. There is a balance between attractive and repulsive van der Waals forces between atoms, at a certain distance specific for each type of atom. This distance is called the *van der Waals radius* of an atom in a molecule.

Binding forces are sufficiently low at cell temperatures so that an effective van der Waals bond develops only when several atoms in a molcule are bound to several atoms in another molecule. The energy of interaction is much greater than the tendency to dissociate in response to thermal movements, under these conditions. The strongest bonds develop when the shape of one molecule "fits" the shape of another molecule, like a lock and key.

All these types of relatively weak bonds provide a wider spectrum of flexibility as molecules associate and disassociate during chemical interactions in cells. Hydrogen bonds have a bonding energy of about 3 to 7 kcal/mole, but they form only when covalently-bonded oxygens and nitrogens share a hydrogen nucleus. These bonds clearly cannot occur unless the interacting regions include the specific interacting atoms. Ionic bonds require electrically charged components of opposite sign, and have a bonding energy of about 5 kcal/mole. Van der Waals bonds are the least energetic, having a bonding energy of about 1 to 2 kcal/mole, but they are also the least specific of all three types of biologically important, weak chemical bonds and can therefore link a greater variety of

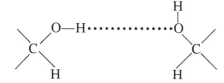

**Figure 2.2**
Hydrogen bonding involves sharing of a covalently-bonded hydrogen between two covalently-bonded oxygens, nitrogens, or one of each. The hydrogen usually is closer to one of the two atoms which share the proton.

interacting molecules. As organic compounds are made and degraded, assembled and disassembled, and interact in numerous ways in cells, each type of chemical bond provides some particular advantage in relation to a particular event or system in cellular chemistry.

## WATER

Life depends on water. From 60 to 95 percent of most cells consists of water, and even dormant seeds and spores contain from 10 to 30 percent water. There are a number of important properties of water that make it distinctive and uniquely suitable as the medium of cellular activities.

### Properties of Water

Covalent bonds are **nonpolar** if the negative charge of symmetrically distributed electrons is shared equally by the bonded atoms, or **polar,** if the electrical charge is closer to one atomic nucleus than to another in the electron-sharing unit. Water molecules exist as polar H—O—H units, whose shared electrons are closer to the oxygen than to the hydrogen nuclei. Furthermore, water molecules are **dipoles,** since there is a net positive charge at the end of the molecule carrying the two hydrogens and a net negative charge at the oxygen end of the aggregate.

Under physiological conditions, very few water molecules ionize to form H⁺ and OH⁻ ions. Instead, the hydrogen and oxygen atoms of the polar water molecules form strong hydrogen bonds. These bonds are directed tetrahedrally, so that each water molecule tends to have four nearest neighbors (Fig. 2.3). The arrangement of molecules is fixed in ice, since the bonds are very rigid between neighbors. In the liquid state, between 0°C and 100°C, there is enough energy of thermal motion to break hydrogen bonds, which allows continual change of nearest neighbors. At any given instant, however, most water molecules are held by four strong hydrogen bonds, even in the liquid state.

Water will dissolve an appreciable quantity of almost any molecule that carries a net charge or that has polar, especially dipole, groups. Any molecule with an

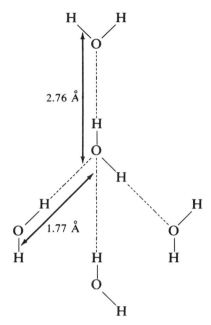

**Figure 2.3**
Hydrogen bonding around a water molecule in ice, showing the bond lengths involving the central oxygen atom and the oxygens of water molecules with hydrogens in two different regions of the tetrahedral arrangement.

asymmetrically distributed electron cloud, or permanent dipole such as O—H, is likely to be soluble in water because of attractive forces between it and the permanently dipolar water molecules. Hydrogen or other weak bonds usually form between the water molecules in the **solvent** phase and the **solute** molecules that are added to the water. The relative degree of solubility depends on how many water-water bonds can be disrupted when solute-water bonds replace them. Since water-water bonds usually are energetically more favorable (at a lower-energy state) than solute-water bonds, most compounds have only limited solubility in water. Compounds that readily form bonds with water molecules are called **hydrophilic.**

Organic compounds that cannot readily form hydrogen bonds are poorly soluble in water; that is, they are **hydrophobic.** When a nonpolar molecule, such as benzene, is mixed with water, there is a rapid separation of the water and benzene molecules. The water molecules form hydrogen bonds among themselves while the benzene molecules bind to each other by van der Waals forces or by hydrophobic interactions. The hydrophobic molecules are effectively separated in space from the water molecules, each compound existing in its own space and not mingling with the others.

If some electrically charged group such as a phosphate residue is added to a hydrophobic molecule, then that molecule becomes more readily soluble in water. The increased solubility is due to greater capacity for formation of hydrogen bonds and other weak bonds between the electrically charged residue and the dipole water molecules. Since water molecules are dipolar either or both of the charged regions can undergo bonding, depending on the particular charge of the solute molecule or region. It also follows that water is an excellent solvent for electrically charged ions or inorganic charged molecules, even when these are in the crystalline state.

### The pH Scale

Water can enhance the dissociation of substances such as weak acids and bases, that already exist in partially dissociated or ionized form. In addition, water itself can undergo slight dissociation to ionized components. Although it is not strictly true, for practical purposes we refer to $H^+$ and $OH^-$ ions as the **dissociation products of water.**

Dissociation of water is an equilibrium process and at constant temperature this can be expressed by

$$K_{eq} = \frac{[H^+][OH^-]}{[H_2O]} \qquad [2.1]$$

where $K_{eq}$ is the equilibrium constant and concentrations of water molecules and their ionized components are expressed in moles per liter (signified by enclosure in brackets [ ]). Since the molar concentration $(M)$ of water in pure water is 55.5 $M$ (number of grams of water in a liter divided by gram molecular weight of water, or 1000/18), and $H^+$ and $OH^-$ ion concentrations in pure water are very low ($1 \times 10^{-7} M$ at 25°C), we can simplify the equilibrium constant to 55.5 ×

$K_{eq} = [H^+][OH^-]$. The term 55.5 × $K_{eq}$ is called the **ion product of water** or the constant $K_w$.

$$K_w = [H^+][OH^-] \qquad [2.2]$$

$K_w$ is the basis for the **pH scale,** which is a means of designating the actual concentration of $H^+$ ions (and thus of $OH^-$ ions as well) in any aqueous solution in the biologically significant acidity range between 1.0 $M$ $H^+$ and 1.0 $M$ $OH^-$ ions (Table 2.1). The term **pH** is defined as

$$pH = -\log_{10}[H^+] \qquad [2.3]$$

It is convenient to use a logarithmic scale for pH values because of the wide variations in $H^+$ ion concentrations. The negative logarithm to the base 10 is used so that a positive scale of readings can be obtained. In a precisely neutral solution at 25°C, $[H^+] = [OH^-] = 1 \times 10^{-7} M$, and the pH of such a solution is

$$pH = -\log_{10}[H^+] = -\log_{10}(10^{-7}) = 7.0 \qquad [2.4]$$

The value of pH 7.0 for a neutral solution is thus derived from the ion product of water at 25°C and not from some arbitrary standard. In an acidic solution, the pH is less than 7.0 since the $H^+$ ion concentration is high, whereas in an alkaline solution, the pH is greater than 7.0 because the solution has a low $H^+$ ion concentration.

Because the pH scale is logarithmic, there is a tenfold difference in $H^+$ ion concentration between one pH unit and the next, a 100 times difference in $H^+$ ion concentration between two whole pH units, a 1000 times difference in $H^+$ ion concentration for a span of three pH units, and so forth. Measurements of $H^+$ ion concentration are made rapidly and routinely using a pH meter.

Cellular activities are extremely sensitive to even slight changes in internal pH, mostly because *enzyme activity* is affected by $H^+$ ion concentration. Enzymes have a maximal activity at some characteristic pH,

**TABLE 2.1 The pH scale and the molar concentrations of H$^+$ and OH$^-$ ions**

| H$^+$ ions (M) | pH | OH$^-$ ions (M) |
|---|---|---|
| 1.0 | 0 | $10^{-14}$ |
| 0.1 | 1 | $10^{-13}$ |
| 0.01 | 2 | $10^{-12}$ |
| 0.001 | 3 | $10^{-11}$ |
| 0.0001 | 4 | $10^{-10}$ |
| $10^{-5}$ | 5 | $10^{-9}$ |
| $10^{-6}$ | 6 | $10^{-8}$ |
| $10^{-7}$ | 7 | $10^{-7}$ |
| $10^{-8}$ | 8 | $10^{-6}$ |
| $10^{-9}$ | 9 | $10^{-5}$ |
| $10^{-10}$ | 10 | 0.0001 |
| $10^{-11}$ | 11 | 0.001 |
| $10^{-12}$ | 12 | 0.01 |
| $10^{-13}$ | 13 | 0.1 |
| $10^{-14}$ | 14 | 1.0 |

called the **optimum pH,** and their activities decline sharply above and below this optimum value. The striking effects of pH on enzyme activity almost certainly reflect electrical changes in surface groups of the enzyme molecules. Since such changes lead to alteration in enzyme molecule shape, which is critical to enzyme action, lowered reactivity of the enzyme results in solutions that have inappropriate pH values.

Variations in fractions of a pH unit may be damaging or even lethal to some cells. These fluctuations in pH are controlled by powerful *buffering action* of coupled **H$^+$ ion donors** (acids) and **H$^+$ ion acceptors** (bases) that are present in intracellular and extracellular fluids of living organisms. Buffered systems tend to resist changes in pH when H$^+$ and OH$^-$ ions are added. The principal buffering system in blood plasma of vertebrates is the donor-acceptor pair $H_2CO_3$—$HCO_3^-$. The pH of human blood plasma is closely regulated to a pH of about 7.40. Irreparable damage may occur if plasma pH falls below 7.0 or rises above 7.8. The difference between pH 7.4 and pH 7.8 in blood reflects a change in H$^+$ ion concentration of

only $3 \times 10^{-8}\ M$. The small magnitude of this change emphasizes the importance of pH-regulating mechanisms as precise modulators of acidity and alkalinity of cellular fluids.

## THE CHEMISTRY OF CARBON ATOMS

Modern organic chemistry is the chemistry of carbon. After Wöhler and other chemists of the early nineteenth century showed that organic compounds could be made in the laboratory as well as in living cells, it became clear that the uniqueness of organic compounds depended on special properties of the carbon atom and not on whether the chemical was produced in living cells. Modern biochemistry deals with the chemical dynamics of living systems, while organic chemistry today is concerned with all classes of carbon-based compounds, only some of which are biologically important.

Organic compounds are very versatile, existing in virtually unlimited numbers and having widely varied properties. They are quite stable, as shown by their relatively sluggish reactions with one another and with water or molecular oxygen. These trademarks of versatility and stability are due to particular features of carbon atom interactions, both with other carbon atoms and with the relatively few other elements that make up the cellular materials.

### Carbon Atom Interactions

Carbon atoms interact with other carbon atoms or with hydrogen, oxygen, nitrogen, phosphorus, sulfur, and the few other elements generally found in cellular organic molecules, by forming covalent bonds with these atoms. The basic framework of organic molecules is the carbon "skeleton" of atoms. These covalently bonded atoms can exist in various arrangements, including chains, rings, networks, or combinations of these forms.

Structural diversity of organic molecules is due to the fact that carbon has a **valency of four;** that is, there are only four electrons in the outermost electron orbital, which can hold up to eight electrons. Carbon can form covalent bonds with one to four other carbon atoms, as well as with electrically charged atoms of H, O, N, P, and S. Since carbons interact with each other, organic molecules can occur in various sizes and lengths. Because carbons can also interact with various other elements, many different combinations of a few kinds of atoms become possible in organic molecules of widely varying sizes. In addition to this basis for versatility, carbon atoms can form double and triple as well as single covalent bonds. An organic molecule or region of a molecule will have different properties if nitrogen is single-, double-, or triple-bonded to a carbon atom, even though it may be the same combination of carbon and nitrogen. The same holds true for other elements with a valency greater than one.

Since the carbon atom is **tetravalent** its outer electron orbital becomes completely filled when four covalent bonds are formed. The filled electron orbital is the energetically most favorable configuration, that is, the most stable state. Stability of the entire carbon framework of an organic molecule is therefore due in large measure to the energetically-satisfying configuration of its individual carbon atoms. The covalent interaction between carbon and atoms of oxygen, nitrogen, and hydrogen is strong, as shown by bond energies between 70 and 100 kcal/mole (Table 2.2). Still higher amounts of energy are required to break double and triple bonds involving carbon atoms. Organic molecules with double-bonded construction are therefore even more stable than carbon chains made of single covalent bonds. The tetrahedral distribution in

space of the valency of four in the carbon atom confers a basic geometrical symmetry to the atom (Fig. 2.4). This symmetry also contributes to greater stability of bonding with other atoms.

### Isomers

The tetrahedral nature of the carbon valency can lead to **asymmetry** in many organic molecules. When a carbon atom is bonded to four different atoms or groups of atoms, two spatial arrangements of the molecule can be constructed with one alternative being the mirror image of the other (Fig. 2.5). The structure of such asymmetric molecules is usually drawn so that two sides of the tetrahedron face the viewer and the other two sides lie away from the viewer. The alternative configurations of the same compound are called **isomers** or stereoisomers. If the two members of a mirror-image pair are exposed in solution to polarized light, each may be able to rotate the plane of polarized light equally but in opposite direction. In this case the mirror-image pair of a compound are called **optical isomers.**

The potential asymmetry of *each* carbon atom in an organic molecule is significant because the same compound may occur in a number of isomeric forms, each having different properties. This feature also

**TABLE 2.2  Values for some covalent bond energies\***

| Single Bonds | | Double Bonds | | Triple Bonds | |
|---|---|---|---|---|---|
| O—H | 110 | | | | |
| H—H | 104 | | | | |
| C—H | 99 | | | | |
| C—O | 84 | C=O | 170 | | |
| C—C | 83 | C=C | 146 | C≡C | 195 |
| C—N | 70 | C=N | 147 | C≡N | 212 |

\*Energy (kcal/mole) required to break the bond.

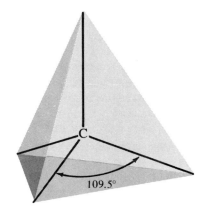

**Figure 2.4**
The tetrahedral nature of carbon is a result of the tetravalency of the atom disposed in space in particular angles and lengths.

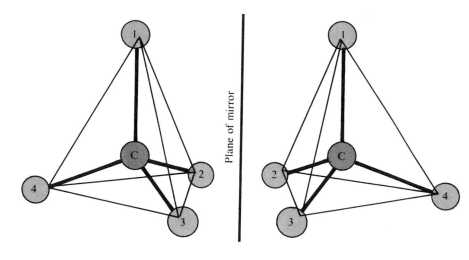

**Figure 2.5**
Isomers of a compound are found when a carbon atom is bonded to 4 different atoms or groups of atoms. The alternative forms of the compound are mirror images of one another.

contributes to the versatility of carbon-based compounds. In biological systems, most organic molecules assume only a fraction of the number of possible isomer configurations.

### Biologically Important Carbon Compounds

The four major classes of organic compounds that are important in biological systems are carbohydrates, lipids, proteins, and nucleic acids. Except for lipids, many of the molecules in these classes are **polymers** made up of repeating **monomer** subunits, or building blocks. Some kinds of carbohydrates consist of only one kind of monomer; for example, starch is built from numerous glucose units. Proteins, on the other hand, may be constructed from twenty different kinds of amino acid units. Molecular weight of a polymer may range up to hundreds of millions of **daltons** (units of molecular weight; each dalton is approximately equal to the weight of one hydrogen atom).

All four classes of cellular organic compounds contribute to cell structure, function, and regulation of cellular activities. Each class makes unique contributions, however, so that all these groups must be present if the cell as we know it is to exist. In the following sections we will look at each of these four classes and note their characteristics and their importance in living systems.

### CARBOHYDRATES

Carbohydrates are compounds of the general formula $(CH_2O)_n$. Some biologically important carbohydrate derivatives also contain nitrogen and sulfur atoms. Among all carbohydrates in living systems, the most widespread molecule is glucose. Glucose is a simple sugar containing 6 carbon atoms as its basic skeleton. Glucose and other simple sugars that are single units are called **monosaccharides.**

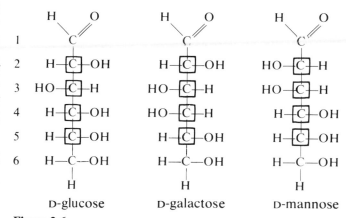

**Figure 2.6**
Three naturally occurring isomers of the hexose sugar D-glucose. There are asymmetric carbon atoms at positions 2, 3, 4, and 5 in the molecule.

### Monosaccharides

Monosaccharides are classified according to the number of carbon atoms in the molecule; a **hexose**, like glucose, has 6 carbons; a **pentose** has 5, a **triose** has 3, and so on. Because at least some of their carbon atoms are asymmetric, monosaccharides can occur in a number of isomeric forms. Glucose has 4 asymmetric carbons, leading to 16 possible stereo-isomers. Only 3 of these isomers, however, are found in nature (Fig. 2.6).

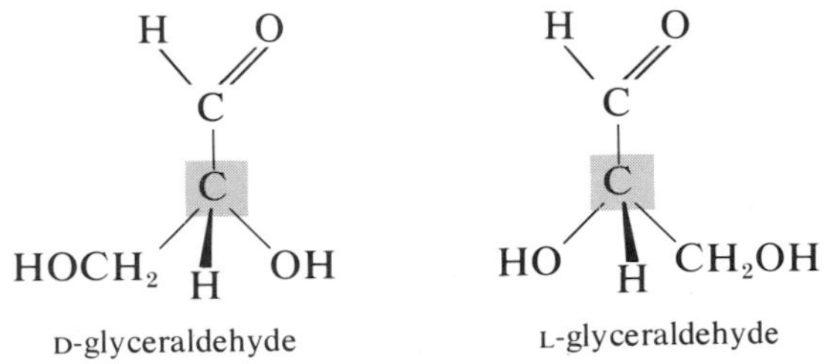

Monosaccharides are identified as D- or L-forms according to a convention based on the configuration of the triose D-glyceraldehyde. A monosaccharide is of the D-variety if the hydroxyl group on the bottommost asymmetric carbon atom is on the right, when the carbonyl group, $>C=O$, is at the opposite end of the formula, as in D-glyceraldehyde (Fig. 2.7). If this hydroxyl group is on the left, then the molecule is the L-form. Glucose in its naturally-occurring form is a D-hexose according to this convention.

Pentose and hexose sugars in solution exist largely in a ring form in equilibrium with a small amount of the linear form of the molecule (Fig. 2.8). When the oxygen bridge is located between carbons 1 and 5, it is

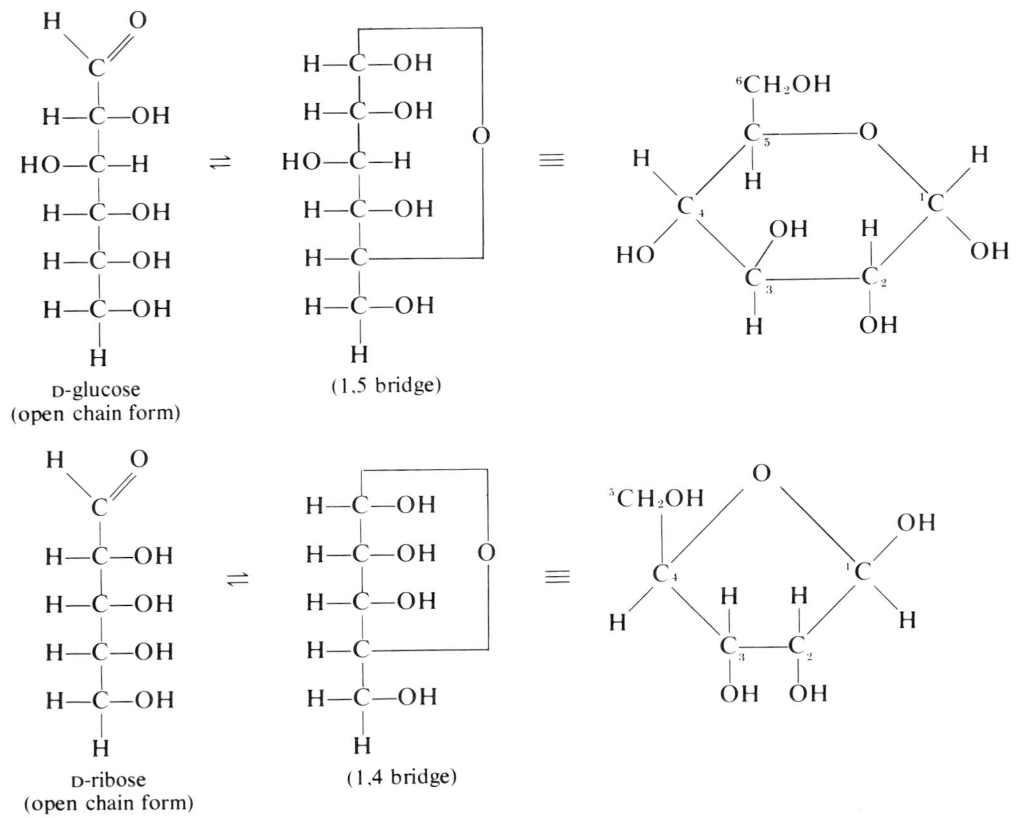

**Figure 2.7**
The D- and L-isomers of gylceraldehyde are shown in relation to the asymmetric carbon atom of the molecule.

**Figure 2.8**
The linear and predominant ring forms of pentose and hexose sugars exist in solution in an equilibrium mixture. The conventional hexagon and pentagon formulations refer to the ring form of the molecules.

conventional to depict the molecule as six-sided, or hexagonal. When there is a 1,4 oxygen bridge, the molecule is drawn as a pentagon. These notations show a heavier line across the bottom of the pentagon or hexagon figure to indicate that this is the part of the molecule nearest the viewer, and the plane of the molecule is perpendicular to the plane of the paper. The hydrogens and hydroxyls are oriented up or down and are shown by vertical lines at each of the carbons (Fig. 2.9). These conventional representations, called Haworth formulae, make it very simple to specify and recognize a particular isomer at a glance. Monosaccharide rings can have only 5 or 6 carbon atoms, because of spatial restrictions. A 7-membered carbon chain would be subject to excessive strain if it were bent into ring form.

   With formation of the oxygen bridge, the ring form of the molecule gains asymmetry at carbon atom 1. The hydroxyl of carbon-1 is either adjacent to the hydroxyl of carbon-2 or is rotated 180° relative to it (Fig. 2.10). These isomers are called the $\alpha$- and $\beta$-

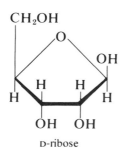

D-glucose

D-ribose

**Figure 2.9**
Haworth formulae depicting 5- and 6-membered ring forms of sugars.

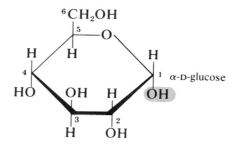

$\alpha$-D-glucose

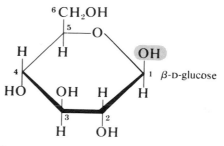

$\beta$-D-glucose

**Figure 2.10**
$\alpha$- and $\beta$-isomers of D-glucose are distinguished on the basis of the position of the hydroxyl group at carbon-1 relative to the hydroxyl group at carbon-2 of the molecule.

forms of the cyclic monosaccharides. Polymers made up of $\alpha$-units are profoundly different from polymers constructed from $\beta$-units of the same monosaccharide building block.

### Polysaccharides

   Polysaccharides, which have the general formula $(C_6H_{10}O_5)_n$, are polymers that form on condensation of smaller units. Each link between units is called a **glycosidic bond.** These may be $\alpha$- or $\beta$-glycosidic bonds, depending on whether the hydroxyl group of carbon-1 is in the $\alpha$- or $\beta$-conformation in the monosaccharide units. Monomers can be joined more directly by $\alpha$-linkages whereas a $\beta$-glycosidic bond requires a rotation of 180° for the hydroxyl of one unit to come into an appropriate spatial relationship to the hydroxyl of its neighbor monomer (Fig. 2.11). These linkages are physiologically important for at least three reasons:

1. They provide a means for joining two or more subunits in construction of a variety of larger molecules with different functions and specificities.

2. Different enzymes attack $\alpha$- and $\beta$-glycosidic links allowing the cell additional ways of discriminating among compounds used in structure and function.
3. The molecular potential in the cell is different for each type, with $\alpha$-glycosides being readily mobilized for metabolism and $\beta$-glycosides contributing to formation of many structural, stable molecules.

Polysaccharides usually contain hundreds or thousands of monosaccharide residues even though as few as 9 or 10 of these subunits are enough to define a molecule as the polymer form. Compounds containing from 2 to about 8 monosaccharide units are called **oligosaccharides** (Gr. *oligos:* few). The most important of these are the two-sugar compounds, or **disaccharides,** such as sucrose (table sugar), lactose (milk sugar), and maltose (a degradation product of starch, Fig. 2.12).

The major functions served by polysaccharides in living systems are concerned with food storage and with structure. Two major food reserve polysaccharides in eukaryotes are **starch** and **glycogen.** Both of these foods are hydrolyzed to their constituent glucose monomer units by specific enzyme actions. Starch is deposited in large granules in the chloroplasts of some green cells or in colorless plastids (leucoplasts) of root,

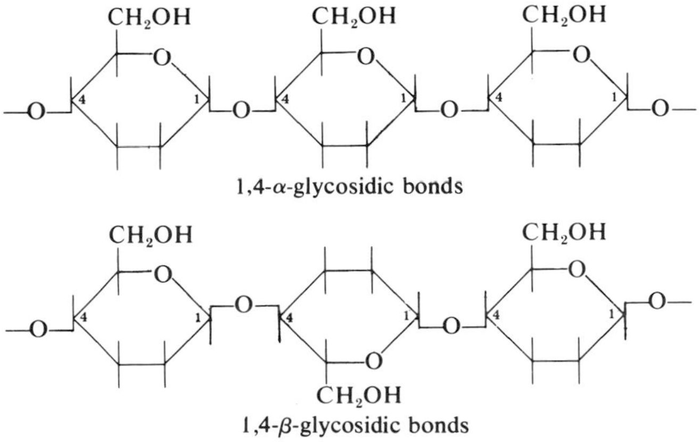

**Figure 2.11**
Polysaccharide fragments showing $\alpha$- and $\beta$-glycosidic bonds; in these examples glycosidic bonds occur between $C_1$ and $C_4$ of adjacent monomer units.

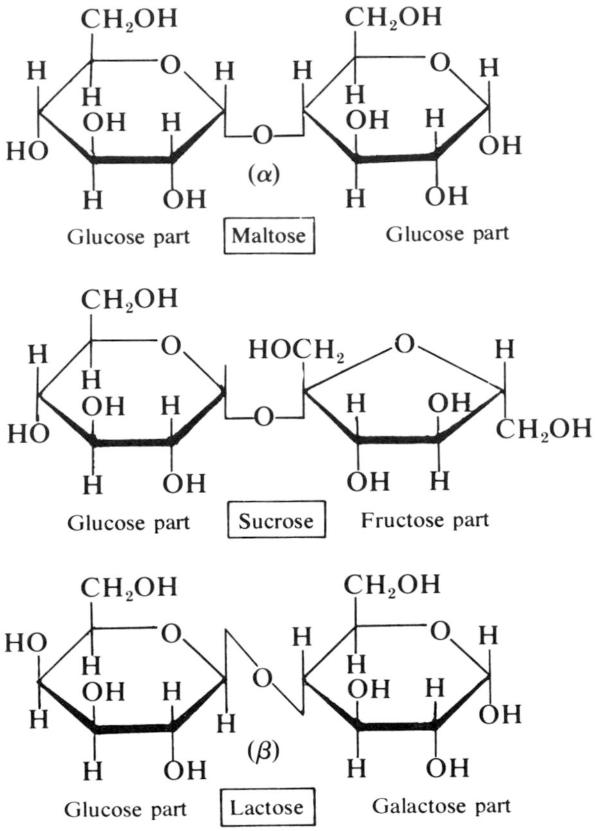

**Figure 2.12**
Three common disaccharide sugars.

stem, and other plant tissues. These starch granules can be seen with the light microscope and identified in several ways, including a simple staining test using an iodine solution. There are two types of molecules in the starch granule, both of which are constructed using $\alpha$-glycosidic bonds. One kind of molecule is made of 250–300 glucose units in an unbranched chain. The other type of starch molecule has approximately 1000 glucose residues in a long chain with some branching along its length.

While starch is a major form of stored food in algae and land plants, glycogen is the principal food reserve in animals and fungi. Glycogen is deposited in the cytoplasm rather than within an organelle as is starch. Glycogen occurs in particularly large amounts in the liver of animals. The liver glycogen polymer is a long branched chain containing about 30,000 glucose units connected by $\alpha$-glycosidic bonds.

Structural polysaccharides in eukaryotes include **cellulose** and **chitin,** which have $\beta$-glycosidic links between monomer subunits. Cellulose usually occurs along with other materials in the cell wall of most algae and all higher plants, as well as in certain fungi and protists. About 8,000 glucose units are arranged in the long unbranched polymer chain. These polymers fold by hydrogen bonding in such a way as to form long fibrils aggregated into bundles easily seen with the electron microscope (Fig. 2.13). Chitin is a nitrogen-containing polysaccharide that is constructed of $N$-acetylglucosamine residues joined by $\beta$-glycosidic bonds (Fig. 2.14). These monomers are derivatives of glucose. The chitin found in cell walls of many fungi and in the rigid exoskeleton of insects, crustacea, and certain other invertebrate animals is chemically identical.

The polysaccharide chains in bacterial cell walls are long unbranched molecules made of sugars joined by $\beta$-glycosidic linkages. These polysaccharide chains are connected by cross-links made of 4–5 amino acid units, so that the basic structure is a sheet composed of these different kinds of monomers in a specific organizational framework. There is considerable variation among bacterial species in the amino acid cross-links, but the carbohydrate portions of the cell-wall framework are essentially the same throughout this prokaryotic group. There are other kinds of molecules in bacterial cell walls in addition to the basic sheet described above, all of which contribute to the complexity of organization and function of cell walls in bacteria.

## LIPIDS

Lipids are a diverse group of substances that have the common property of solubility in nonpolar, organic solvents, such as ether and alcohol. Among the many

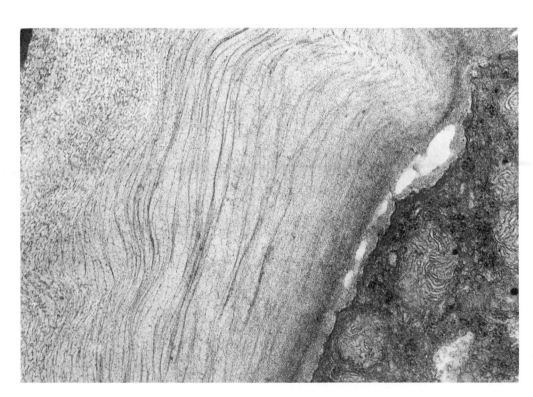

**Figure 2.13**
Electron micrograph showing parallel bundles of cellulose fibers in the cell wall of wheat *(Triticum aestivum).* $\times$ 37,000. (Courtesy of M. C. Ledbetter)

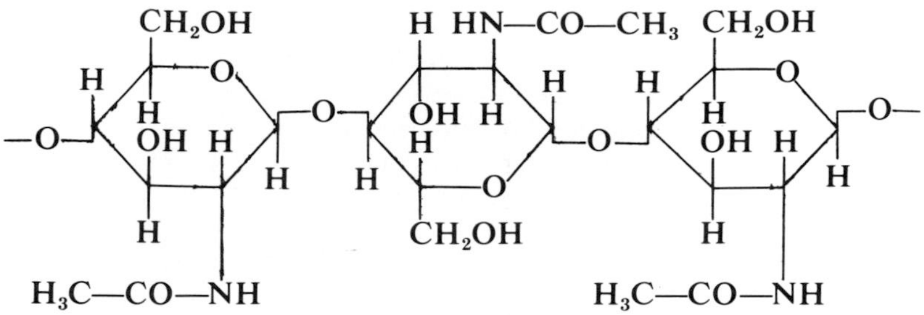

**Figure 2.14**
Portion of a chitin molecule showing 1,4-β-glycosidic links between *N*-acetylglucosamine monomer units that make up the polymer.

N-acetylglucosamine unit

types of lipids we can include fats, fatty acids, waxes, steroids, phospholipids, glycolipids, sphingolipids, and terpenes, as selected examples of biologically important substances.

### Fatty Acids

Naturally-occurring **fatty acids** are unbranched hydrocarbon chains with a carboxyl group at one end (Fig. 2.15). Since fatty acids are synthesized from 2-carbon precursor molecules, they usually have an even number of carbon atoms, the commonest numbers being 16 and 18. When all carbon atoms of a fatty acid chain are joined by single covalent bonds the compound is "saturated" (with hydrogens at sites not involved in the —C—C—chain). An "unsaturated" fatty acid has one or more double bonds between carbons in the backbone of the chain (Fig. 2.16).

The carboxyl end of the fatty acid chain is water-soluble and highly polar, while the hydrocarbon portion of the chain is water-insoluble and highly nonpolar. When fatty acids interact with water, the soluble carboxyl end is contained within the water as a layer while the hydrocarbon tails of these molecules remain outside the water surface (Fig. 2.17). Although fatty acids occur in trace amounts in cells and tissues they are important as building blocks of several classes of lipids.

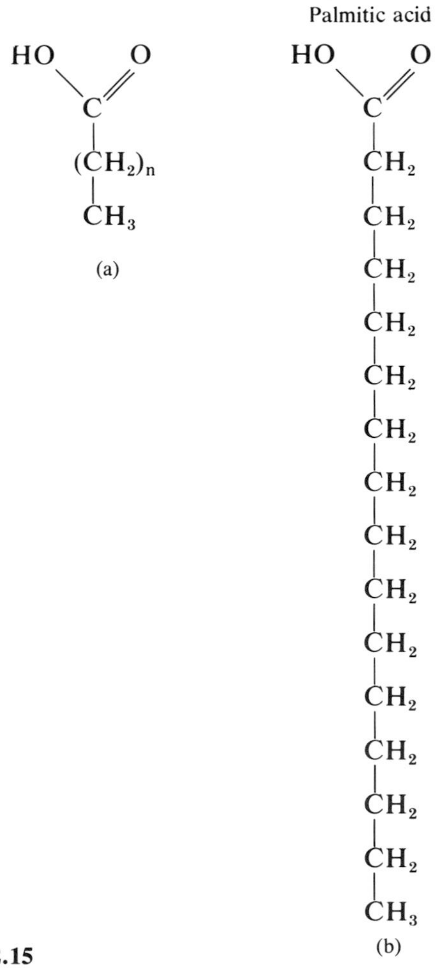

### Neutral Fats

When the 3-carbon alcohol **glycerol** interacts with three fatty acid molecules in a dehydration reaction, **neutral fats,** or triglycerides, are formed (Fig. 2.18). One, two, or three different kinds of fatty acids can combine with the hydroxyl groups of glycerol to form

Palmitic acid

**Figure 2.15**
The saturated fatty acid: (a) general nature of the molecule with a carboxyl group at one end and a long chain of carbon atoms fully saturated with hydrogens; and (b) the molecule of palmitic acid, whose formula is $C_{16}H_{32}O_2$, or $COOH—(CH_2)_{14}—CH_3$.

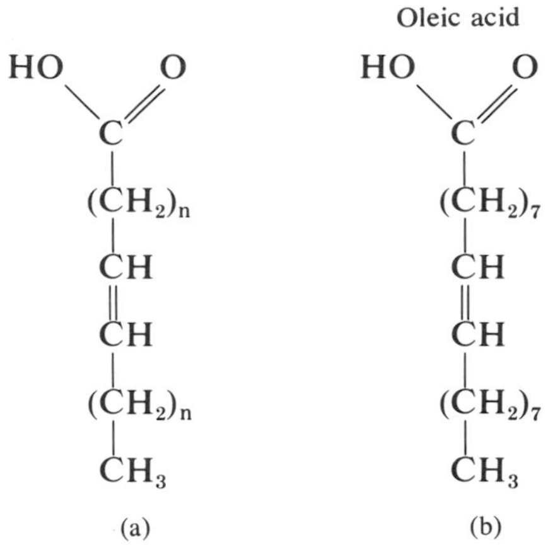

Oleic acid

(a)          (b)

**Figure 2.16**
The unsaturated fatty acid: (a) general type of molecule having only one double bond in the hydrocarbon chain; and (b) oleic acid, whose formula is $C_{18}H_{34}O_2$, or $COOH-(CH_2)_7-CH=CH-(CH_2)_7-CH_3$.

boiled in the presence of alkali, a practical method for home manufacture still continued in some societies today.

Oils liquefy at room temperature. The particular melting point depends on the degree of saturation of the constituent fatty acids, with higher melting points for higher levels of saturation; that is, more —C—C— bonds and fewer —C=C— bonds in the hydrocarbon chains. Vegetable oils are saturated to convert them to the "hard" fat form of margarine.

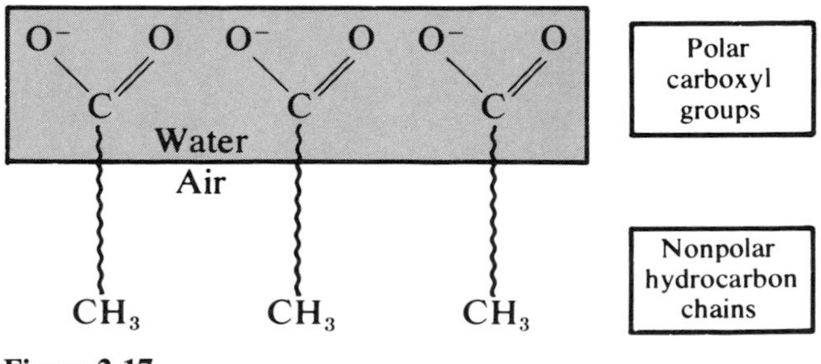

a molecule of fat. These fats are a major storage form of lipids in both plants and animals. They may be formed after suitable metabolic reactions from excess carbohydrate, protein, or lipid in cells and tissues.

Neutral fats are nonpolar molecules and insoluble in water, since the carboxyl groups of the fatty acid components are no longer accessible for molecular interactions. Neutral fats are digested in cells by the action of enzymes called **lipases,** or outside living systems they may be hydrolyzed when boiled with acids or bases. Animal fats can be converted to soap if

**Figure 2.17**
When fatty acids interact with water, the polar carboxyl end of the fatty acid undergoes hydrophilic interaction with water molecules, while the nonpolar hydrocarbon chain is sequestered away from water because of its hydrophobic nature.

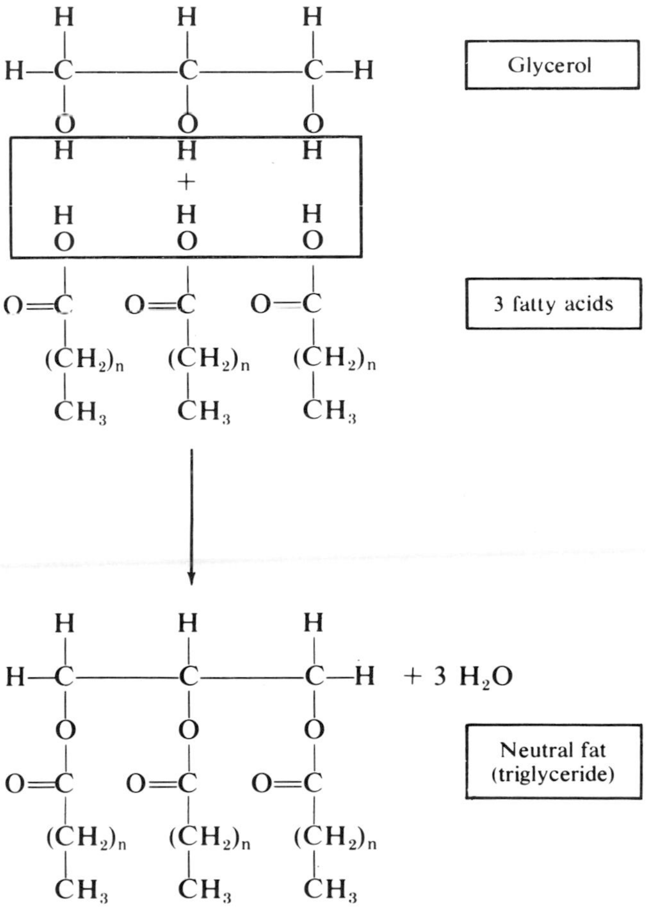

**Figure 2.18**
Neutral fats form by a dehydration reaction between the alcohol glycerol and the carboxyl groups of fatty acid chains, producing a fatty acid ester of glycerol. This molecule is a triglyceride.

### Phospholipids

Phospholipids are a major constituent of cellular membranes. Fatty acids are bonded to two of the hydroxyl groups of glycerol while a phosphoric acid residue is present, instead of a fatty acid, at the third hydroxyl of the glycerol group (Fig. 2.19). Phospholipids share the common property of having a hydrophobic "tail" consisting of two fatty acid chains, and a hydrophilic "head" made up of a negatively charged phosphoric acid residue to which a positively charged group is usually bonded. Phospholipids are **amphipathic** molecules, having both hydrophobic and hydrophilic regions in the same molecule. Phospholipids can therefore serve as vital structural links between the watery and nonwatery phases outside and inside the cell. They also play a functional role in the activities of some enzymes.

Some major phospholipids of membranes are listed in Table 2.3. When some phospholipids interact with water in a water-air system, they aggregate spontaneously into a single layer of molecules or, more usually, into two layers each of which is one molecule thick. In these bimolecular sheets the hydrophilic ends of the molecules are in the water and the hydrophobic tails are in the air phase of the system (Fig. 2.20). These aggregates resemble cell membranes in some ways. They are even more like cellular membranes if proteins are added to the phospholipids and then combined with water. The amphipathic properties of phospholipids are therefore important in *membrane conformation* as well as *membrane functions* in interaction with water, proteins, and lipids.

### Sphingolipids and Glycolipids

Sphingolipids and glycolipids resemble phospholipids in having an amphipathic construction based on hydrophobic "tail" and hydrophilic "head" residues. These two classes of lipids also occur in membranes, but they are more restricted in cellular distribution

**Figure 2.19**
The phospholipid compound ethanolamine phosphoglyceride, or phosphatidyl ethanolamine, a major component of cell membranes. The three carbon atoms of the glycerol region are shown in gray.

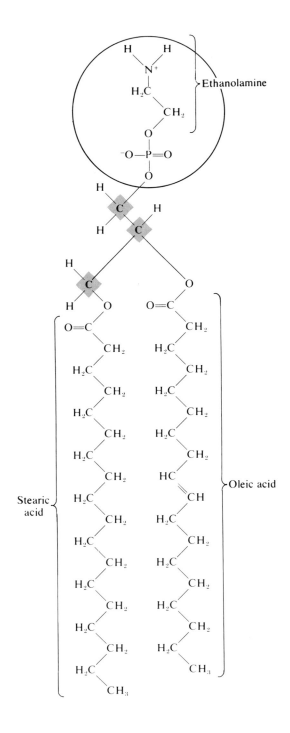

**TABLE 2.3  Major types of lipids in cells**

| Lipid Group | Some Major Types | Important Cellular Location |
|---|---|---|
| Fatty acids | Oleic acid, palmitic acid, stearic acid | Cytosol, mitochondria, glyoxysomes of fatty seeds |
| Glycerides (neutral fats) | Coconut oil, beef tallow | Fat storage depots |
| Phosphoglycerides (phospholipids) | Ethanolamine phosphoglyceride, choline phosphoglyceride, cardiolipin | Membranes |
| Sphingolipids | Sphingomyelin | Membranes |
| Glycolipids | Cerebrosides, gangliosides | Membranes |
| Steroids | Cholesterol | Membranes |
| Terpenes | Essential oils, carotenoids | Plant cytosol, chloroplasts |

than are phospholipids. Sphingolipids are found in plant and animal membranes, but are especially prominent components of cell membranes in brain and nerve tissues; **sphingomyelin** is the most abundant compound in this group. All sphingolipids lack a glycerol component, and they contain a hydrophobic **sphingosine** residue in place of one of the two fatty acid chains (Fig. 2.21).

Glycolipids are distinguished by having polar hydrophilic carbohydrate groups in the "head" region of the molecule, usually D-glucose or D-galactose. Two important classes of these compounds are **cerebrosides** and **gangliosides.** These compounds contain both carbohydrate "head" residues and sphingosine "tail" residues, and are therefore often referred to as glycosphingolipids. Cerebrosides are found particularly in the myelin sheath of nervous tissue. Gangliosides are abundant in the outer surface of the plasma membrane of nerve cells, and are important factors in cell-surface phenomena, including immune responses.

### Steroids and Terpenes

Lipids that cannot be converted to soaps (nonsaponifiable) include steroids and terpenes. Both types are derived from common 5-carbon building blocks and are therefore related groups of compounds. The most familiar steroids are male and female sex hormones in vertebrates, bile acids, and adrenocortical hormones. Most steroids occur in trace amounts, but **sterols** are a relatively abundant class of

these compounds. Sterols are steroids that occur as free alcohols or as long-chain compounds containing fatty acids. The most abundant of these substances in animal tissues is **cholesterol,** which is found particularly in the plasma membrane (Fig. 2.22). Plants and fungi contain other kinds of sterols; sterols have not been found in bacteria.

Terpenes are especially evident constituents of certain plant species, and are responsible for char-

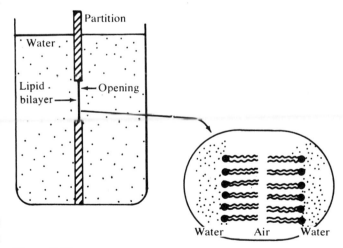

**Figure 2.20**
When phospholipid molecules interact with water, a spontaneous aggregation into a bimolecular layer occurs. The hydrophilic heads of the amphipathic phospholipid interact with water, and the hydrophobic tails of the molecules are sequestered away from water and within the air phase of the system.

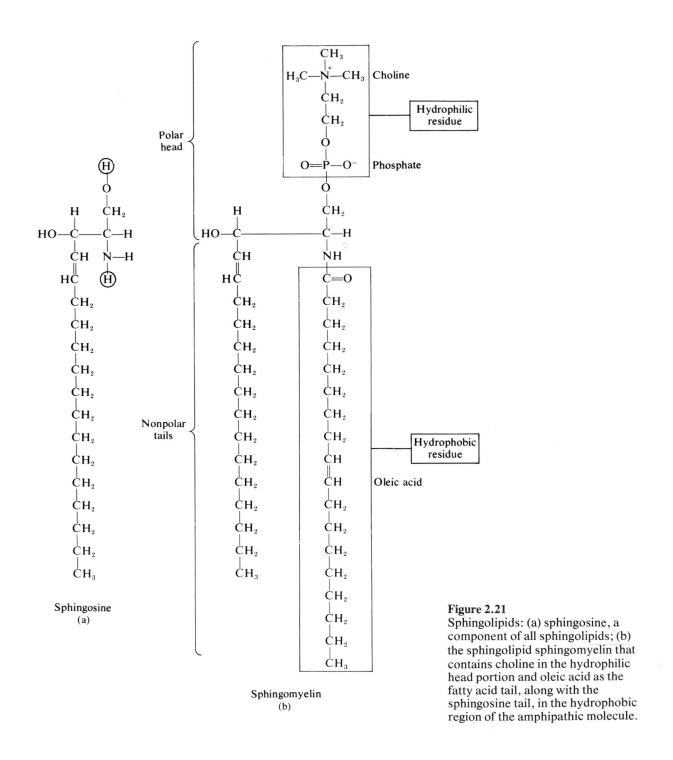

Sphingosine
(a)

Sphingomyelin
(b)

**Figure 2.21**
Sphingolipids: (a) sphingosine, a component of all sphingolipids; (b) the sphingolipid sphingomyelin that contains choline in the hydrophilic head portion and oleic acid as the fatty acid tail, along with the sphingosine tail, in the hydrophobic region of the amphipathic molecule.

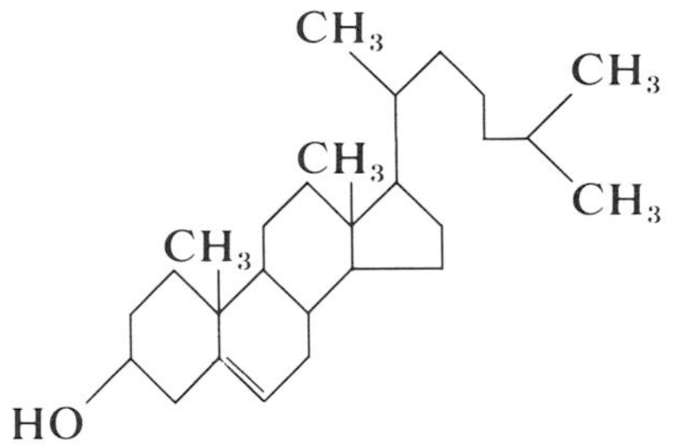

**Figure 2.22**
Cholesterol, a sterol member of the steroid group of lipids and an important consituent of animal cell plasma membrane.

acteristic odors and flavors. They are a major component of "essential oils" derived from such plants; for example, the terpenes camphor, limonene, and menthol occur in oil of camphor, lemon, and mint, respectively. A terpenoid alcohol called phytol is part of the molecule of the green pigment chlorophyll. Some other terpenoid components in plants are natural rubber, and carotenoid pigments that absorb light energy in photosynthesis and also contribute to yellow and orange colors of carrots, autumn foliage, and other materials. The fat-soluble vitamins A, D, E, and K are synthesized from precursors of the same 5-carbon unit as are the terpene compounds. Carotenes from plants are precursors of Vitamin A, which is found only in animals.

## PROTEINS

The enormous variety of proteins serve as major building materials and as regulatory molecules that control the many activities of living systems. Major types of fibrous proteins that act as *structural* elements include **actin** and **myosin** of muscle and other contractile systems, **collagens** that form connective ligaments within the body, **keratins** that form protective coverings such as skin, hair, claws, horns, feathers, and other structures of land vertebrates, and a number of other compounds. Proteins that *regulate* the numerous processes and activities of the organism include **enzymes** that modulate chemical reactions, **antibodies** that provide immunity against infection, **hor-**mones, and various other substances that make each life form respond appropriately to the constantly changing internal and external environments. Regulatory proteins are globular in shape, in contrast with fibrous structural proteins. These conformations have an important bearing on the character of the molecule and its activity or function, as we shall see shortly.

### Amino Acids

There is a common plan of construction for the thousands of kinds of proteins in living systems. The 20 kinds of naturally-occurring **amino acid** monomers are strung together in unbranched, linear polymer chains of proteins. These are the 20 amino acids specified in the genetic code that is universal to all organisms. Some other kinds of amino acids are also found in cells, but they are either degradation products or residues that have been modified from one of the 20 commonly occurring amino acids after the latter has been inserted into the polymer chain. Hydroxyproline is a major amino acid constituent of collagen in connective tissue, but proline residues are initially included in the protein and are converted to hydroxyproline after polymerization. Hydroxyproline is not one of the encoded amino acids, but proline is. Many proteins contain fewer than 20 kinds of amino acids. The relative proportions and the absolute numbers of the amino acid repertory vary from one protein to another, as a reflection of the specific information in genes, which are the blueprints for protein construction.

All 20 amino acids have the same basic structure (Fig. 2.23). There is a carboxyl group (—COOH) and an amino group (—$NH_2$) joined to the first, or $\alpha$-carbon atom, and a hydrogen atom as a third unit bonded to this carbon in all amino acids. Except for glycine, which has a second hydrogen joined to the $\alpha$-carbon, the other 19 amino acids have a fourth group that differs from the other three joined to the first carbon atom. Except for glycine, therefore, the amino acids have an asymmetric $\alpha$-carbon atom, and they also are optically active molecules. The L-isomer of these 19 amino acids occurs in almost all natural proteins, al-

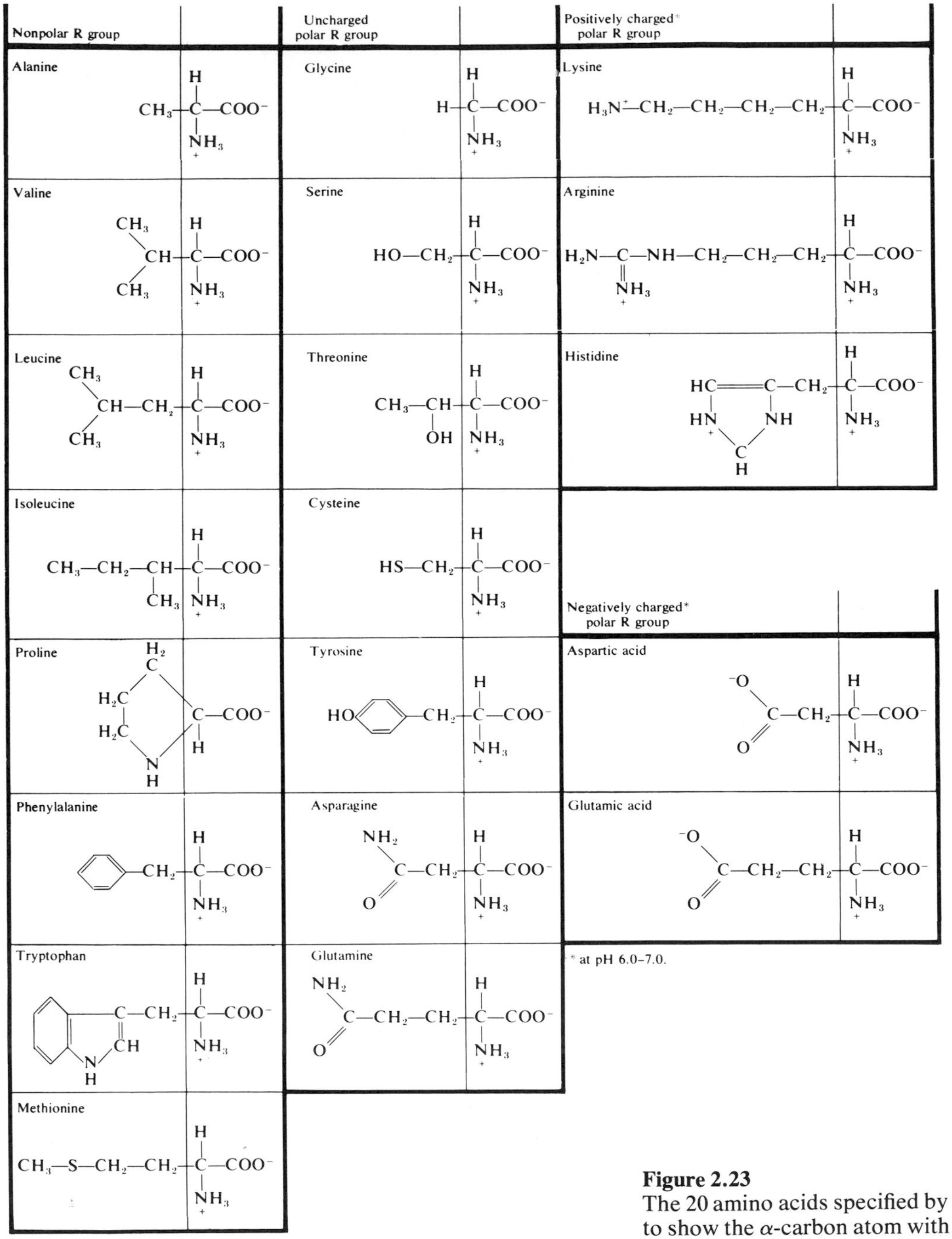

**Figure 2.23**
The 20 amino acids specified by the genetic code, arranged to show the $\alpha$-carbon atom with its variable R groups and common residues at the other three valency positions.

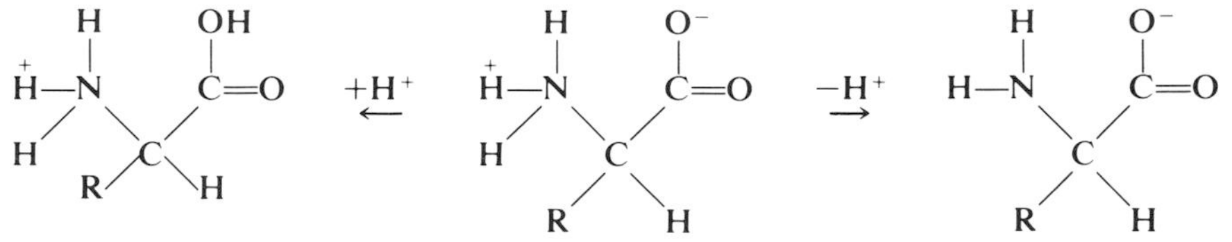

**Figure 2.24**
Amino acids have "zwitterion" properties (groups with positive and negative charges are present simultaneously) and can neutralize added $H^+$ ions in acid solution or $OH^-$ ions in basic solution.

though some D-amino acids have been found in certain molecules from plants and bacteria.

The side-chain differences are responsible for the varying properties of different amino acids, and for particular properties of the proteins in which they occur. When individual amino acids are in solution at pH 7.0, added alkali or acid is neutralized by the amino acid. The amino acid is a "zwitterion" with simultaneous negatively- and positively-charged groups (Fig. 2.24). These effects disappear when amino acids condense to form a **peptide** unit because of the nature of the bond that joins amino acid monomers into a linear chain in the dehydration reaction. Formation of a **peptide bond** or **amide** involves the linking of an amino group of one amino acid with the carboxyl group of the adjoining amino acid (Fig. 2.25). Since this linkage involves one charged group of each amino acid, the "zwitterion" property no longer exists for individual amino acids in the chain.

An amino acid polymer may still display acidic or basic properties because of the presence of acidic or basic side-chains in its constituent amino acid units. At pH 7.0, aspartic acid and glutamic acid residues confer acidic properties on a polymer region, while the positively charged polar groups of histidine, arginine, and lysine contribute basic properties to a protein.

When acid is added to protein in solution, a net positive charge develops because —COO⁻ changes to —COOH. Addition of base causes —NH₃⁺ to change to —NH₂, leaving the protein with a net negative charge. At a particular intermediate pH called the **isoelectric point** there are equal numbers of positive and negative charges in the protein, and the net charge is zero (Fig. 2.26). Positively- and negatively-charged proteins migrate in an electrical field, but proteins at their isoelectric point do not move toward either charged terminal. Proteins are most easily precipitated from solution by appropriate solvents when in solution at their isoelectric pH, the pH at which the protein is at its minimum solubility. Most proteins have an isoelectric point on the acid side and therefore carry a net negative charge at physiological pH. Basic proteins, such as **histones** found in chromosomes, carry a net positive charge under normal cellular conditions. Relatively high amounts of lysine and arginine contribute to the histone positive charge and to chro-

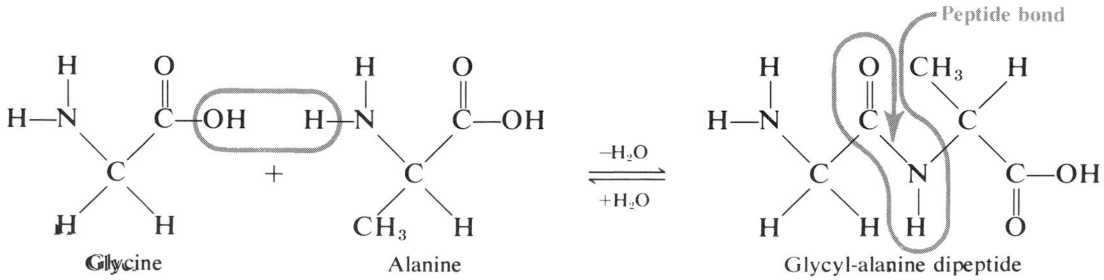

**Figure 2.25**
Amide linkage formation between adjacent amino acids in a peptide occurs as the carboxyl group of one amino acid joins with the amino group of its neighbor amino acid in a dehydration reaction (toward the right).

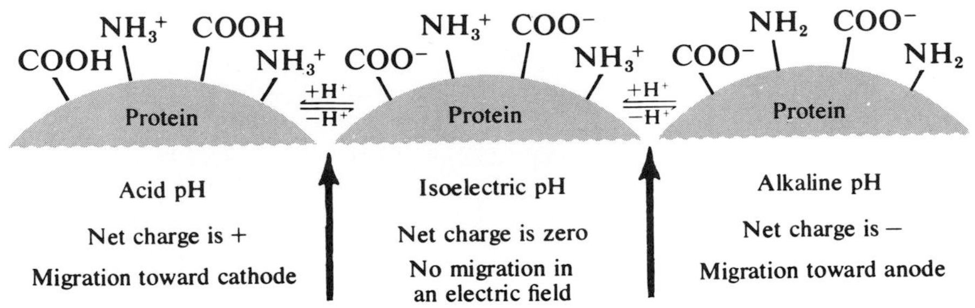

**Figure 2.26**
Migration of proteins in an electrical field depends on the net charge of the molecule. The net charge is zero at the isoelectric pH, even though the number of positive and negative charges is at the maximum for the molecule.

mosome structure, since positively charged histone proteins readily bind with negatively charged DNA to form highly stable nucleoprotein complexes.

## Polypeptides

The formation of a peptide bond is one part of the very complex sequence of events during protein synthesis at ribosome surfaces. These processes are discussed in Chapter 8. A dipeptide is formed when two amino acids are joined by amide linkage, a tripeptide involves three amino acids, and a **polypeptide** contains a large number of amino acid monomers, perhaps as many as 1000. The same peptide linkage that joins two amino acids is responsible for the addition of each

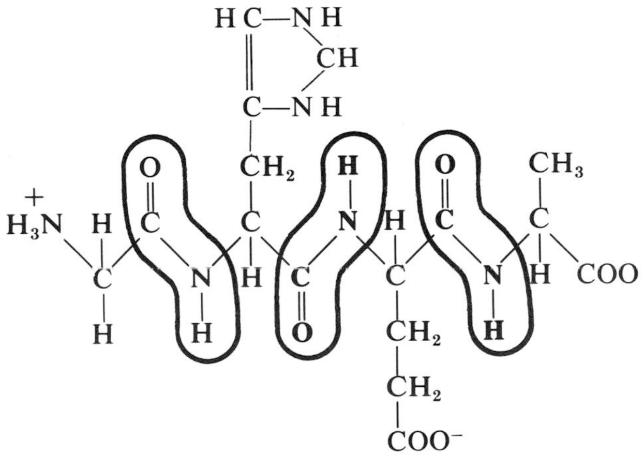

glycyl—histidyl—glutamyl—alanine
(at pH 7)

**Figure 2.27**
The amide link or peptide bond is the same throughout the length of a peptide, whether there are two or two hundred amino acids in the molecule.

adjoining amino acid in the linear chain (Fig. 2.27). Most polypeptides are long-chain molecules, but some biologically important peptides may have only 8–10 amino acids per molecule. The hormone **oxytocin** consists of only 8 amino acids. This molecule influences muscle contractions during the labor stages of birth and during suckling in mammals.

Proteins may be composed of one or more polypeptide chains held together by various forces in the functional protein molecule. In some cases the individual polypeptides are the same, while other proteins may be composed of two or more kinds of polypeptides. While proteins may range into millions of daltons molecular weight, their constituent polypeptide chains have molecular weights in an average range of 15,000 to 100,000 daltons. Insulin is an unusually small protein, with a molecular weight of about 6,000 daltons, yet it is constituted of two polypeptide chains having 21 and 30 amino acid units, respectively. The enzyme ribonuclease is larger, with a molecular weight of 13,700 daltons, but contains only one polypeptide chain of 124 amino acids. Insulin and ribonuclease were the first two proteins whose amino acid composition and exact sequence were described in the 1950s. The historic dissection of the insulin molecule by Frederick Sanger, reported in 1953, opened the way to detailed analysis of other proteins and to the realization of how important this knowledge was in understanding protein structure in relation to function and gene action.

## Protein Structure

Each kind of polypeptide consists of a unique sequence of amino acids, called the **primary structure** of the molecule. This primary structure is important for at least two reasons. First, the primary structure determines the three-dimensional conformation of the protein and in that way, the cellular role of the molecule. Second, the primary structure of a protein is a *co-linear translation* of the sequence of nucleotides in DNA and therefore provides crucial information about genetic input to protein synthesis.

Sometimes there is a drastic effect if even one amino acid is changed in protein primary structure. This is true for hemoglobin which is made up of two $\alpha$-chains and two $\beta$-polypeptide chains per molecule. People with the inherited sickle cell anemia disease have $\beta$-chains in which one of the 146 amino acids is substituted. Instead of glutamic acid, valine has been substituted at position number 6 of the amino acid chain of 146 monomers, leading to changes in amino acid interactions within the molecule and to altered shape of the protein. The change in molecular shape interferes with binding between an active group in the protein and molecular oxygen in the bloodstream.

This sort of profound effect from one amino acid substitution is not typical. For example, as many as 40 amino acids out of 104 may be different in a respiratory enzyme found in yeast and in human beings, yet the protein functions equally well in both species. Fewer than 40 substitutions have occurred in this same protein during evolution in most eukaryotes, but the protein remains functionally identical in all species. The amino acid substitutions that characterize this protein in different species have arisen by mutations during evolution, but they do not involve the critical region of the protein that is responsible for its respiratory activity. The region retains its required shape. These and similar studies have shown that protein shape depends on interactions between some but not necessarily all of the amino acid residues in a molecule.

The standard dimensions of the polypeptide backbone are determined by the bond lengths and bond angles (Fig. 2.28). The restrictions imposed by the zigzag, rigid polypeptide backbone, whose amide groups are planar in shape, lead to restrictions on the manner in which this linear chain can fold into three-dimensional structures in space. From physical studies

it is clear that proteins are rigid, compact molecules which must be folded, since the molecule is much shorter than we would expect from the lengths of the constituent polypeptides. Interactions between neighboring amino acid residues contribute to **secondary structure** of polypeptide chains; interactions between residues at some distance from one another in a chain contribute to **tertiary structure.** The three-dimensional character of a polypeptide results from both these spatial considerations.

A principal mode of secondary structure, called the $\alpha$-**helix,** was first postulated by Linus Pauling and

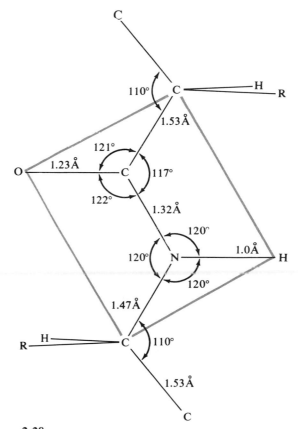

**Figure 2.28**
Bond angles and bond lengths of the peptide linkage in a polypeptide chain are shown. The plane is somewhat rigid because there are some double-bond characteristics to the C—N atoms that lie at its center.

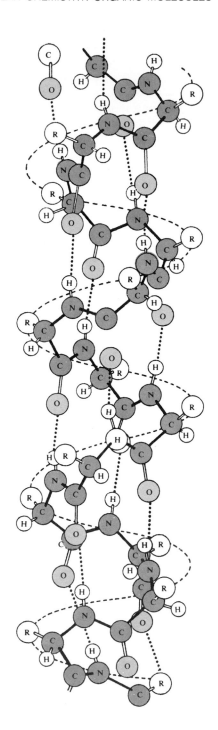

Robert Corey in 1951 (Fig. 2.29). The helix is a natural conformation for *regular* linear polymers, since it places each monomer in an identical orientation within the molecule. This leads to formation of the same group of secondary bonds for every monomer in the polymer, and to a more stable state since each set of bonds is as strong as any other set in the molecule's secondary bonding (secondary to covalent bonds of the polymer backbone). Most polymers, particularly proteins, contain different kinds of monomer units and are therefore not regular in construction. Even though the polypeptide backbone is regular, the attached side groups are irregular, that is, different from their neighbor side groups. Since a helical structure may be energetically satisfactory for the backbone groups but unsatisfactory for side group bonding, some compromise is usually achieved between the tendency of the regular backbone to form a regular helix and the tendency of the amino acid side groups to twist the backbone into a configuration that maximizes the strength of the secondary bonds formed by the side groups. Most proteins assume a three-dimensional shape dictated more by side group interactions (tertiary structure) than by interactions between neighboring groups of the backbone (secondary structure), but both kinds of interactions occur in virtually all proteins.

Fibrous proteins exhibit much greater regularity of secondary structure throughout the molecule, leading to highly ordered three-dimensional conformations of these important cellular building materials. Side group interactions predominate in globular proteins, such as enzymes and antibodies, thereby contributing to much greater folding. Globular proteins usually have a considerable amount of unordered **random coil** regions, along with occasional regions of ordered $\alpha$-helix and other types of secondary structure.

Proteins are extremely stable in the watery envi-

**Figure 2.29**
A polypeptide chain folded into an $\alpha$-helix configuration. Highly ordered secondary structure is maintained by hydrogen bonds between the carbonyl group of one residue and the amino group of the fourth amino acid residue along the polymer chain.

ronment of the cell. One of the major reasons is that the polypeptide polar backbone forms strong hydrogen bonds with surrounding water molecules. This contributes to an energetically satisfactory state and considerable molecular stability. But hydrogen bonds are considerably weaker than covalent bonds within the molecule, so that an ample amount of flexibility exists in protein modulation of chemical reactions.

When two or more polypeptide chains are present in a protein molecule, their spatial organization imposes a **quaternary structure** on the compound (Fig. 2.30). The smaller polypeptide units are held together in the protein aggregate by weak chemical bonds, as shown by their ready dispersion in the presence of reagents that are known to break noncovalent bonds but not covalent bonds. The number of polypeptides and the number of different kinds of polypeptides varies in different proteins. Hemoglobin contains two $\alpha$-chains and two $\beta$-chains, each having a molecular weight of about 16,000. Ferritin, a protein which functions to store iron atoms in mammals, has a molecular weight of about 480,000. It contains 20 identical, smaller polypeptide chains of about 200 amino acids each. While most of the large proteins are constructed of polypeptide subunits, even many of the smaller proteins also have quaternary structure (Fig. 2.31). Disruption of protein secondary, tertiary, or quaternary structure leads to disorganization of the molecule and to protein malfunction.

## NUCLEOTIDES AND NUCLEIC ACIDS

Nucleotides are involved in at least two major cellular functions: (1) they are monomeric units from which DNA and RNA polymers are constructed, and (2) they act as agents in certain energy-transferring reactions during metabolism. A **mononucleotide** is made up of one nitrogen-containing organic base, one

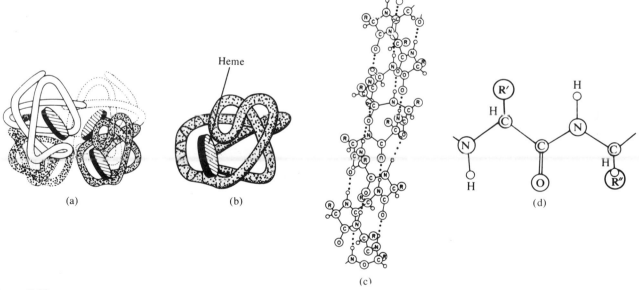

**Figure 2.30**
The four levels of structural organization in proteins as exemplified by the hemoglobin molecule: (a) *quaternary*, aggregation of two or more polypeptide chains (four in hemoglobin); (b) *tertiary*, folding of a polypeptide chain in space produces a globular three-dimensional shape; (c) *secondary*, neighboring interactions as in the $\alpha$-helix portion of a polypeptide chain; and (d) *primary*, the sequence of amino acids joined by peptide bonds shown here.

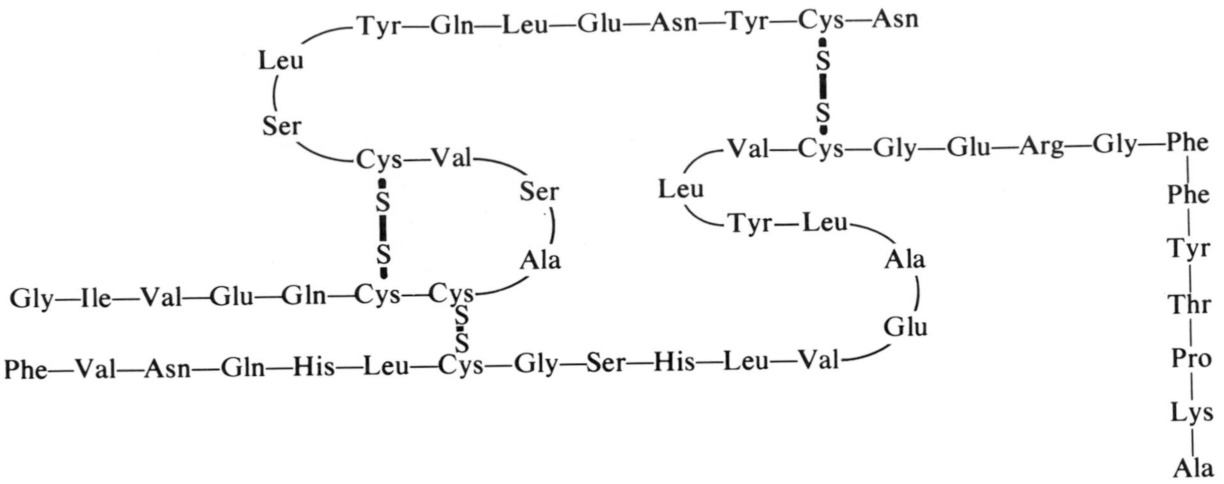

**Figure 2.31**
The tertiary structure of the hormone insulin depends on the formation of disulfide
(—S—S—) bridges that form at specific regions by interaction between sulfhydryl (—SH)
groups of amino acids at some distance from one another in one or both polypeptide chains
of the protein.

pentose sugar, and one phosphate residue derived from phosphoric acid. When there is no phosphate group, the sugar–base combination is called a **nucleoside** (Table 2.4). For this reason, nucleotides (phosphate–sugar–base) are also called **nucleoside phosphates.** Nucleoside mono-, di-, and triphosphates contain one, two, or three phosphate groups, respectively.

The nitrogenous bases commonly found in nucleic acids and their nucleotide building blocks are derivatives of **purine** and **pyrimidine** (Fig. 2.32). The commonly occurring purines **adenine** and **guanine** are found in both DNA and RNA, as is the pyrimidine compound **cytosine.** The second kind of pyrimidine in DNA is **thymine,** while its demethylated form, **uracil,** occurs in RNA. Since each kind of nucleic acid

**TABLE 2.4 Nomenclature of nucleic acids and their constituent units**

| Base | Nucleoside | Nucleotide | Nucleic Acid |
|------|-----------|-----------|-------------|
| Purines: | | | |
| Adenine | Adenosine | Adenylic acid | RNA |
| | Deoxyadenosine | Deoxyadenylic acid | DNA |
| Guanine | Guanosine | Guanylic acid | RNA |
| | Deoxyguanosine | Deoxyguanylic acid | DNA |
| Pyrimidines: | | | |
| Cytosine | Cytidine | Cytidylic acid | RNA |
| | Deoxycytidine | Deoxycytidylic acid | DNA |
| Thymine | Thymidine | Thymidylic acid | DNA |
| Uracil | Uridine | Uridylic acid | RNA |

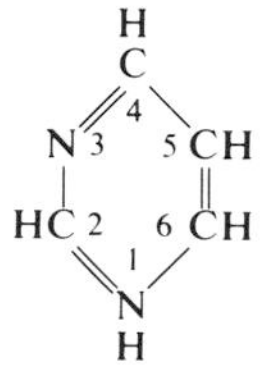
Pyrimidine

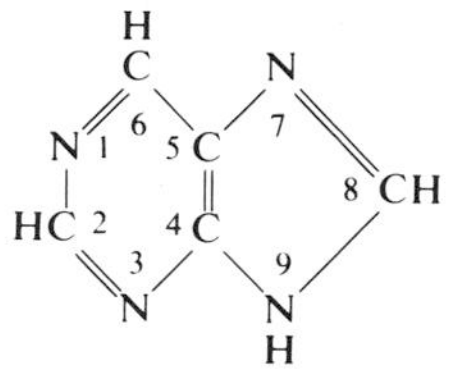
Purine

| Found in | Pyrimidines | Purines | | Pentoses | Phosphates |
|---|---|---|---|---|---|
| DNA | 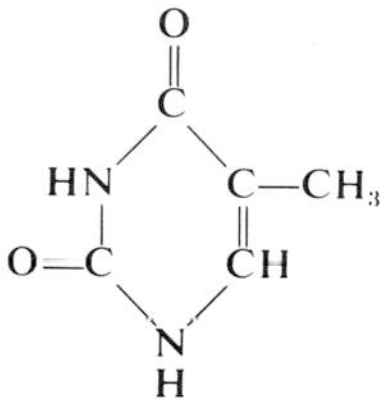 Thymine | | | 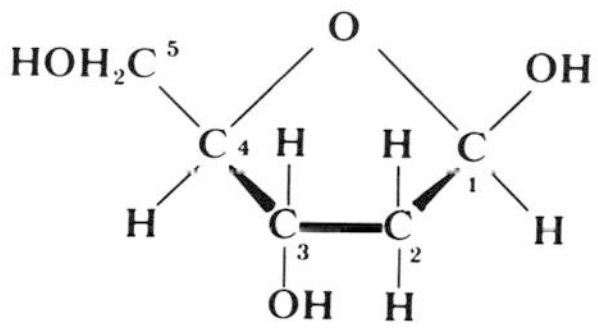 2-deoxyribose | |
| RNA | 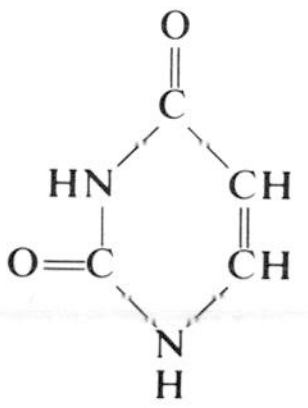 Uracil | | | 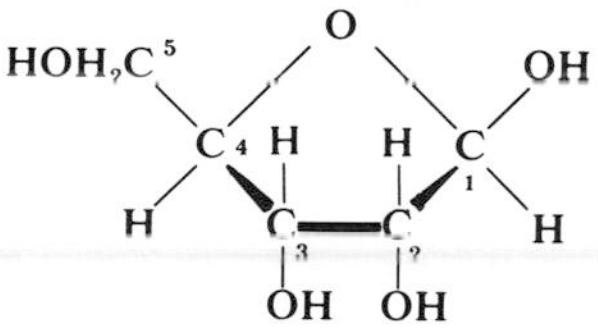 Ribose | |
| DNA and RNA | 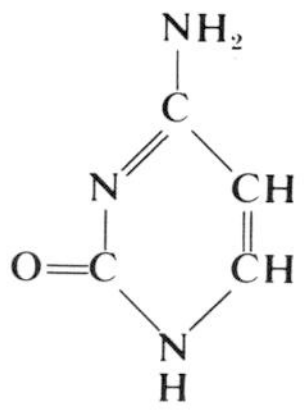 Cytosine | 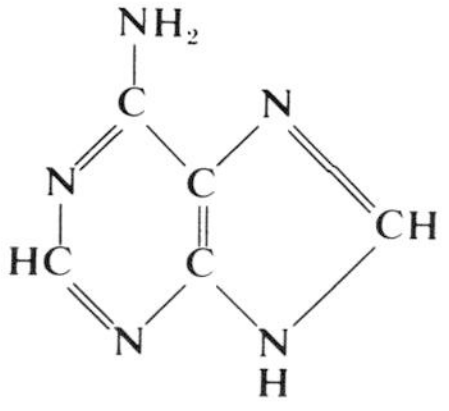 Adenine | 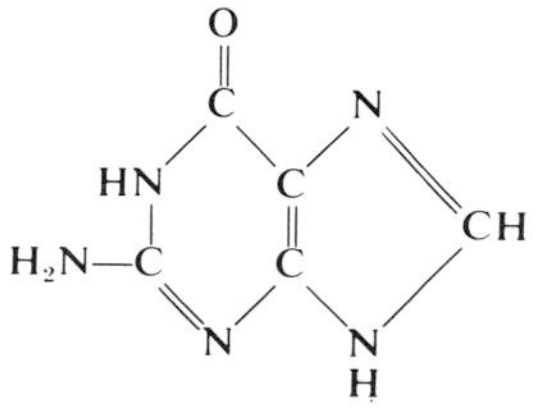 Guanine | | 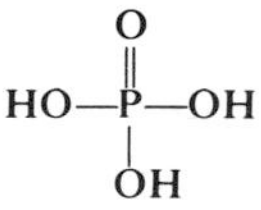 |

**Figure 2.32**
The building blocks of DNA and RNA.

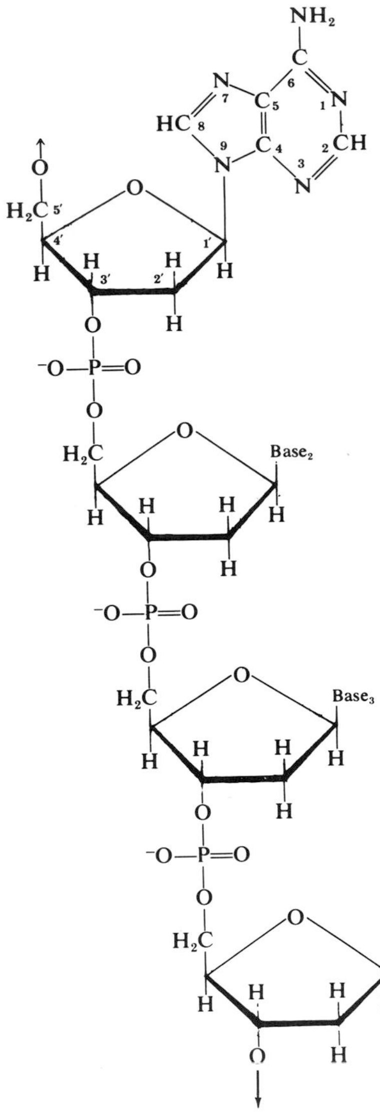

**Figure 2.33**
Region of a DNA molecule showing four nucleotide residues linked by a 3′, 5′-phosphodiester bridge between each residue and its neighbors in the chain. The sugar—phosphate "backbone" of the nucleic acid is therefore a repeating sequence of sugars linked between carbon-3 of one and carbon-5 of another sugar via phosphate bonding. There is no restriction on the particular base bonded to a sugar in the polymer.

contains one unique pyrimidine, it is convenient to study synthesis and activity of DNA or RNA using isotopically-labeled precursors containing one or the other of these bases. Usually the nucleosides **uridine** or **thymidine,** or their nucleotide forms, are added to the biological system under study.

The only difference in the pentose sugars of nucleotides is the presence of a hydroxyl group at carbon atom 2 of **D-ribose** in RNA, but a hydrogen at carbon-2 of **2-deoxy-D-ribose** in DNA monomers and polymers. This seemingly simple difference is partly responsible for profound differences in stabilities, pairing potential, and functions of DNA and RNA.

Polynucleotides of both the DNA and RNA varieties are built from mononucleotides that are linked covalently via **phosphodiester bridges** between the 3′ position of one unit and the 5′ position of the next (Fig. 2.33). Since there is no restriction on the *vertical* sequence of adjacent mononucleotides in either DNA or RNA, a considerable variety of molecules is possible even though only 4 kinds of nucleotide monomers (one for each of the four kinds of bases in the combination with sugar and phosphate) are used in polymer construction. The theoretical variety is calculated as $4^n$, where 4 is the number of different kinds of nucleotides, and $n$ is the number of monomers in the polymer. For a molecule made of only 75 monomeric units, as in some of the smallest RNAs, there may be $4^{75}$ different arrangements of the constituent units. Each arrangement theoretically constitutes a molecule of different specificity. Where the average gene may include about 500 nucleotides in a DNA sequence, $4^{500}$ different sequences are theoretically possible and, therefore, that many different and specific genes. Despite the apparently meager number of monomer types, an astronomically high number of possible genes can be constructed. Such variety can easily account for all past and present life forms.

### The DNA Double Helix

DNA molecules usually have regular helical configurations because most DNA molecules consist of two *complementary* polynucleotide strands. The two strands are held together by hydrogen bonds between complementary pairs of purines and pyrimidines (Fig. 2.34). Adenine always binds with thymine while guanine always binds with cytosine. This repeated hydrogen bonding within the double helix structure and bonding between virtually all the surface atoms in the sugar and phosphate groups with water molecules serve to stabilize the structure.

Since the purine-pyrimidine pairs are found in the center of the molecule, their flat surfaces can stack on top of each other and thereby limit their contact with water. In double-helical molecules, a regular structure is possible because the complementary base pairs are exactly the same size. Single polynucleotide chains could not have a regular backbone structure because pyrimidines are smaller than purines, which would cause the angle of helical rotation to vary with the sequence of bases.

DNA double helix molecules are very stable at physiological temperatures because: (1) disruption of the double helix breaks hydrogen bonds and brings hydrophobic purines and pyrimidines into contact with water, which is energetically unsatisfactory; and (2) there are many weak bonds within the DNA molecule, arranged so that most of them cannot break without many others breaking at the same time. Even though some hydrogen bonds may be broken by thermal motion, hydrogen bonds in the rest of the molecule remain intact and the molecule does not fall apart. In fact, when held together by more than ten nucleotide pairs the double helices are quite stable at room temperature. The cooperative result of a number of weak bonds is stability of molecular shape, in proteins as well as in nucleic acids. At abnormally high temperatures there is more frequent breakage of weak

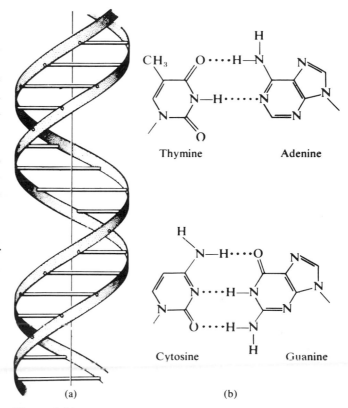

**Figure 2.34**
The DNA double helix consists of (a) two polynucleotide chains held in register by hydrogen bonding between the nitrogenous bases of partner chains; (b) the constant diameter of a duplex molecule and the specificity of interaction is due to specificity of complementary base-pairing, adenine with thymine and guanine with cytosine.

bonds, which become less stable as temperatures rise above physiological levels. Once a significant number of weak bonds have been broken, a protein or nucleic acid molecule usually loses its original form (in the process of denaturation) and changes to an inactive, or denatured, form.

### SUMMARY

1. Atoms are aggregated into molecules and are held together by chemical bonds. Strong covalent bonds, which involve sharing of electrons between atoms, are the primary bonds of organic molecules. Weaker bonds, such as hydrogen bonds, ionic bonds, and van der Waals bonds, are secondary to covalent bonding. Weak bonds involve attractions between atoms but do not involve electron sharing. Covalent bonds permit great molecular stability at cell temperatures while weaker bonds permit increased flexibility in chemical reactions involving organic compounds.

2. Water molecules are dipoles and can therefore bond with either electronegative or electropositive atoms in organic molecules. Hydrogen bonds between water molecules help maintain an energetically favorable state, and also partially account for water as a solvent in cells. Compounds that are soluble in water can form ionic bonds or hydrogen bonds with water molecules, whereas water-insoluble molecules cannot bond with $H_2O$. Hydrophilic and hydrophobic molecules therefore interact differently with $H_2O$.

3. The pH scale from 0 to 14 (1.0 $M$ $H^+$ ions to $10^{-14}$ $M$ $H^+$ ions) is based on $K_w$, the ion product of water. More $H^+$ ions in solution produce a lower or more acid pH, while lower concentrations of $H^+$ ions lead to higher pH (more alkaline). The pH scale is logarithmic and consists of positive readings in units that differ by a factor of ten from one to the next. Strong buffer pairs stabilize the pH of fluids, resulting in resistance to changes when $H^+$ ions are increased or decreased in concentration after addition of ionized compounds.

4. The unique chemistry of biologically significant organic compounds is largely due to special properties of carbon atoms and their interactions. Carbon is a tetravalent atom and can form covalent bonds with other carbons or with H,O,N,P,S, and the few other elements found in cellular molecules. The stability and versatility of organic compounds are trademarks based on strong covalent bonding and on tetravalency, respectively, of carbon atoms.

5. Isomers are alternative structural forms of the same compound, and four different isomers are theoretically possible for each asymmetric carbon atom in a molecule. Asymmetry results when each valence of a carbon atom is occupied by a different atom or group of atoms. Isomers increase the diversity of organic compounds without introducing new atoms or groups into the molecule construction.

6. Organic molecule variety is further enhanced by constructing larger molecules from smaller building units, which are monomers of the larger polymers. Many different polymers can be made from a finite, common set of monomer units. Three of the four classes of biologically significant organic compounds (carbohydrates, proteins, and nucleic acids) make numerous polymer molecules from relatively few kinds of monomers.

7. Carbohydrates have the general formula $(CH_2O)_n$. The commonest carbohydrate in cells is glucose, a hexose sugar or monosaccharide. Polysaccharides form on condensation of smaller units. Starch and glycogen are food storage polysaccharides, cellulose and chitin are structural polysaccharides, and many carbohydrates serve as metabolites in cell chemistry. Structural carbohydrates have $\beta$-glycosidic bonds stabilizing molecule construction, while easily mobilized molecules are held together by $\alpha$-glycosidic bonds that are more readily made or broken during metabolism.

8. Lipids are a diverse group of compounds with the common property of being soluble in nonpolar, organic solvents. Among various functional types of lipids, some are food storage forms (neutral fats) and some are parts of membrane structure (phospholipids, sphingolipids, and glycolipids predominantly). Membrane lipids are amphipathic, having both hydrophilic and hydrophobic regions in the same molecule. These lipids can interact with both the watery and nonwatery regions of cell environments and, therefore, provide a vital link between the outside and inside environments of cells. Some lipids, such as cholesterol, are found only in animal cell membranes, while others, such as essential oils, are only found in plants. Most lipids occur in all kinds of cells belonging to prokaryote and eukaryote groups.

9. Proteins are polymers made up of more than twenty kinds of amino acid monomers, but only twenty amino acids are included in the genetic code. All amino acids have a carboxyl group and an amino group linked to the first or $\alpha$-carbon atom of the molecule. A hydrogen atom occurs at a third valence while some fourth group occupies the fourth carbon valence, making for asymmetry and isomer potential.

Amino acids are strung together in peptide or polypeptide chains by a common peptide bond, which is formed by repeated joining of the amino group of one monomer with the carboxyl group of an adjacent monomer along the unbranched polymer chain. Net electrical charge of polypeptides results from properties of amino acid side-chains included in the molecule. Proteins are made up from one or more polypeptide chains.

10. Proteins (and other polymers) are characterized by primary, secondary, tertiary, and perhaps quaternary organization. The primary structure is based on the kinds and sequence of amino acids in the chain; secondary and tertiary structure are based on three-dimensional shape due to interactions between neighboring or more distant monomers, respectively; quaternary structure depends on interactions between two or more polypeptides in a protein molecule. Disruption of three-dimensional organization can lead to molecule malfunction, since structure, function, or regulatory properties are largely based on three-dimensional themes in molecules.

11. Mononucleotides form nucleic acid polymers when joined covalently by phosphodiester bridges. Nucleic acid variety stems from lack of restriction on the sequence of adjacent mononucleotides, where $4^n$ different molecules can be constructed from $n$ monomers of four different kinds (thymine or uracil, guanine, cytosine, and adenine in nucleoside-phosphate monomer units). DNA and RNA have a repetitive "backbone" of sugar and phosphate residues, with one of the four bases linked to each sugar throughout the length of an unbranched nucleic acid chain. DNA occurs usually as a two-stranded helical molecule, with hydrogen bonding between complementary bases along the molecule's length. Nucleic acids tend to be relatively stable at physiological temperatures, but they are made more flexible by hydrogen bonding between bases.

### STUDY QUESTIONS

1. What is a chemical bond? What are the important chemical bonds in biological systems? How do we characterize these types of chemical bonds? Where do we find each kind of chemical bond most often in cellular chemicals? What is a "strong" chemical bond?

2. What are the special properties of water? Why is water considered to be essential for life as we know it (even on Mars)? What is a solvent? a solute? What kinds of interactions occur between water molecules and organic compounds? What is the basis for the pH scale? How can we

calculate the pH of a particular solution after we discover the $H^+$ ion concentration of that solution? What is a buffer?

3. What do we mean by saying that organic compounds are "versatile" and "stable"? What are the consequences of tetravalency of carbon atoms? What is an isomer? What is a monomer? a polymer?

4. Why do we classify glucose as a carbohydrate? as a monosaccharide? How does the linear form differ from the ring form of a pentose or hexose sugar molecule? How do polysaccharides form? What bonds do we find in (a) polysaccharides that are easily metabolized and (b) polysaccharides that are not easily metabolized by most organisms? What are two common types of polysaccharides that (a) function as food storage forms and (b) function as structural molecules of the cell?

5. What is the common feature shared by all or most lipids? What is the difference between a "saturated" and an "unsaturated" fatty acid? What distinguishes a neutral fat from a phospholipid? What is the biological significance of phospholipids, sphingolipids, and glycolipids?

6. How is a polypeptide constructed? What functional types of proteins are known? What aspect of its amino acids leads to acidic or basic properties of protein molecules? What is the difference between a polypeptide and a protein? How do we define or recognize polypeptide or protein primary structure? secondary structure? tertiary structure? quaternary structure?

7. What are the functions of nucleic acid monomer units? What is the difference between a nucleoside and a nucleotide? What are the components of ribonucleotides? of deoxyribonucleotides? How are monomer units linked together in a nucleic acid molecule? Why is there so much potential variety in nucleic acid molecules? What is the nature of the interactions between two strands of a duplex DNA molecule? What features of the DNA duplex contribute to its stability at cell temperatures? Why is the DNA duplex a helical molecule with a constant diameter all along its length? What is the significance of such regularity (constancy) of DNA duplex construction?

### SUGGESTED READINGS

Albersheim, P. The walls of growing plant cells. *Scientific American* **232:**80 (Apr. 1975).

Dickerson, R. E., and Geis, I. *The Structure and Action of Proteins.* Menlo Park, Calif.: W. A. Benjamin (1969).

Doty, P. Proteins. *Scientific American* **197:**173 (Sept. 1957).

DuPraw, E. J. *The Biosciences: Cell and Molecular Biology*. Stanford: Cell and Molecular Biology Council (1972).

Lambert, J. B. The shapes of organic molecules. *Scientific American* **222**:58 (Jan. 1970).

Pauling, L., Corey, R. B., and Hayward, R. The structure of protein molecules. *Scientific American* **191**:51 (July 1954).

Perutz, M. F. The hemoglobin molecule. *Scientific American* **211**:64 (Nov. 1964).

Rodley, G. A., Scobie, R. S., Bates, R. H. T., and Lewitt, R. M. A possible conformation for double-stranded polynucleotides. *Proceedings of the National Academy of Sciences* **73**:2959 (1976).

Scholander, P. F. Tensile water. *American Scientist* **60**:584 (1972).

Stryer, L. *Biochemistry*. San Francisco: W. H. Freeman (1975).

Watson, J. D. *Molecular Biology of the Gene*, 3rd ed. Menlo Park, Calif.: W. A. Benjamin (1976).

# Chapter 3

# Cellular Metabolism: Energy and Enzymes

In this chapter we will scan some fundamental characteristics that distinguish living systems from their nonliving environments. To see how cellular chemistry, or metabolism, operates we need some basic information about: (1) the nature of *energy* needed to do work in the unique environment within a cell, (2) the coordinated systems responsible for *energy transfers* between energy-consuming and energy-releasing reactions, and (3) the *catalysts* that modulate reaction rates and direct the ebb and flow of metabolism. We can then examine a metabolic pathway to illustrate how these cellular features interact.

## CELLULAR ENERGETICS

Energy is broadly defined as the capacity to do work. There are different forms of energy and different kinds of work. One form of energy can be transformed into another, and the transformed energy can then be applied to do work. For example, thermal energy of steam can be transformed into mechanical energy by a steam engine, and the mechanical energy can be used to perform mechanical work. The application, movement, and transformation of energy ultimately underlie all physical and chemical processes. The area of physical science that deals with exchanges of energy in collections of matter is called **thermodynamics,** a term handed down from the earliest studies when heat was the focus of measurement. The equivalent term applied more specifically to the study of energy transformations in living systems is **bioenergetics.**

In thermodynamic studies we distinguish the collection of matter, or **system,** from the **surroundings,** or all other matter in the **universe** apart from the system under study. It is possible to measure the total energy content of a system in its **initial state,** before the reaction begins, and in its **final state** of equilibrium at the conclusion of the reaction. It is simpler, however, to measure the *difference* (shown by the delta sign, $\Delta$) in energy content, that is, the amount of energy exchanged between the system and its surroundings as the reaction takes place. For biological systems the only significant form of useful energy is **free energy,** or Gibbs free energy, designated as $G$. A change in free energy, or $\Delta G$, refers to the difference in free energy between organic compounds and their reaction products, and it is this energy which can be used to perform work in the system. The free-energy difference can be measured in the laboratory by recording the kilocalories of heat released to the surroundings as 1 mole of a substance is burned completely in air. For example, 686 kilocalories of heat are released when 1 mole of glucose is oxidized in air to its end products of $CO_2$ and $H_2O$. The difference in free energy content between glucose and $CO_2 + H_2O$ in this reaction can be stated as $\Delta G = -686$ kcal/mole. A negative free-energy change indicates energy release from the system during a reaction, while a positive free-energy change indicates that energy must be added to the system if the reaction is to proceed. The synthesis of glucose from $CO_2$ and $H_2O$ as initial reactants in the reverse reaction has a free-energy change stated as $\Delta G = +686$ kcal/mole. Energy-releasing reactions, signified by a negative free-energy change, are called **exergonic;** reactions which require free energy to be added to the system are signified by a positive $\Delta G$ and are called **endergonic** reactions.

In the example cited above there was heat energy released from the system. Although heat is a useful form of energy for performing work in many manufactured devices, it cannot be used to do work in biological systems. Heat can do work only if it acts through a temperature differential from warmer to cooler, and living cells have little or no temperature differential between their parts. We can, however, assume that oxidation of glucose to $CO_2 + H_2O$ in cells also will show a difference of 686 kcal/mole between the initial reactant and its end products. But the only fraction of the released energy which will be useful in cells is $\Delta G$; the rest of the energy is dissipated as heat, which is not used to do work in cells. The relative *efficiency* of a reaction is based on the percentage of free energy relative to the total energy involved. If 343 kcal were to be released as free energy during glucose oxidation in cells, the reaction would be 50% efficient ($343/686 \times 100$), and we would state that $\Delta G = -343$ kcal/mole for the reaction.

### Measurement of Free Energy

The **First Law** of thermodynamics states the principle of conservation of energy in the universe (system + surroundings): *energy can neither be created nor destroyed.* Energy can be transformed (changed from one form to another), however, so the energy content of the universe remains constant. In biological reactions we usually are interested only in energy change in the system, so we refer to reactions which lead to an increase or a decrease in free energy within the system. It is implied, according to the First Law, that the free-energy change in the system is compensated by energy increase or decrease in the surroundings such that the energy content of the universe remains constant.

The Second Law of thermodynamics tells us that *all systems tend toward an equilibrium state;* that is, all systems tend to minimize their free energy content. When this minimum is reached the system is in equilibrium; that is, $\Delta G = 0$. The reaction can only proceed in a specific direction when energy exchange takes place, or when $\Delta G < 0$ or $> 0$. When a system moves toward a state of higher energy, energy must be added to the system or the reaction cannot proceed in that direction. Such energy-requiring endergonic reactions run "uphill" (against the natural tendency to minimize free energy content), while energy-releasing exergonic reactions can be said to run "downhill" (toward their states of minimum free energy) spontaneously. The *rate* of reaction depends on other factors. As we will see in a later section, catalysts are important factors which influence reaction rates.

In a reaction such as A → B, which proceeds toward equilibrium, or A ⇌ B, we can determine the **equilibrium constant,** $K_{eq}$, as the ratio [B] / [A] when [B] and [A] are such that the reaction is at equilibrium, or

$$K_{eq} = \frac{[\text{final product at equilibrium}]}{[\text{initial reactant at equilibrium}]} \qquad [3.1]$$

(Molar concentrations are indicated by brackets, [ ].)

At equilibrium there is no further free-energy change since $\Delta G = 0$, or

$$\Delta G = \Delta G^0 + RT \ln K_{eq} \qquad [3.2]$$

where $R$ is the gas constant (1.987 cal/mole-degree), $T$ is the absolute temperature (°Kelvin), $\ln K_{eq}$ is the natural logarithm of the equilibrium constant, and $\Delta G^0$ is a thermodynamic constant called the **standard free-energy change** of the reaction. The relationship of $K_{eq}$ to $\Delta G^0$ is shown by

$$\Delta G^0 = -RT \ln K_{eq} \qquad [3.3]$$

Through analytical determination of $K_{eq}$ it is possible to calculate the standard free-energy change of a reaction. These two constants, $K_{eq}$ and $\Delta G^0$, are important values in characterizing a reaction. The relationships of the two sets of values are shown in Table 3.1. When $K_{eq} > 1$ ($\Delta G^0$ is negative), we know that the reaction proceeded with a decline in free energy and we know the amount of free energy that would be released under specified standard conditions. When $K_{eq}$ is $< 1$ ($\Delta G^0$ is positive), we know that the reaction proceeded toward equilibrium with energy added to the system, and we know the amount of free-energy increase at equilibrium.

These points can be illustrated in the following

**TABLE 3.1  Relationship between the equilibrium constant and the standard free-energy change at 25°C and pH 7.0**

| $K_{eq}$ | $\Delta G^0$ (kcal/mole) |
|---|---|
| 0.001 | +4.09 |
| 0.01 | +2.73 |
| 0.1 | +1.36 |
| 1.0 | 0 |
| 10.0 | −1.36 |
| 100.0 | −2.73 |
| 1000.0 | −4.09 |

example: During sugar oxidation in cells, glucose 1-phosphate is converted to glucose 6-phosphate in the presence of a certain enzyme. If we add this enzyme to 0.020 $M$ glucose 1-phosphate in a system at pH 7 and 25°C (= 298°Kelvin), we can determine the concentrations of the initial reactant and the final reaction product at equilibrium. We find that glucose 6-phosphate has risen in concentration from zero to 0.019 $M$ while glucose 1-phosphate has decreased to 0.001 $M$. The equilibrium constant is then

$$K_{eq} = \frac{[\text{glucose 6-phosphate}]}{[\text{glucose 1-phosphate}]} = \frac{0.019}{0.001} = 19 \quad [3.4]$$

and the standard free-energy change can be calculated as

$$\begin{aligned}
\Delta G^0 &= -RT \ln K_{eq} \\
&= -1.987 \times 298 \times \ln 19^* \\
&= -1.987 \times 298 \times 2.303 \times \log_{10} 19^* \qquad [3.5] \\
&= -1363 \times 1.28 \\
\Delta G^0 &= -1745 \text{ cal/mole, or } -1.745 \text{ kcal/mole}
\end{aligned}$$

These measurements were made outside the cell in a controlled laboratory system. We know that the

---

*It is often convenient to change from base $e$ natural logarithms (ln) to base 10 logarithms (log or $\log_{10}$) in working with equations. The conversion is $2.303 \log x = \ln x$.

standard-state free-energy change rarely is achieved in living systems, since the specified standard conditions are rarely present. The *actual* free-energy change of a reaction usually varies according to the actual concentrations of reactants and products. But for the sake of consistency in characterizing reactions and making predictions about reactions, $\Delta G^0$ is used routinely in describing biological systems.

### Open Systems and Steady States

In classical thermodynamics one deals with a **closed system** which does not exchange matter with its surroundings. In living cells, however, we find an **open system** which does exchange matter with its surroundings. Furthermore, living cells exist in different **steady states** in which the rate of input equals the rate of output at each given moment, but a different pair of rates may lead to a different steady state from moment to moment. Because of this feature, cells do not usually achieve thermodynamic equilibrium states of $\Delta G = 0$. Two significant features which describe the living cell emerge in relation to the nonequilibrium open system existing in steady states:

1. These systems can perform work precisely because they do not attain equilibrium; equilibrium is a state of no work.
2. Only systems that are away from equilibrium can be subject to control and regulation of their activities.

## COUPLED METABOLIC REACTIONS

Metabolism consists of energy-releasing and energy-requiring reactions. There is a problem of obtaining energy for "uphill" reactions which are otherwise thermodynamically unfavorable. Similarly, "downhill" reactions of food breakdown in cells must be exploited in such a way that the released free energy can be made to do work of all kinds. The solution to these problems in living systems is to *couple* energy-consuming reactions to energy-releasing reactions. In this way, energy released during food breakdown can be tapped to drive biosynthesis reactions which can only take place if adequate energy is pro-

vided. Coupled reaction systems predominate in the living world, quite unlike the nonliving world where such systems are rare. The energy siphoned from breakdown reactions must be free energy, stored in the form of chemical energy for immediate or future use in performing cellular work. Cellular efficiency is based in large measure on the ability to trap and use energy that might otherwise go to waste as unusable heat.

The common denominator of "energy currency" in cells is adenosine triphosphate, or **ATP.** Energy-releasing reactions may be coupled to the energy-requiring formation of ATP, with the released energy captured and stored in ATP molecules. Energy-requiring reactions may be coupled to energy-releasing breakdown of ATP, with the free energy difference between reactants and products driving an otherwise unfavorable, "uphill" reaction to completion.

### The Role of ATP in Energy Transfer

When ATP is hydrolyzed, the terminal **phosphoryl group** $\left(-\text{P}\!\!\!\begin{array}{c}/\text{O}^-\\ =\!\!\text{O} \\ \backslash\text{O}^-\end{array}\right.$, or $-\text{PO}_3^{2-}$) is released, and **ADP** (adenosine diphosphate) is formed (Fig. 3.1). The *difference in free-energy content* between the initial and final components, or the $\Delta G^0$ of the reaction, is about $-7.3$ kcal/mole. Conversely, when a phosphoryl group is added to ADP in a **phosphorylation** reaction producing ATP, the $\Delta G^0$ of the reaction is about $+7.3$ kcal/mole. If ATP hydrolysis is coupled to some energy-requiring work, such as biosynthesis or movement of molecules from one place to another, 7.3 kcal of free energy is available to sponsor this work for every mole of ATP converted to lower-energy ADP. Energy transfer is accomplished as the terminal phosphoryl group is transferred from ATP to some molecule, which in turn becomes phosphorylated and, therefore, exists at a higher energy level than before. Such phosphorylated molecules can then engage in reactions or other work because they now have a higher energy level and can proceed to the next step in a "downhill" direction. Such phosphorylated metabolites are **intermediates** in reaction sequences. Each step in the sequence takes place under energetically favorable conditions only because some molecule is temporarily raised to a high enough energy level to participate in the next reaction step. As it participates in the next reaction step it becomes dephosphorylated. In some cases, dephosphorylation is coupled to ATP synthesis from ADP, thus regenerating ATP for other rounds of work in the same or in different cellular activities.

What we have really said is that ATP acts as a *common intermediate* in biological-energy transfer. The ATP-ADP cycle acts as a link between energy-releasing and energy-requiring reactions, through the mechanism of phosphoryl group transfer. For example, consider two reactions in which ADP and ATP interact with compounds X and Y and their phosphorylated forms X—P and Y—P (P signifies a phosphoryl group here):

$$X—P + ADP \rightarrow X + ATP;$$

$$\Delta G^0 = -5 \text{ kcal/mole} \quad [3.6]$$

and

$$ATP + Y \rightarrow ADP + Y—P;$$

$$\Delta G^0 = -1 \text{ kcal/mole} \quad [3.7]$$

In each case we are really summing up a pair of reactions:

| | | |
|---|---|---|
| X--P → X + P | ; | $\Delta G^0 = -12$ kcal/mole |
| ADP + P → ATP | ; | $\Delta G^0 = +7$ kcal/mole [3.8] |
| X--P + ADP → X + ATP ; | | $\Delta G^0 = -5$ kcal/mole |

and

| | | |
|---|---|---|
| ATP → ADP + P | ; | $\Delta G^0 = -7$ kcal/mole |
| Y → Y--P | ; | $\Delta G^0 = +6$ kcal/mole [3.9] |
| ATP + Y → ADP + Y--P ; | | $\Delta G^0 = -1$ kcal/mole |

In reaction [3.8] we can see that dephosphorylation of X--P is *coupled* to the reaction of ADP accepting the phosphoryl group to form ATP. In reaction [3.9] the

formation of Y--P is coupled to ADP formation since the phosphoryl group from ATP is transferred to compound Y. Each pair of coupled reactions is thermodynamically possible because an energetically unfavorable reaction (e.g., Y → Y--P, or ADP → ATP) is coupled with an energy-releasing reaction that drives them both.

Suppose the sets of coupled reactions were part of a biochemical pathway in which energy transfers depended on reaction [3.8] preceding reaction [3.9], linked by an ADP-ATP cycle. We might then indicate the relationships in the pathway as

$$\begin{array}{ccc} X\text{--}P & ADP & Y\text{--}P \\ & \times \quad \times & \quad [3.10] \\ X & ATP & Y \end{array}$$

In reaction system [3.10] we have shown that coupled reactions are involved in transfer of phosphoryl-bond energy from a higher-potential compound, via a com-

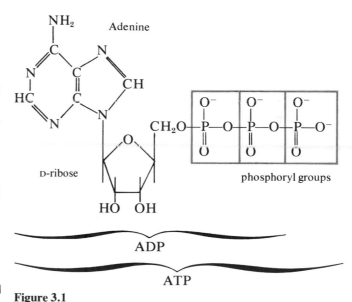

**Figure 3.1**
The energy transferring molecule adenosine triphosphate (ATP) is a nucleoside triphosphate. It is converted to adenosine diphosphate (ADP) when its terminal phosphoryl group is transferred to some acceptor molecule. ATP is synthesized when ADP is phosphorylated. There is a free-energy difference of 7 kcal/mole between ATP and ADP.

mon ADP-ATP cyclic link. By "phosphoryl-bond energy" we really mean the free-energy difference between the reactants and the products of the reaction, usually indicated by $\Delta G^0$. In other words, we are not referring to energy localized to the chemical bond itself although references to bond energy may seem to imply this. We use bond energy as a colloquial expression in energetics, but actually mean the free-energy difference between free energy of the reactants in the initial state and the free energy of the products in the final state of the reaction. The free-energy difference, of course, is the energy available to do work in the cell.

There are two unique features of ATP in relation to energy flow in the cell:

1. The standard free-energy change in the hydrolysis of ATP is intermediate between phosphorylated compounds of high and of low energy potential (Fig. 3.2).

2. ATP and ADP participate in almost all phosphoryl group transfers in metabolism.

Because ATP occupies an intermediate position in the thermodynamic scale, it connects reactions involving high-potential and low-potential compounds. ATP is formed when ADP accepts phosphoryl groups from high-potential compounds. Subsequently ATP donates its terminal phosphoryl group to particular acceptor molecules. These acceptors have a higher energy potential once they have been transformed, and reactions continue in the general "downhill" direction dictated by the standard free energies of the reactants and the products.

Because ATP and ADP take part in nearly all phosphoryl group transfers, ATP serves as a general energy carrier in the cell; that is, it is the energy "currency" for a multitude of unrelated reactions and activities in cells. Enzymes which function in energy transfer catalyze phosphoryl-group transfer from higher-

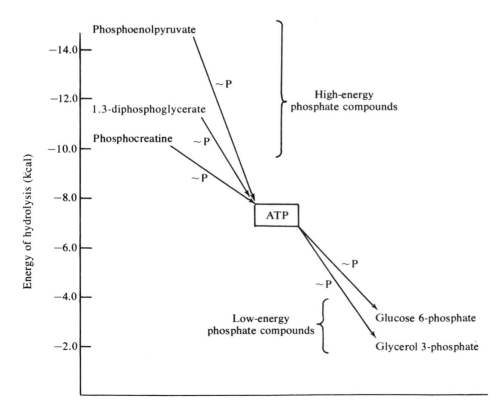

**Figure 3.2**
Energy transfer from compounds of high free-energy content to those of low free-energy content occurs through mediation of ATP, whose free-energy level is intermediate between those two groups of molecules. (Adapted with permission from Albert L. Lehninger, *Bioenergetics: The Molecular Basis of Biological Energy Transformations*, Second Edition, copyright © 1971, W. A. Benjamin, Inc., Menlo Park, California.)

energy compounds to ATP, and from ATP to lower-energy compounds (see Fig. 3.2). No enzymes have been discovered which transfer phosphoryl groups directly from some high-energy component to low-energy substances; all operate in conjunction with the ATP-ADP cycle.

### Oxidation-Reductions in Energy Transfer

Energy transfer can be accomplished by coupled reactions involving *gain and loss of electrons,* as well as by phosphoryl group transfer. In some cases it is the electron itself which is involved in transfer, in other cases **electron transfer** is achieved when hydrogen atoms (including their electron) are gained or lost. Reactions involving loss of electrons, or **oxidations,** are coupled with reactions in which electrons are gained, or **reductions.** In such coupled **oxidation-reduction reactions,** the substance accepting electrons is an **oxidizing agent,** or **oxidant,** and a **reducing agent,** or **reductant,** loses electrons. Oxidation-reductions are also known as **redox reactions,** an abbreviation which further implies that such reactions are coupled.

The ability to donate or accept electrons varies among oxidants and reductants. The tendency to lose electrons can be quantitatively compared and expressed as a positive or negative value in a **redox series,** arranged according to the oxidation-reduction potential **(redox potential)** of various substances. These values are obtained from measurements of electrode potential made against the standard of hydrogen, according to the reaction

$$H_2 \rightleftarrows 2H^+ + 2e^- \qquad [3.11]$$

In this reaction molecular hydrogen is in equilibrium with hydrogen ions (protons) and electrons, under standard conditions.

When electrons of a substance are donated to the standard hydrogen electrode, the potential has a negative value ($< 0$); when the substance accepts electrons its potential registers as positive ($> 0$). The standard electrode potential of a substance, or $E_0$, is its potential relative to a hydrogen electrode, expressed in volts (V). Stronger oxidizing agents have a higher positive potential, or a greater affinity for electrons than weaker oxidants, or than reductants which have negative potentials.

The redox series represents a range of increasing electron affinity, going from negative to positive $E_0$ values (Table 3.2). The redox series has great predictive value for determining which reactions are theoretically possible, just as the $\Delta G^0$ series permits predictions for direction of energy flow and thermodynamically possible reactions. An oxidant can be reduced by a substance with a lower $E_0$ value than its own, and conversely, a reductant can be oxidized by a substance with a higher $E_0$ value than its own. In coupled reactions, therefore, electron transfer occurs spontaneously in predictable directions according to $E_0$ values.

Standard redox potentials for biochemical systems, at the physiological pH of 7.0, are usually determined by potentiometric titrations in which an oxidizing agent is added to the completely reduced form of a system. The difference in potential is then determined at various stages of oxidation by reference to a standard electrode. The process can be carried out in reverse, by adding a reducing agent to the completely oxidized form of a substance. The inflection point of the titration curve is the point at which 50 percent of the reaction is complete, and this point is taken as the standard redox potential, or $E_0'$ (Fig. 3.3).

The standard free-energy change ($\Delta G^0$) is related to the difference in redox potentials ($\Delta E_0$) between the oxidation and reduction half-reactions, according to

$$\Delta G^0 = -nF\Delta E_0 \qquad [3.12]$$

where $n$ is the number of electrons that move, $F$ is the Faraday (23,040 cal/volt), and $\Delta E_0$ is the redox potential difference (in volts). The interconvertibility of $\Delta G^0$ and $\Delta E_0$ is illustrated in the following hypothetical scheme for the reaction $AH_2 + B \rightleftarrows BH_2 + A$, shown as oxidation and reduction half-reactions:

$$AH_2 \rightleftarrows A + 2H^+ + 2e^- \qquad E_0 = 0.20V \qquad [3.13]$$

$$B + 2H^+ + 2e^- \rightleftarrows BH_2 \qquad E_0 = -0.10V \qquad [3.14]$$

**TABLE 3.2  Redox potentials of redox couples and substrate couples**

| Redox Couple (red/ox) | $E_0'$ | Substrate Couple (red/ox) | $E_0'$ |
|---|---|---|---|
| NADH/NAD⁺ + H⁺ | −0.32 | α-ketoglutarate/succinate | −0.67 |
| FMNH₂/FMN | −0.12 | Ethanol/acetaldehyde | −0.20 |
| Cytochrome $c$, red/ox | +0.26 | Lactate/pyruvate | −0.19 |
| Cytochrome $a$, red/ox | +0.29 | Malate/oxaloacetate | −0.17 |
| Fe²⁺/Fe³⁺ (nonprotein) | +0.77 | Succinate/fumarate | +0.03 |
| | | Hydrogen peroxide/oxygen | +0.30 |
| | | Water/oxygen | +0.82 |

Reaction [3.13] proceeds by being coupled to reaction [3.14], so that

$$\Delta E_0 = 0.20V - (-0.10V) = 0.30V \qquad [3.15]$$

and according to equation [3.12]

$$\Delta G^0 = -2 \times 23,040\ cal/V \times 0.30V$$

$$[3.16]$$

$$= -13,824\ cal,\ or\ -13.8\ kcal.$$

### The Role of NAD⁺/NADH in Energy Transfer

Just as the ADP-ATP system operated in energy transfer involving phosphoryl groups, redox couples act in energy transfer reactions involving electrons (or hydrogen atoms containing these electrons). A **redox couple** is the oxidized and reduced form of some electron-transferring compound. One of the more widely-used of these compounds is **nicotinamide adenine dinucleotide (NAD)**, which exists as the NAD⁺/NADH redox couple (Fig. 3.4). Redox couples interact with substrate couples (oxidized and reduced forms of a substrate) according to the general reaction

reduced substrate + NAD⁺

$$\Updownarrow \qquad [3.17]$$

oxidized substrate + NADH + H⁺

Each NADH is formed when NAD⁺ accepts two electrons from a substrate; one of these electrons is a part of the hydrogen atom. The proton remaining after removal of the electron from the second hydrogen

atom is released into the medium as free H⁺ ion. When protons are required in two-electron transfers producing NAD⁺ and reduced substrate, any H⁺ ion in the medium can provide the proton to be added to the substrate, along with a hydrogen atom and an electron from NADH.

The NAD⁺/NADH system can also transfer electrons from one substrate couple to another

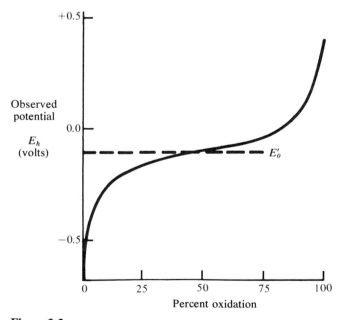

**Figure 3.3**
Titration curve of a biological substance, showing the point that determines the standard electrode potential ($E_0$) of that substance.

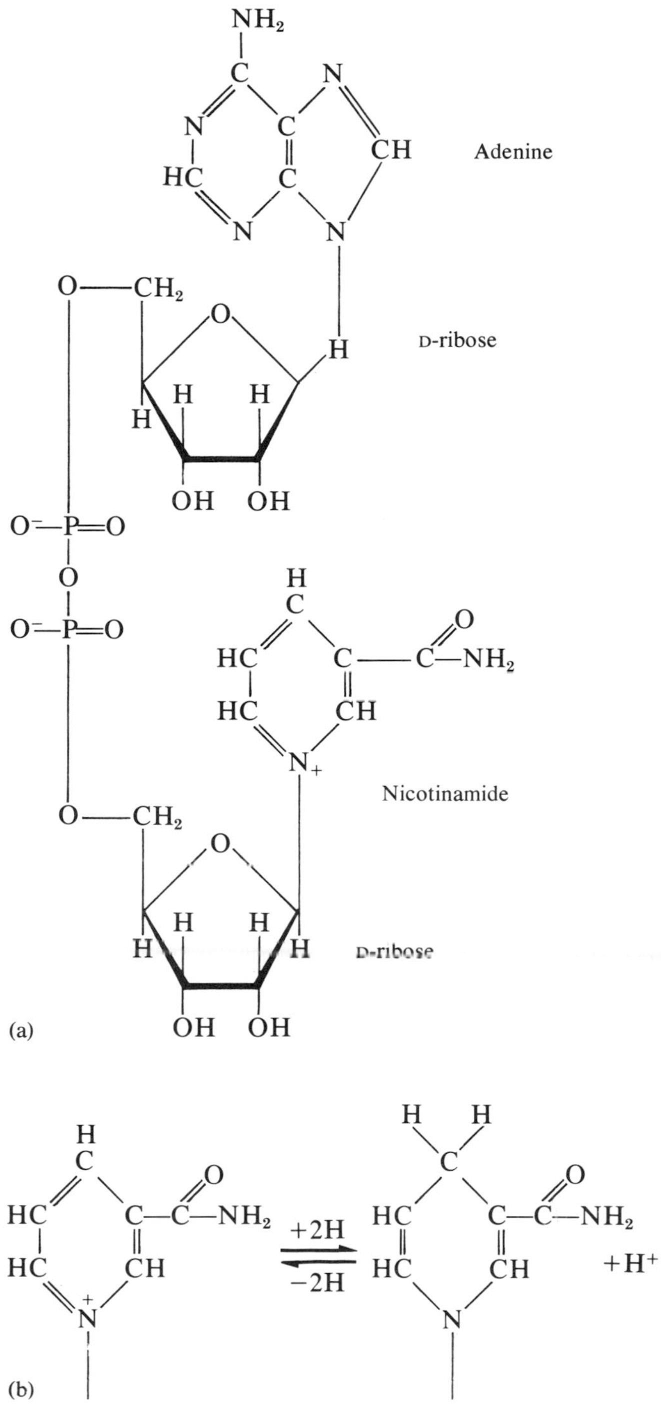

Adenine

D-ribose

Nicotinamide

D-ribose

(a)

(b)

substrate couple, since $NAD^+/NADH$ can act as a *common intermediate* shared by two NAD-linked, coupled reactions, each catalyzed by a specific enzyme. For example,

$$H_2X \diagdown \qquad NAD^+ \diagdown \qquad H_2Y$$
$$X \diagup\diagdown NADH + H^+ \diagup\diagdown Y \qquad [3.18]$$

In most cases NAD is bound to an enzyme protein as its **coenzyme.** Coenzymes are usually nonprotein, organic molecules essential for the activity of some enzymes. NAD-linked enzymes generally are involved in breakdown of fuel molecules and other energy-releasing reactions.

## ENZYMES: CATALYSTS OF LIFE

A **catalyst** is a substance that modulates or influences reaction rates without altering the equilibrium point of a reaction. Nonliving systems use inorganic catalysts, such as iron or hydrogen ions. Living systems also use these same kinds of inorganic catalysts, but they depend primarily on *protein* catalysts called **enzymes.** Enzymes are more uniquely suited than inorganic catalysts because enzymes possess regulatory properties as well as catalytic properties. Organic chemical reactions are notoriously slow under the mild conditions of temperature and pressure of the average cell. Enzymes not only speed up these reaction rates, but they also exert a fine control over the actual rate at which a reaction will proceed. Enzymes themselves are subject to regulation, both by external factors and by particular qualities of their own.

Since there are about 3000 different enzymes in an average cell's repertory, regulation of enzyme activity and regulation of rates of reactions in progress are important aspects of the orderliness that distinguishes

**Figure 3.4**
(a) The energy-transferring electron carrier NAD. (b) The pyridine ring in the nicotinamide portion of the NAD molecule undergoes oxidation-reduction, producing $NAD^+$ and NADH, respectively.

the open system of the cell from its disordered surroundings. The activities taking place at different times in a cell can undergo profound and rapid changes under the catalytic direction of enzymes. The distinctiveness of the chemistry of different cells or of different compartments of a cell can be traced to differential enzyme content and activity. These features, in turn, are the outcome of gene action and of the regulation over the activity of enzymes which are produced according to genetic instructions in certain cells at certain times.

Although first formulated in the 1850s, the concept of catalysis was not expanded to include proteins as biological catalysts until the 1930s. The first enzyme to be purified in crystalline form was **urease,** which catalyzes the conversion of urea to carbon dioxide and ammonia in a hydrolysis reaction.

$$H_2N \diagdown$$
$$C=O + H_2O \rightarrow CO_2 + 2NH_3;$$
$$H_2N \diagup$$
$$\Delta G^0 = -13.8 \text{ kcal/mole} \qquad [3.18]$$

Crystalline urease was first reported in 1926 by James B. Sumner. During the next 10 years there were reports of other enzymes that were purified or crystallized, all of which were shown to be proteins.

Naming enzymes is a formal procedure based on an internationally accepted system. We usually use their informal names, however, rather than the more cumbersome (but more informative) formal nomenclature. The suffix **-ase** identifies almost all enzymes, as with urease. Sometimes the informal name refers to the chemical acted upon (urease and urea), and sometimes to the nature of the catalyzed reaction (a hydrolase catalyzes hydrolysis reactions). A few enzymes are still called by their original names because these names are very familiar and occurred often in the extensive older literature of chemistry. Enzymes such as pepsin and trypsin, which digest proteins, are examples of enzymes whose older names have been

retained in common usage even though the suffix *-ase* is not included.

**Enzyme Activity**

All the complexity of cellular biochemistry is based ultimately upon simple chemical reactions that are catalyzed at each step by a specific enzyme. Enzyme-catalyzed reactions follow the same basic rules of chemistry as any other chemical reaction that may take place in our world. The hydrolysis of urea can take place spontaneously in the absence of any catalyst or in the presence of some catalyst, yielding the same end products and energy release. The spontaneous, uncatalyzed reaction rate can be speeded up if $H^+$ ions are provided as a catalyst. The rate of hydrolysis is made considerably faster when the enzyme urease is provided to the system. In all three instances there is a difference only in the rate of reaction, but not in the end products, the energy balance sheet, the direction of reaction, or the point at which equilibrium is reached.

The basis for the differences in rate of urea hydrolysis, as for other reaction rates, is the relative ease with which the reactant molecules pass over the **barrier of activation energy** (Fig. 3.5). In order for urea to react with water molecules there must be sufficient free energy to form an activated, or higher energy, complex. The activation-energy barrier is reduced to the greatest extent in an enzyme-catalyzed reaction system. For this reason, the reaction occurs at a faster rate when urease is present than when $H^+$ ions are present or in the total absence of a catalyst (Fig. 3.6).

The main function of an enzyme therefore is to lower the activation-energy barrier to a chemical reaction. Reactions will proceed spontaneously in the direction of lower energy potential (higher $\Delta G$) whether or not a catalyst is present because molecules possess thermal energy. Some molecules have more thermal energy than others, and there is a statistical probability of some molecules interacting after random collisions, thus overcoming the energy barrier, at all times. If thermal energy is added to the system by raising the temperature, molecules gain additional energy and more easily hurdle the activation-energy barrier, since interactions occur more frequently between the reactants. The reaction rates therefore accelerate at higher temperatures. Cells experience little change in temperature, however.

Enzymes act to lower the activation-energy barrier and thus increase the probability of molecules passing this barrier and going on to completion of a reaction. The enzyme does this by combining with the reactant, or **substrate** (molecule changed in the reaction), in *temporary* association. Since the enzyme-substrate complex has a lower activation-energy requirement, more substrate molecules can pass beyond the energy of activation barrier per unit time. Some enzymes can process millions of substrate molecules every second, repeatedly dissociating from the temporary enzyme-substrate complex. A relatively few enzyme molecules can therefore handle millions of substrate molecules. While inorganic catalysts such as iron or hydrogen ions are relatively nonspecific in interactions with substrates, enzymes show a high level of *specificity* for particular substrates. Since a particular enzyme has a

highly specific affinity for a particular substrate, it is highly *efficient* in catalyzing the particular reaction in question. Each enzyme handles its own share of the workload, and many different reactions can therefore proceed simultaneously in the confines of an organelle or a cell.

Like any chemical reaction, increasing the temperature of the reactants will lead to an increase in reaction rate for enzyme-catalyzed systems. But because enzymes are proteins, they are subject to **denaturation** (loss of original active form) when many weak bonds are broken at the elevated temperature. Many enzymes are inactivated at 45°C, and most are rapidly denatured at 55°C and above.

### Substrate Specificity

The mechanism of enzyme action first proposed by Emil Fischer in 1894 was described as a "lock and key" relationship. Many enzymes are so discriminating between substrates that virtually identical molecules may be handled by different enzymes in

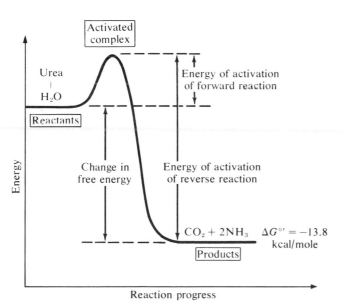

**Figure 3.5**
Diagram showing the energy relations of a reaction in which urea is hydrolyzed to form carbon dioxide and ammonia, and the reverse reaction.

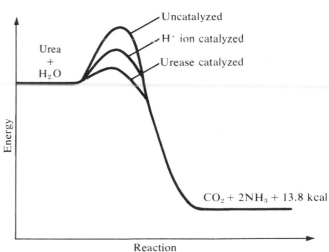

**Figure 3.6**
Relative effects on reducing the activation-energy barrier in hydrolysis of urea when catalyzed by $H^+$ ions or by the enzyme urease.

chemical reactions. For example, different enzymes catalyze reactions involving optical isomers of the same compound; different enzymes break β-glucose and β-galactose linkages even though the only apparent difference is in the position of the hydroxyl group on carbon atom 4 of these hexose sugars; and many similar examples exist. The analogy of lock and key seemed appropriate in describing such exquisite discrimination by enzymes, due to spatial and conformational relationships between enzyme and substrate molecules in their interactions.

Some enzymes, however, are able to interact with a specific group of molecules rather than a single compound. For example, trypsin and chymotrypsin disrupt peptide links in many proteins. Even in such cases, however, the different substrates have a similar molecular site with which the enzyme interacts (Fig. 3.7).

Specificity is believed to be determined largely by the catalytic site, or **active site,** of the enzyme. It is this part of the enzyme which combines specifically with the substrate molecule. The active site of an enzyme contains the particular functional groups that can bind the substrate (or its relevant portion) and then bring about the catalytic event. A great deal of our knowledge about the nature of enzyme specificity has come from studies of enzymes whose amino acid sequence is known and whose three-dimensional conformation has been constructed. From such molecular models it is clear that the critical side groups on the

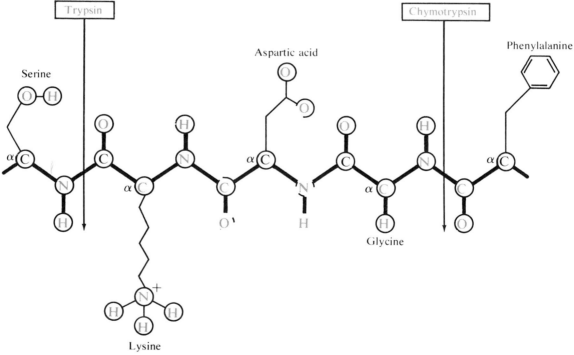

**Figure 3.7**
Trypsin digests peptide bonds that are followed by either lysine or arginine, which have positively charged side chains, while chymotrypsin acts only at peptide bonds followed by large, hydrophobic side chains, such as phenylalanine. These and other proteases act on the same peptide bonds, but each of the enzymes cleaves only specific links depending on the side chains projecting from the polypeptide "backbone."

protein backbone which participate in catalysis are not necessarily very close together along the length of the protein chain. Folding brings these groups into closer proximity (Fig. 3.8).

The substrate molecule is oriented to the enzyme by the active site(s) of the enzyme, and making and breaking of chemical bonds is enhanced by enzyme-substrate complexing (Fig. 3.9). Interestingly, the geometry of the active site is closely related to the shape of the particular substrate molecule and to the way in which this molecule is processed in the catalytic reaction. Some enzymes have an active site that is shaped like a crevice or cleft in the molecule. Such enzymes

seem to cut bonds which are located in relatively exposed positions in the substrate molecule. The substrate chain appears to fit into the cleft containing the active site of the enzyme, very much like a thread that can be snipped anywhere in the middle of its length by a pair of scissors (Fig. 3.10). Other enzymes have the active site in a pit or depression. In such cases the enzyme can fit up against the substrate molecule and cut away at the bond near the end of the substrate chain. In many situations the substrate molecule is tightly folded and exposed strands are rare, so that it would be difficult to interact with the chain except for the occasional exposed ends. An ac-

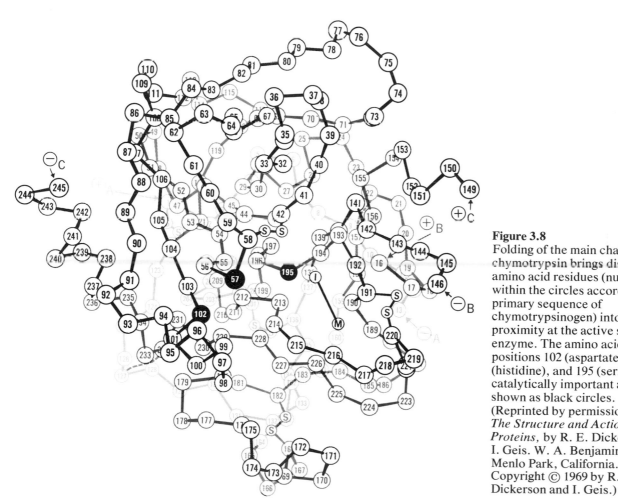

**Figure 3.8**
Folding of the main chain of α-chymotrypsin brings distant amino acid residues (numbered within the circles according to primary sequence of chymotrypsinogen) into proximity at the active site of the enzyme. The amino acids at positions 102 (aspartate), 57 (histidine), and 195 (serine) are catalytically important and are shown as black circles. (Reprinted by permission from *The Structure and Action of Proteins,* by R. E. Dickerson and I. Geis. W. A. Benjamin, Inc., Menlo Park, California. Copyright © 1969 by R. E. Dickerson and I. Geis.)

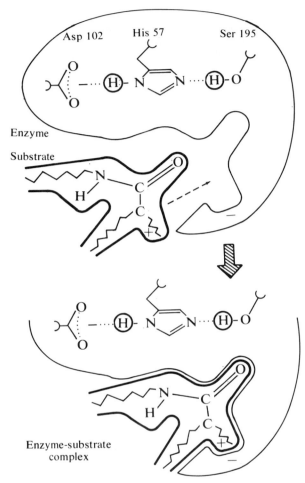

**Figure 3.9**
The complexing of the enzyme chymotrypsin with its
peptide substrate occurs at the active site of the enzyme
Three catalytically important amino acid residues in the
enzyme are shown (see Fig. 3.8)

tive site shaped like a crevice would be ineffective in
this case, whereas the end of such a folded substrate
would be more easily accommodated and brought into
proper juxtaposition if the enzyme's active site were
shaped like a shallow pit.

The active site occupies a relatively small area of
the surface of the enzyme molecule, which means that
only a small portion of the molecule actually engages
in catalysis. Other surface areas may be involved in

interactions with molecules that regulate enzyme
activity, and in subunit associations that contribute to
enzyme quaternary structure. Most of the enzyme's
components, however, are involved in development of
secondary and tertiary structure of the globular
molecule.

### Self-Regulation

We will restrict ourselves in this discussion to those
characteristics of enzymes themselves which
contribute to regulation of the overall rate of reactions
during their interactions in cellular metabolism. An important phenomenon in the category of enzyme self-

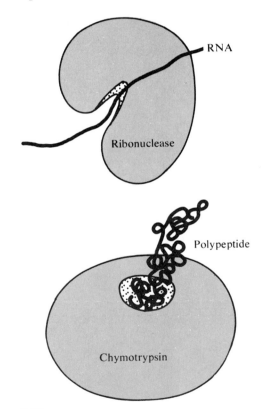

**Figure 3.10**
Diagram illustrating the relationship between active site
geometry and conformation of the substrate in relation to the
way the substrate is processed. Ribonuclease, a crevice
enzyme, snips exposed regions in the middle of the RNA
chain, whereas chymotrypsin, a pit enzyme, attacks the
accessible ends of the polypeptide chain that is its substrate.

regulation systems is called **feedback inhibition.** The situation usually involves a multi-step reaction sequence, with a specific enzyme catalyzing each step or reaction in the series. The end product of such a reaction sequence will often inhibit the formation of the initial product catalyzed by the first enzyme of the series, that is, the first enzyme in the pathway is inhibited by the end product of that pathway (Fig. 3.11). This enzyme is called a **regulatory enzyme,** and the inhibitory metabolite (a reactant or reaction product of one step in a metabolic pathway) is called a **modulator.** Since the inhibitor and substrate of the regulatory enzyme are not structurally similar molecules, the term **allostery** (meaning "different structure") has been coined to describe the modification of an enzymatic reaction by a compound that is different in shape from the true substrate. Since there is strict specificity of fit between the substrate molecule and the active site of an enzyme, **allosteric modulators** must bind to the enzyme at some other site than its catalytic region.

An allosteric modulator may be an activator or an inhibitor of enzyme action, which is why the neutral term is more appropriate than either of the other two, unless we discuss some specific activating or inhibiting modulator molecule. The action of an allosteric modulator has been shown to be due to an induced change in the three-dimensional shape of the enzyme. However the conformational change may take place, and there is some difference of opinion on the subject, modulator molecules are not consumed in reaction with the enzyme and are therefore still available in the cell for metabolism once released from the complex with the enzyme. Using the same set of components, enzymes and modulators, the cell regulates its metabolism through this system of "on-off" switching of catalysis.

Another way in which enzymes regulate their activity in relation to the cellular pool of metabolites has been called **cooperativity.** Here, too, we are dealing with change in enzyme shape upon interaction with substrate molecules in metabolism. In binding with a substrate molecule, one subunit of an enzyme may undergo a conformational change that makes the next subunit more receptive to the substrate, and therefore more readily bound to it. A kind of "domino" effect leads to more efficient enzyme—substrate interactions and higher enzyme activity at relatively lower substrate concentrations than in noncooperative systems. In other cases, binding of substrate to enzyme apparently induces a conformational change in an enzyme subunit that makes the next subunit less receptive to the substrate molecules. Enzyme activity in this situation remains at lower levels even when very high concentrations of substrate exist in the cell.

Regulation of enzyme activity by protein subunit cooperativity helps explain the different degrees of sensitivity of enzymes to the actual concentrations of substrates in the cell. Positively cooperative enzymes are extremely sensitive to minor fluctuations in substrate concentrations, whereas negatively cooperative enzymes appear rather insensitive to such fluctuations. Each type of cooperativity provides an advantage to particular catalytic situations, especially in relation to the amount of available substrate at any one time in the cell.

Systems involving modulators are different but complementary to systems involving cooperativity.

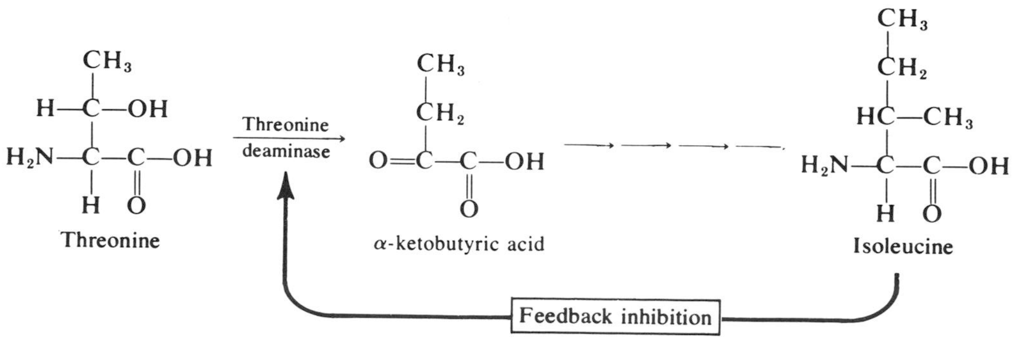

**Figure 3.11**
The feedback inhibition of activity of the regulatory enzyme threonine deaminase by the allosteric modular L-isoleucine, the end product of the five-step reaction sequence.

Positive modulators (activators) turn on enzymes and negative modulators (inhibitors) turn them off. Cooperativity increases or decreases the sensitivity of these enzymes to environmental fluctuations in metabolite concentrations. While modulators play a major role in metabolic regulation, cooperativity adds a further dimension of fine-tuning of the catalytic system. All or most of these regulations depend on the ability of an enzyme to undergo changes in shape, some of which involve a displacement in space of only a few angstroms.

### Genetic Regulation Of Enzyme Synthesis

Regulation of enzyme activity provides a fine control over cell metabolism, and is effective at the level of the enzyme protein itself. The important but coarser controls over catalysis are those determining the *kind* of enzyme and the *amount* of enzyme synthesized in the cell. These aspects of regulation are influenced by the genes. The gene codes for enzyme protein, thus determining the kind of enzyme that can be synthesized in particular cells with appropriate nucleotide sequences. If the genetic information specifying a certain enzyme is absent from a cell or organism, that protein will not be manufactured. If the gene is mutant and either fails to direct synthesis of the

enzyme or directs synthesis of a defective enzyme, the catalyst is not available for its specified reaction.

If a particular gene is present, the enzyme it specifies may be synthesized. But the amount of enzyme manufactured is controlled by processes that involve the gene directly rather than the enzyme for which it codes. DNA of the gene contains the blueprints for protein synthesis, but these instructions must be copied into intermediary RNA molecules, called **messenger RNA.** The messenger RNA travels to the sites of protein synthesis and directs polymer construction from specified amino acid building blocks in a sequence dictated by the coded nucleotide sequences in the messenger copy of DNA (Fig. 3.12). The process of copying DNA instructions into messenger RNA is called **transcription.** The process of **translation** involves protein biosynthesis according to the genetically coded instructional sequence.

If the proper genetic instructions are present, a cell has the potential for synthesizing particular proteins, many of which are enzymes. This tells us the basis for manufacturing different *kinds* of enzymes in cells, or at least the potential for their manufacture. The *amount* of each enzyme synthesized in a cell is regulated by controls operating at the levels of transcription and, to a lesser-known degree, of translation. Two particular categories of enzymes have been studied most inten-

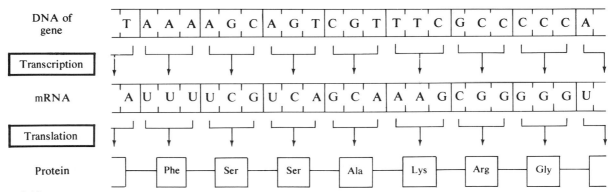

**Figure 3.12**
The *kind* of enzyme synthesized in the cell depends on the genetic blueprint in DNA. The *amount* of enzyme synthesized depends on controls over transcription of genetic information into messenger RNA and translation of that message into the primary structure of the enzyme protein.

sively. Both fluctuate rapidly and substantially in response to the presence and absence of substrates. These are **inducible** and **repressible** enzymes.

Inducible enzymes are formed in response to the introduction of the substrate on which they act. For example, when glucose is not available the cell will begin to make enzymes needed to metabolize an alternate energy source that is available. If there is no glucose but there is lactose or some other $\beta$-galactoside present, the cell begins to synthesize the enzyme $\beta$-galactosidase. Only a few molecules of this enzyme usually are found in cells grown in glucose-containing solutions, but up to 5000 molecules can be found in some cells within minutes after adding lactose to the glucose-free solution. When lactose is removed or used up, synthesis of $\beta$-galactosidase stops abruptly (Fig. 3.13).

Synthesis of repressible enzymes, on the other hand, stops when the product of the reaction is present in excess. For example, the amino acid tryptophan is needed in relatively small amounts, since it is not an abundant constituent of proteins. When tryptophan is synthesized in excess of its need in protein synthesis, further manufacture of this amino acid stops because the enzymes in the reaction pathway are repressed. The actual synthesis of new enzyme molecules stops. This phenomenon is called **end product repression** of enzyme synthesis, since the last product of a reaction sequence usually shuts down the system. In addition, any of this enzyme existing in the cell will be inhibited from further activity. Shutting down existing enzyme activity is feedback inhibition of *enzyme activity;* shutting down manufacture of new enzyme molecules is end product repression of *enzyme synthesis*. The two phenomena together act by shutting down catalysis when substrate is in excess.

Inducible enzymes generally catalyze breakdown reactions. They are usually found in trace amounts, but may exist in higher concentrations when their substrates are present. Repressible enzymes usually catalyze biosynthesis reactions. They are present and active in growing cells in relation to the amount of reaction product being made and used in biosynthesis. Enzymes are inducible or repressible; the responses to

metabolites are referred to as **enzyme induction** and **enzyme repression,** or induction and repression of enzyme synthesis.

The above discussion sketched in the general outlines of differential enzyme synthesis, but we must now look at the parts of the control system that regulate the amount of enzyme being synthesized at any one time in the cell. Enzymes will be synthesized if messenger RNA has been transcribed from the coded genes. Control over manufacture of the messengers will therefore determine whether or not RNA transcripts are available for translation into proteins. Transcription of messenger RNA is turned "off" and "on" according to interactions between two kinds of molecules, **repressors** and **inducers.** Repressors are proteins made according to information in regulatory genes. Inducers are the substrates or reaction products involved in a catalytic pathway.

Synthesis of inducible enzymes will take place if inducers (substrates of these enzymes) are present. Repressor protein, manufactured at all times from information in regulatory genes, blocks transcription of the gene coding for inducible enzyme. When the inducer is present, however, it combines with the repressor and leads to detachment of the molecular

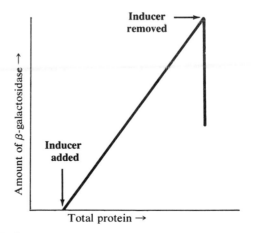

**Figure 3.13**
Synthesis of the inducible enzyme $\beta$-galactosidase begins on addition of the inducer substrate and declines very rapidly when substrate is depleted or removed from the cell.

block from the DNA. Once the repressor is removed, the enzyme that catalyzes messenger RNA synthesis (called RNA polymerase) is free to move along the gene, and DNA information is copied into newly made messenger RNA. The messenger can be translated into enzyme protein afterward. In the case of $\beta$-galactosidase, repressors block transcription in the absence of lactose. Once lactose is provided, the inducer binds with repressor thereby causing repressor to be released, and the block along DNA is removed (Fig. 3.14).

Synthesis of repressible enzymes will *not* take place if inducers are present. In this case the repressor product of the regulatory gene for the system cannot prevent transcription all by itself. When the inducer is present, it binds to the repressor and the repressor-inducer complex is then able to bind to DNA and block

transcription of messenger RNA. The inducer in these systems is usually referred to as a **co-repressor,** since it acts together with the repressor protein to block transcription (Fig. 3.15).

Both induction and repression of enzyme synthesis are regulated by negative control systems. In each case, the synthesis of enzyme takes place only when the repressor is not blocking transcription of messenger RNA from DNA coded for the enzyme protein. The two systems are only different sides of the same coin.

We will now look at a metabolic pathway and see how the stored energy in a food molecule can be extracted and salvaged for ultimate use in cellular work. These events depend on specific enzymes acting in conjunction with energy transfer systems, such as those described earlier.

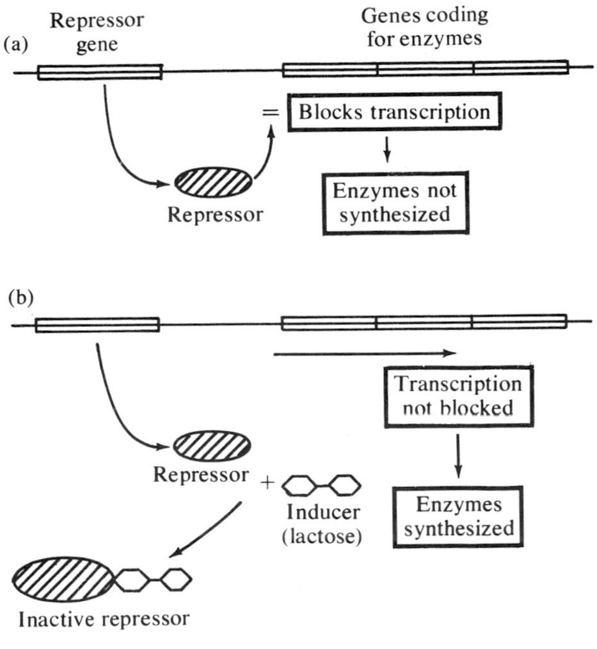

**Figure 3.14**
Diagram illustrating the interaction between the repressor product of the regulator gene and transcription of the genes coding for inducible enzymes in the (a) absence and (b) presence of the inducer metabolite lactose.

## GLUCOSE BREAKDOWN BY GLYCOLYSIS

The most widely used fuel molecule in cells is the hexose sugar glucose. In addition to obtaining glucose directly in foods, glucose can be derived from a variety of stored foods and from various carbohydrates after suitable chemical processing. The pathway we will examine is a **fermentation** sequence, that is, a group of sequential reactions that can take place in the absence of molecular oxygen. In fermentations, oxidation-reductions take place, with the end product of the sequence acting as the last electron acceptor in the group. In other reaction systems that can incorporate $O_2$ in the reaction sequence, $O_2$ usually can act as the terminal electron acceptor. The particular fermentation we will discuss is known as **glycolysis** ("dissolution of sugar"), which takes place without the intervention of molecular oxygen. Glycolysis is among the earliest and best-studied metabolic pathways. In addition, the same glycolytic pathway has been shown to occur in the simplest bacteria up through the complex mammalian animals, such as ourselves. In honor of Gustav Embden and Otto Meyerhof, two German chemists who provided important information and insight into glycolysis during the 1920s and 1930s, the reaction sequence is also known as the Embden-Meyerhof pathway.

An overall summation of glycolysis may be written as the anaerobic breakdown of glucose into two molecules of the 3-carbon compound lactic acid (lactate), or

$$C_6H_{12}O_6 \rightarrow 2CH_3-CHOH-COOH;$$
glucose    2 lactic acid

$$\Delta G^0 = -47.4 \text{ kcal/mole}$$

[3.19]

The reaction is thermodynamically favorable and therefore can proceed spontaneously with a large decline in free energy, if a suitable pathway is available to catalyze it. This equation is not described as accurately as it might be. We know that inorganic phosphate (referred to as $P_i$), ADP, and ATP also participate in the sequence. A more appropriate equation than [3.19] would therefore be

$$\text{glucose} + 2P_i + 2\text{ADP}$$
$$\downarrow$$
$$2 \text{ lactate} + 2\text{ATP} + 2H_2O$$
$$\Delta G^0 = -32.4 \text{ kcal/mole}$$

[3.20]

Two coupled processes are summed up in equation [3.20]. In one process, glucose is broken down to two molecules of lactic acid and, in the other, two molecules of ADP and two of phosphate are used to make two ATP. Glucose breakdown and ATP formation cannot take place independently of one other; each depends on the other.

We should also note that there are 15 kcal less in the free-energy decline in reaction [3.20] than in [3.19]. This difference approximately reflects the 7.3 kcal needed to produce each mole of ATP from ADP and phosphate. This means that a significantly large part of the free energy lost during glucose breakdown to lactic acid is *conserved* in the form of ATP. About 32 percent (15/47.4 × 100) of the free-energy difference between the reactants and the products has been retained by the cell. In fact, this is the major rationale for fuel breakdown during glycolysis and many other degradative processes.

Although some of the released fuel energy has been conserved in ATP, the vastly greater potential energy of glucose is still tied up in the lactic acid products. If we were to burn glucose completely to carbon dioxide and water, in the presence of molecular oxygen, the $\Delta G^0$ of the entire process would be −686 kcal/mole. About 7 percent of the energy in glucose is released in glycolysis (47.4/686 × 100), which is a far less efficient process than aerobic breakdown of fuels. We will discuss the further processing of glycolytic breakdown products in Chapter 5, when we describe activities of the eukaryotic mitochondrion. For anaerobic cells that depend on glycolysis, much more glucose is consumed per unit of time per unit of weight than is the case for aerobic cells. As much as ten times more glucose may be consumed by some anaerobic cells as by aerobic cells, to accomplish the same amount of cellular work.

**The Sequence of Reactions**

There are eleven different enzymes involved at specific steps in the pathway, each of which catalyzes one of the consecutive glycolytic reactions leading from glucose to lactic acid. The reactions can be

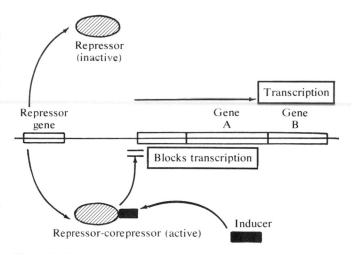

**Figure 3.15**
Diagram illustrating the negative control of repressible enzyme synthesis in prokaryotes. Transcription of mRNA is blocked when co-repressor binds with repressor at the DNA. Transcription proceeds when the co-repressor metabolite is absent, since repressor alone cannot bind to the DNA in such a system.

divided into two general stages (Fig. 3.16). In the first stage, the glucose molecule is energized by phosphorylation at the expense of two molecules of ATP. The phosphorylated glucose is then split to yield 2 molecules of the 3-carbon sugar, glyceraldehyde 3-phosphate. This molecule is converted to lactic acid during the second stage of glycolysis. There is a net gain of two ATP per glucose, since four ATPs are formed but two are used up earlier in phosphorylating glucose at the beginning of the glycolytic sequence.

The initial step in degrading glucose is modifying the molecule to a form which can be metabolized, or "priming the pump", so to speak. The conversion of glucose to glucose 6-phosphate costs the cell some of its stored chemical energy, in the form of ATP. Further cost is incurred when another ATP phosphoryl group transfer takes place to yield fructose 1,6-diphosphate. This is a cost that is repaid twice over to the cell, for a simple reason. When fructose diphosphate is cleaved to form *two* phosphorylated 3-carbon intermediates, *each* of these can participate in the second stage reactions and *each* can be coupled to two ATP-making steps. The cell gets two ATPs from each glucose this way. If fructose 6-phosphate were split in two, instead of a diphosphate, only one-half of the products would be able to enter into stage-2 reactions because the other product would not be phosphorylated and could not go on to "downhill" reactions. With only half the glucose continuing on to lactic acid, undergoing two coupled reactions to make ATP, the net yield from each glucose would be only one ATP (one ATP used in stage 1, two made in stage 2; net yield is 1 ATP per glucose). Such lower efficiencies are known to occur in glucose breakdown in some bacteria.

In all fermentations of glucose under anaerobic conditions, glucose is broken down to two or more fragments, one of which is oxidized by another. Some of the energy released in the oxidation-reduction process is conserved as ATP. In stage 2 of glycolysis we can see that glyceraldehyde 3-phosphate is oxidized to 1,3-diphosphoglycerate. The ultimate electron acceptor of the two electrons removed from glyceraldehyde 3-phosphate is pyruvate, another

3-carbon intermediate in the sequence. Pyruvate accepts these electrons and becomes reduced to lactate

$$\text{glyceraldehyde 3-phosphate} + P_i + \text{pyruvate}$$
$$\downarrow \qquad [3.21]$$
$$\text{1,3-diphosphoglycerate} + \text{lactate}$$

One 3-carbon fragment of glucose has been oxidized by another 3-carbon fragment; glyceraldehyde 3-phosphate is oxidized to 1,3-diphosphoglycerate, and pyruvate is reduced to lactate. NAD acts as a common intermediate in electron transfers in these oxidation-reductions. $NAD^+$ accepts electrons from glyceraldehyde 3-phosphate and is thus reduced to NADH, which then donates electrons to pyruvate, and $NAD^+$ is regenerated (see Fig. 3.16).

It is because this enzymatic oxidation of glyceraldehyde 3-phosphate by $NAD^+$ proceeds together with uptake of phosphate ($P_i$) that a high-energy derivative is formed (1,3-diphosphoglycerate), which in turn donates its phosphoryl group at carbon-1 to ADP to form ATP. The reactions are:

$$\text{glyceraldehyde 3-phosphate} + P_i + NAD^+$$
$$\downarrow \qquad [3.22]$$
$$\text{1,3-diphosphoglycerate} + NADH + H^+$$
$$\text{and}$$
$$\text{1,3-diphosphogylerate} + ADP$$
$$\downarrow \qquad [3.23]$$
$$\text{3-phosphoglycerate} + ATP$$

The basic point to be illustrated by these reactions is that oxidation is accompanied by the formation of a

**Figure 3.16**
The processing of glucose to lactate occurs in two stages during glycolysis. Glucose is phosphorylated before being rendered into two $C_3$ compounds in the first stage, with 2 ATP expended in these events. Each half of the original glucose molecule is then processed in the same pathway—an economical metabolic device—to produce lactic acid at the end of the second stage. There are 4 ATP produced per glucose, but the net yield is only 2 ATP, since 2 ATP were used in first stage phosphorylations. See text for details.

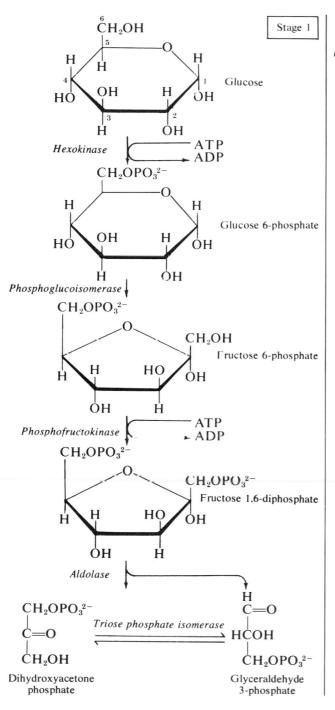

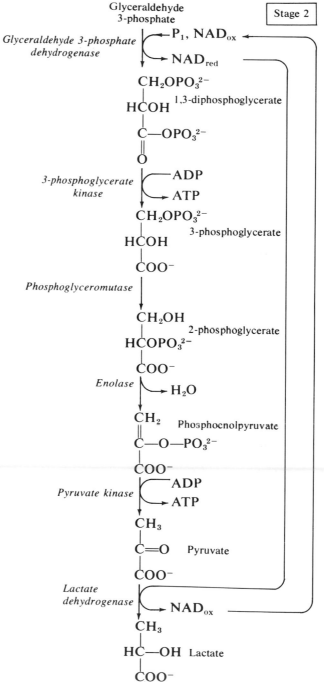

high-energy derivative of the oxidation product, which ultimately leads to donation of a phosphoryl group to ADP, making ATP. The common intermediate in reactions [3.22] and [3.23] is 1,3-diphosphoglycerate, which is a high-energy oxidation product. This general feature is typical of many of the known energy-conserving oxidation-reduction reactions in the cell.

Since it is the oxidation-reduction reactions that are responsible for generating high-energy phosphorylated compounds, which can then donate their phosphoryl groups to ADP in enzymatically catalyzed reactions, it may seem that the second site of ATP formation in the glycolytic pathway does not follow this basic rule. Although it may seem that oxidation-reduction is not involved in the formation of phosphoenolpyruvate from its precursor 2-phosphoglycerate, there is a shift of electrons within the molecule so that it can be regarded as an intramolecular oxidation-reduction reaction. The internal modifications of phosphoenolpyruvate cause a shift in its internal energy, so that phosphoenolpyruvate becomes a high-energy compound ($\Delta G^0 = -14.8$ kcal/mole) and donates its phosphoryl group to ADP in the next reaction

$$\text{phosphoenolpyruvate} + \text{ADP}$$
$$\Updownarrow \qquad\qquad [3.24]$$
$$\text{pyruvate} + \text{ATP}$$

Except for the two ATP-forming steps we have just discussed for glycolysis, no other reactions in the pathway have enough standard free-energy change to sponsor ATP formation in a coupled reaction. Each of these other steps in glycolysis proceeds with a relatively small $\Delta G^0$, as compared with about 7 kcal/mole required for ATP formation from ADP.

If we check the balance sheet of glycolysis by adding up all the reactants entering the scheme and all the products emerging, we should find that the overall reaction is the same as we proposed earlier in reaction [3.20]. One glucose molecule enters the pathway and is primed by two ATP molecules eventually forming fructose 1,6-diphosphate, which is cleaved to two 3-carbon fragments that are interconvertible. It is two 3-carbon fragments in the form of glyceraldehyde 3-phosphate that become oxidized to 3-phosphoglycerate; this requires two molecules of ADP, two molecules of $P_i$, and two molecules of $NAD^+$. Later, two more molecules of ADP enter the pathway to accept phosphoryl groups from two molecules of phosphoenolpyruvate. The two molecules of NADH that were formed when glyceraldehyde 3-phosphate was oxidized are now used in the reduction of pyruvate to lactate.

The complete equation for glycolysis, including all that entered and all that emerged is

$$\text{glucose} + 2\text{ATP} + 2P_i + 4\text{ADP} + 2\text{NAD}^+ + 2\text{NADH}$$
$$\downarrow \qquad\qquad [3.25]$$
$$2\text{ lactate} + 2\text{ADP} + 4\text{ATP} + 2\text{NADH} + 2\text{NAD}^+ + 4\text{H}_2\text{O}$$

If we cancel the components appearing on both sides of the equation, we have the net equation

$$\text{glucose} + 2P_i + 2\text{ADP}$$
$$\downarrow \qquad\qquad [3.26]$$
$$2\text{ lactate} + 2\text{ATP} + 2\text{H}_2\text{O}$$

$$\Delta G^0 = -32.4 \text{ kcal/mole}$$

This agrees with equation [3.20], and it is the overall statement for anaerobic glycolysis.

### Regulation of the Rate of Glycolysis

All eleven enzymes catalyzing the glycolysis reaction steps are found in the soluble portion of the cytoplasm (cytosol), and apparently are neither grouped nor a part of any known structure within the cell. Reactants and products diffuse toward and away from the enzyme molecules without significant hindrance.

In common with other multienzyme systems, there is at least one regulatory enzyme that not only catalyzes a specific reaction in the pathway but also is sensitive to some modulator molecule, which may or may not be its own substrate. One of the regulatory enzymes of glycolysis is *phosphofructokinase,* which catalyzes the phosphorylation of fructose 6-phosphate to fructose 1,6-diphosphate in stage 1 of the pathway

$$\text{fructose 6-phosphate} + \text{ATP} \rightarrow$$

$$\text{fructose 1,6-diphosphate} + \text{ADP.} \quad [3.27]$$

The rate of the reaction is strongly dependent on the relative ratio of ADP to ATP. When more ADP is present, the rate of the catalyzed reaction is increased. ADP acts as a positive modulator, or activator, of the enzyme. When there is an excess of ATP relative to ADP concentration, the reaction rate decreases. ATP acts as a negative modulator, or inhibitor, of phosphofructokinase. Since the *sum* of the concentrations of ADP and ATP in the cell is constant, an increase in ATP due to phosphorylation of ADP will result in a high ATP/ADP ratio. This favors the inhibition of the enzyme. If more ADP is made because of some cellular energy-requiring reactions using ATP, the increase in ADP and decrease in ATP leads to activation of the enzyme and to a higher rate of glycolysis.

Since ATP is one end product of glycolysis, it is part of a feedback inhibition system, since the whole reaction sequence slows down when ATP, an end product, inhibits an enzyme catalyzing an early step in the pathway.

### SUMMARY

1. The application, movement, and transformation of energy ultimately underlie all physical and chemical processes. Thermodynamics deals with energy relations in general, but bioenergetics is concerned only with energy in living systems. The First and Second Laws of thermodynamics define the principle of energy conservation in the universe and the direction of a process in which energy is exchanged between a system and its surroundings. The energy available to do work, or free energy, is more easily measured as the difference in energy content ($\Delta G$) of a system in its initial state and in its final state at the end of the reaction, and allows us to make important predictions about reactions.

2. Living cells are open systems, exchanging energy and matter with their surroundings, which exist in different steady states. They are not likely to reach the state of thermodynamic equilibrium. Nonequilibrium open systems are capable of performing work because they do not reach equilibrium, and they proceed toward a decline in free energy at a minimal rate in comparison with nonliving systems.

3. Energy-requiring and energy-releasing reactions of metabolism are coupled in living systems, which increases efficiency and allows greater orderliness. Useful free energy is stored as chemical energy in living cells, most commonly as ATP. The ADP-ATP cycle links reactions, acting as common intermediates in sequences in which energy from breakdown reactions becomes available for biosynthesis reactions. Energy transfers in metabolism may proceed by phosphoryl group transfer via the ADP-ATP system, by electron transfer via redox couples in oxidation-reduction reactions, and by other means. Redox couples are the oxidized and reduced forms of a compound, such as the $NAD^+$/NADH couple, and they act as common intermediates in coupled energy-requiring and energy-releasing reactions involving electron transfer.

4. Reaction rates in living systems are modulated by unique protein catalysts called enzymes, which function primarily by lowering the barrier of activation energy of a reaction and thereby increase the otherwise sluggish organic chemical reaction rates. Enzymes combine temporarily with specific substrate molecules, which then undergo changes during metabolism. Enzymes are highly efficient catalysts, since they are highly specific for certain substrate molecules. Specificity of an enzyme's affinity for its substrate depends on the region of the enzyme called the active site, where enzyme and substrate fit together like a lock and key. The geometry of the active or catalytic site corresponds with the particular substrate and the way in which the substrate is processed in the reaction.

5. Two systems of self-regulation of enzyme activity are feedback inhibition and cooperativity. In feedback inhibition, the activity of the first enzyme in a multi-step pathway is inhibited by the pathway's end product, which can then be considered a modulator of enzyme activity. Modulators may stimulate as well as inhibit enzyme activity, but in either case they are allosteric, since modulator shape is different from the shape of the substrate acted upon by that enzyme. Allosteric modulators lead to enzyme activity being switched on and off in catalysis, and this constitutes a coarse control in metabolism. Fine-tune control occurs in systems showing cooperativity, which provides greater or lower sensitivity of the enzyme to substrate concentrations. The

two self-regulation systems depend on the enzymes themselves rather than on gene control; regulation is exerted over enzyme activity in these cases.

6. Regulation of the kind of enzyme and the amount of enzyme synthesized in cells is controlled by the genes. The kind of enzyme made depends on genetic information encoded in DNA, while the amount of enzyme made is regulated at the levels of transcription and translation of the informational DNA. Studies of inducible enzyme synthesis show that induction occurs when the substrate of the enzyme is present. The substrate, or inducer, combines with the repressor product of regulator gene action and releases the repressor from its association with DNA. Upon removal of the repressor, transcription proceeds, and enzyme synthesis can occur in subsequent translation of the messenger RNA copies of DNA information. Repressible enzyme synthesis is another negative control phenomenon, but in this case the repressor made by a regulator gene cannot prevent transcription all by itself. If the substrate or some related metabolite of the enzyme is present, it acts as a corepressor and the repressor-corepressor complex can prevent transcription. Inducible enzyme synthesis takes place when the substrate of that enzyme is present, but repressible enzyme synthesis takes place in the absence of the substrate or other modulator molecule. Repressors are protein products of regulator gene action.

7. The breakdown of glucose during glycolysis involves specific enzymes acting at each step in the sequence. Glycolysis is a type of fermentation, or food breakdown in the absence of oxygen, in which energy transfer from glucose to end products involves phosphoryl group transfer via the ADP-ATP cyclic link and oxidation-reductions involving $NAD^+/NADH$ in coupled reactions. Some of the difference in free energy between glucose and its lactic acid end product is conserved in ATP and NADH, but since the process is less than 100 percent efficient there is also some loss of energy as heat. Regulation by feedback inhibition also characterizes the glycolytic sequence.

## STUDY QUESTIONS

1. In general, what are the roles of (a) energy, (b) energy-transfer systems, and (c) catalysts in cellular chemistry? What is energy? What is the difference between free energy and heat energy in living cells and cellular reactions? What are exergonic and endergonic reactions?

2. What are the statements and principles of the First Law and Second Law of thermodynamics? What is the relationship between $K_{eq}$ and $\Delta G^0$ of a reaction? Which one is easier to measure, and how can we calculate the other after measuring one of the two? What is the significance of the cell existing as an open system in different steady states?

3. What is the significance of coupled reactions in metabolism? How does ATP act in energy-transferring reactions? Can ATP transfer or accept a phosphoryl group from any compound in the cell? Why not? What is the importance of the observation? How do cells manage to accomplish energetically "unfavorable" reactions in biochemical pathways? Why do we consider ATP to be the "energy currency" of the cell?

4. What is an oxidation-reduction reaction? How does oxidation-reduction differ from energy transfer by phosphoryl group transfer? What is the advantage of there being different systems for energy transfer in cellular metabolism? What is a redox series? How is $\Delta E_0$ related to $\Delta G^0$? What is a redox couple? How do redox couples act in metabolism?

5. What is a catalyst? What is the significance of enzymes in metabolism? What is the major function of enzymes? Why does an uncatalyzed reaction proceed more slowly toward equilibrium than a catalyzed reaction? How do enzymes interact with their substrates? What contributes to efficiency of enzyme-catalyzed reactions? Why can so many different reactions proceed without interference in cellular chemistry?

6. What is an active site of an enzyme? How do we relate the active site of an enzyme to its observed specificity for a particular substrate? How does the geometry of the active site relate to substrate shape and the nature of processing of substrate by its enzyme?

7. How is enzyme activity regulated in feedback inhibition systems? What is the relationship between enzyme shape (conformation) and modulator action in regulating enzyme activity? What are the different effects of regulation of enzyme activity by modulator systems and cooperativity systems in metabolism?

8. What determines the kind and amount of enzyme synthesized in cells? What is the relationship between gene transcription and gene translation? How does inducible enzyme synthesis differ from repressible enzyme synthesis? How are they similar? What is the distinction between feedback inhibition and end product repression as systems of enzyme regulation? What is the function of repressor proteins in the regulation of enzyme synthesis? Why do we consider induction and repression of enzyme synthesis to be different versions of a common system of negative control?

9. What are the overall coupled reactions that characterize glucose oxidation in the glycolysis pathway?

What is the significance of ATP synthesis during glycolysis? Why is it important that the $\Delta G^0$ of glucose breakdown to lactate in the cell is $-32.4$ kcal/mole rather than the theoretically possible $\Delta G^0$ of $-47.4$ kcal/mole which can be achieved in a test tube?

10. What is the significant result of stage-1 reactions in glycolysis? Why is it necessary to use up two ATP in stage-1 of glycolysis before stage-2 can take place? What systems of energy transfer do we observe during reactions of the glycolytic pathway of sugar breakdown? Why is there only $\Delta G^0 = -32.4$ kcal/mole for glycolysis when we know there is a free-energy content of 686 kcal per mole of glucose? What is the role of ATP in the regulation of the rate of glycolysis in cells?

## SUGGESTED READINGS

Changeux, J. The control of biochemical reactions. *Scientific American* **212**:36 (Oct. 1965).

Dickerson, R. E. The structure and history of an ancient protein. *Scientific American* **226**:58 (Apr. 1972).

Dickerson, R. E., and Geis, I. *The Structure and Action of Proteins*. Menlo Park, Calif.: W. A. Benjamin (1969).

Koshland, D. E., Jr. Protein shape and biological control. *Scientific American* **229**:52 (Oct. 1973).

Lehninger, A. L. *Bioenergetics*. Menlo Park, Calif.: W. A. Benjamin (1971).

Lieber, C. S. The metabolism of alcohol. *Scientific American* **234**:25 (Mar. 1976).

Phillips, D. C. The three-dimensional structure of an enzyme molecule. *Scientific American* **215**:78 (Nov. 1966).

Roller, A. *Discovering the Basis of Life: An Introduction to Molecular Biology*. New York: McGraw-Hill (1974).

Stroud, R. M. A family of protein-cutting enzymes. *Scientific American* **231**:74 (July 1974).

Stryer, L. *Biochemistry*. San Francisco: W. A. Freeman (1975).

# Chapter 4.

# Cellular Membranes: Dynamic Barriers

All cellular life must maintain membrane integrity to survive and function amidst diverse and potentially disruptive surroundings. The plasmalemma is capable of rapid self-sealing if punctured, but any extensive damage that is not repaired will lead to cell death. This cellular membrane is a dynamic barrier separating the internal order of cell functions from surrounding disorder, since it not only sequesters living material but also regulates the movement of solutes, solvents, and particles into and out of the cell.

Many organelles as well as the cell itself have *selectively permeable* membranes that allow some substances to move through them at different rates from other substances. In addition to monitoring the diffusion of solutes, membranes and their component molecules participate in moving solutes across the membrane against a diffusion gradient. This process is usually accomplished at the expense of energy manufactured within the membrane itself.

## STRUCTURAL MODELS

During the 1890s there were enough physiological studies of various membranes for E. Overton to propose that surface membranes basically were lipid barriers. Lipid-soluble substances penetrated membranes much more readily than water-soluble materials. This observation was most easily interpreted by postulating a lipid membrane which lipid-soluble components could move through easily. Using plasma membranes isolated from red blood cells (erythrocytes), chemical studies indeed showed that lipids were a constituent of this membrane. Very little membrane is present in mature erythrocytes other than the plasmalemma so the analysis was relatively specific for the plasmalemma of these cells. By measuring the surface area of these cells and calculating the amount of lipid per cell, it was concluded that the lipid must occur in a layer that was just two molecules thick. Such a bimolecular layer could account for the fact that there was twice as much lipid per cell than was needed to make a plasma membrane only one molecule thick, based on measured cell surface area.

### Danielli and Davson Model

The surface properties of a postulated lipid-film cell membrane were not the same as the properties of the living cell surface however, so there was a problem in proposing a simple lipid bilayer as a model of the cell membrane. This difficulty was disposed of by James Danielli and Hugh Davson who postulated that a coating of protein over the lipid bilayer could account for the observed differences in the properties of the cell membrane and a film of pure lipid. Their model of membrane structure, proposed in 1935, was the first significant attempt to describe the membrane in molecular terms and to relate the structure to observed biological and chemical properties.

Danielli and Davson proposed a model in which two layers of phospholipid molecules were arranged so that their hydrophobic fatty acid chains faced each other in the interior of the membrane. The hydrophilic portion of each phospholipid layer faced the outer borders of the membrane and were coated with globular proteins. In effect, the proteins sandwiched the phospholipid bilayer (Fig. 4.1). Hydrophobic properties of the membrane were explained on the basis of the lipid center, and hydrophilic properties were accounted for by the protein coatings attached to the phosphate residues of the phospholipid molecules. Davson and Danielli paid more attention to the lipid than to the protein portions of the membrane in relation to membrane functions; this was a reflection of the extent of our knowledge about the two classes of organic compounds in the 1930s.

### The Robertson Unit Membrane Model

J. David Robertson studied the membranes of the cell from electron micrographs of sectioned materials. The preparations involved the usual steps of killing cells in fixing fluids such as solutions of osmium tetroxide or potassium permanganate, dehydrating in solvents such as acetone and alcohol, and embedding in plastics before sectioning. The earlier photographs were of comparatively poor quality, but by 1960 there had been enough improvement in electron microscopy to provide clear images of a three-layered membrane about 7–8 nm thick (Fig. 4.2).

Building on the Danielli-Davson membrane model, Robertson proposed a more detailed structure based partly on electron microscopy and partly on functional

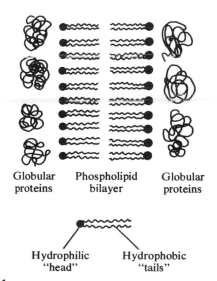

Globular proteins    Phospholipid bilayer    Globular proteins

Hydrophilic "head"    Hydrophobic "tails"

**Figure 4.1**
Model of membrane structure according to Danielli and Davson.

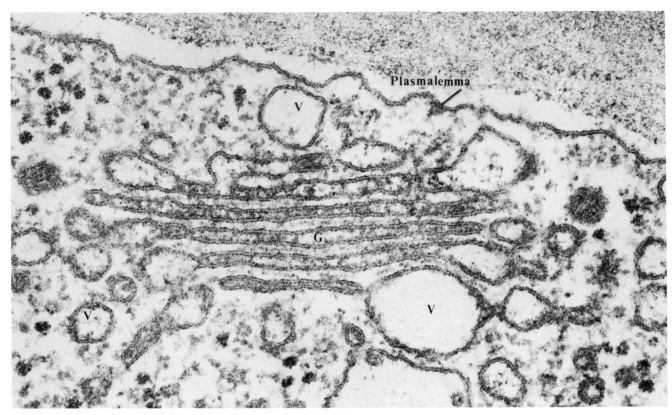

**Figure 4.2**
Electron micrograph of a thin section through a root cell of *Potamogeton natans* showing
the typical dark-light-dark staining pattern of the plasmalemma. Membranes of the Golgi
apparatus (G) and its associated secretion vesicles (V) also show the trilaminate staining
pattern of a "unit" membrane. × 151,000. (Courtesy of M. C. Ledbetter)

criteria. He postulated a single layer of protein
molecules over the inner and outer surfaces of the lipid
bilayer, with extended rather than globular protein
conformations (Fig. 4.3). This model fit the measurements of the three layers of a membrane that could be
seen in electron micrographs, allowing 2 nm each for
the two protein coatings and 3–4 nm for the interior
lipid bilayer. Extended protein conformation fit the 2
nm dimension fairly well, whereas globular proteins
would generally be larger than 2 nm in diameter. Because Robertson found the same dark-light-dark staining pattern and a uniform thickness for the membranes
he examined, he proposed that eukaryotic and prokaryotic membranes had a common structural plan, or
**unit membrane.**

Robertson's model was widely approved, but there
was a continuing undercurrent of dissatisfaction with
the model, mostly because these dynamic structures
had been pictured as static and uniform elements.
Other points of dissatisfaction involved questions
about the presence of artifacts following the harsh
procedures used to prepare materials for electron microscopy, and the difficulties in reconciling the common unit membrane structure with the variety of
membrane functions and specificities. Variety in function is usually accompanied by some variety in structure in living systems, but Robertson's unit membrane
was postulated to be essentially the same in all parts of
all cells.

These initial difficulties in accepting the model became even more troublesome as additional studies showed that membranes varied between 5 and 10 nm in thickness, with particular membranes showing a characteristic thickness. It was also difficult to understand how active globular proteins could remain active, if they assumed an extended conformation over the lipid bilayer. Change in protein shape usually leads to profound change in protein activity, and the unfolding of globular proteins usually leads to inactivation. From a chemical standpoint, the model did not provide a basis for explaining why some proteins were difficult to extract from membranes while others dissociated readily from the lipid components.

The unit membrane did account for a number of observed properties of cell membranes, however, including the following:

1. The densely packed lipid bilayer easily accounted for the presence of 40 percent lipid, by weight, in membranes.
2. Hydrocarbons are poor electrical conductors, so the continuous hydrocarbon phase of the natural membrane would explain the known high electrical resistance of membranes.
3. High permeability of natural membranes to nonpolar molecules could be explained by their ready solubility in the nonpolar lipid phase, and at the same time account for relative impermeability to small ions which do not dissolve readily in this medium.
4. Phospholipids spontaneously form bilayer systems *in vitro* when added to an aqueous environment, and there is no requirement for work input to maintain this minimum-energy conformation of the artificial membrane.
5. The model accounted for the three-layered staining pattern of fixed membranes seen with the electron microscope.

Another positive feature of the unit membrane model was the asymmetry of the structure. Robertson based this in part on differences in appearance of the inner and outer stained layers of the unit membrane, and it fit well with known cellular membranes that contained saccharides and conjugated proteins on one surface and not on the other. Current references to "unit" membranes have come to signify the three-layered stained image and not necessarily acceptance of Robertson's proposed model of the membrane.

### The Fluid Mosaic Model

Objections to the unit membrane model increased during the 1960s and this led to reexamination of lipid—protein interactions and to new models. Studies of mitochondrial and chloroplast membranes especially underscored the differences between observed features of membranes and the uniformity that was required by the unit membrane concept. Mitochondrial and chloroplast membranes contain displays of particulate units in or on the membrane. The plasmalemma did not present the same appearance as mitochondrial or chloroplast membranes. It seemed that different models might be needed to describe different functional membrane types. This unsuitable approach became unnecessary when a mosaic membrane model was proposed. This model could accommodate functionally distinct membranes if different protein or lipoprotein subunits were arranged within the lipid framework of any membrane.

David Green suggested several possible models for a mosaic membrane, and together with Ronald Capaldi

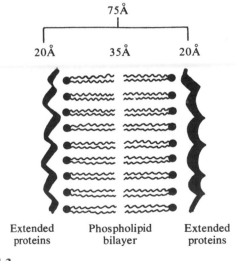

Extended proteins     Phospholipid bilayer     Extended proteins

**Figure 4.3**
Model of the "unit" membrane structure according to interpretations of electron micrographs made by Robertson.

has more recently proposed a model in which the basic membrane structure emerges from protein—protein interactions, with patches of lipid bilayer dispersed among the proteins (Fig. 4.4). Since proteins may account for as much as 75 percent by weight of some membranes, Green and Capaldi emphasized the protein more than the lipid constituents, although both are essential to membrane structure and function (Table 4.1).

In contrast with the Green-Capaldi model, which requires some restriction on protein movements through the thickness of a membrane, S. J. Singer and Garth Nicolson described a **fluid mosaic model** in which the lipid bilayer is the cementing framework of the membrane, and attached or embedded proteins interact with each other and with the lipid but retain the capacity to move laterally in the fluid lipid phase (Fig. 4.5). The *predominant* interactions which hold the membrane together and account for its activities are viewed differently by the two groups. Green and Capaldi stress protein—protein interactions, whereas Singer and Nicolson suggest that protein—protein, lipid—protein, and lipid—lipid interactions contribute to membrane structure and dynamics.

Singer and Nicolson consider two kinds of nonco-

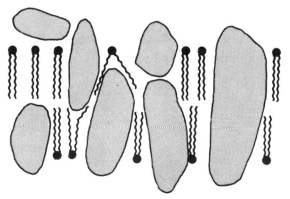

**Figure 4.4**
Model of a fluid mosaic membrane structure proposed by Green and Capaldi. In this model protein-protein interactions predominate. Proteins are distributed at the surfaces and within the patches of phospholipid bilayer, extending from one surface to the other or only part way through the membrane thickness.

valent interactions to be predominant: **hydrophobic** and **hydrophilic.** The stable structure can exist in the minimum free-energy state if both hydrophobic and hydrophilic interactions are maximized. Hydrophobic and hydrophilic interactions are maximized in the aqueous environment inside and outside the cell if the nonpolar fatty acid chains of the phospholipids and the nonpolar amino acid residues of the proteins are kept from contact with water to the greatest possible extent and at the same time, the polar and ionic groups of the membrane proteins, lipids, and carbohydrates are in contact with water (Fig. 4.6). The Robertson unit membrane model does not meet these requirements since the covering protein layers keep the lipid polar groups away from contact with water while the nonpolar residues of these proteins are exposed to water instead of being sequestered away from the aqueous solvent. Both phospholipids and proteins are **amphipathic** molecules, having both polar and nonpolar regions in the same molecule (see Chapter 2), but these regions are not distinguished or accounted for in the unit membrane. Since neither the hydrophobic nor the hydrophilic interactions can be maximized in such a model, Singer has rejected it as being thermodynamically unstable and thus unlikely to exist in the cell.

Both the Singer-Nicolson and the Green-Capaldi models call for two broad categories of proteins: **peripheral** (or extrinsic) and **integral** (or intrinsic). In the most active membranes, a relatively high percentage of protein occurs, much of it in the form of integral proteins, which are tenaciously held in place by strong hydrophobic interactions and thus are difficult to remove from membranes except by harsh and disruptive methods of extraction. Peripheral proteins are located more superficially at the membrane surface and are thus more easily separated during extraction. Some integral proteins span the entire thickness of the membrane and protrude at one or both surfaces from the lipid bilayer, while others project from only one surface with the remainder being embedded within the lipid bilayer. No protein is known to be entirely surrounded by membrane lipid.

The widely differing activities and specificities of

**TABLE 4.1  Composition of mammalian membrane preparations***

| Component | Myelin | Erythrocyte Plasma Membrane | Liver Plasma Membrane | Heart Mitochondria |
|---|---|---|---|---|
| Proteins | 22 | 60 | 60 | 76 |
| Total lipids | 78 | 40 | 40 | 24 |
| Phospholipids | 33 | 24 | 26 | 22 |
| Glycolipids | 22 | trace | 0 | trace |
| Cholesterol | 17 | 9 | 13 | 1 |
| Other lipids | 6 | 7 | 1 | 1 |

*Values taken from various sources, expressed in percentage, by weight.

the various kinds of membranes in cells are assumed to reflect different kinds of proteins that occur throughout the membrane or in some localized regions of the structure. Differences in the content of enzyme proteins have been studied particularly in mitochondria and chloroplasts. In both organelles the enzymes of electron transport and ATP formation are situated within the inner membrane and do not occur in the outer envelope of the organelle.

The photosynthetic complexes involved in the capture and conversion of light energy are also tightly bound within the innermost chloroplast membrane system. These enzyme systems must be integral proteins since they can be extracted from organelle inner membranes only with great difficulty after relatively harsh treatment. Immunological specificities clearly involve peripheral proteins that are located in particular regions of the membrane outer surface and that differ from one cell type to another according to antigen-antibody tests.

Observed differences in membrane thickness can be related to the presence of substantial amounts of peripheral proteins in thicker membranes and to predominance of integral proteins in thinner membranes such as those of the mitochondrion. Measurements of mitochondrial membrane show it to be a little over 5 nm, which is not much thicker than the 5 nm-thick lipid bilayer alone. This membrane has very few pe-

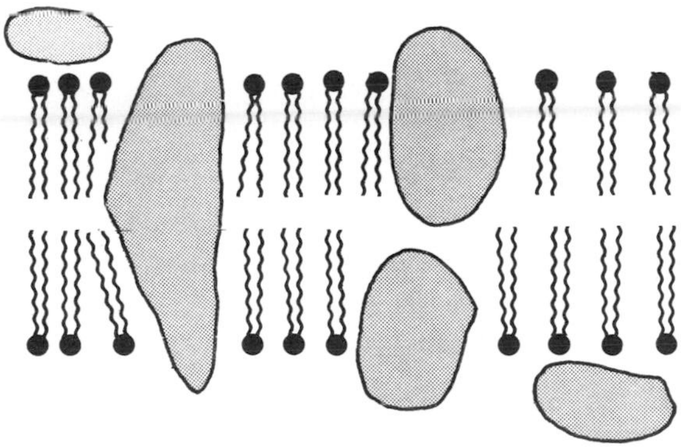

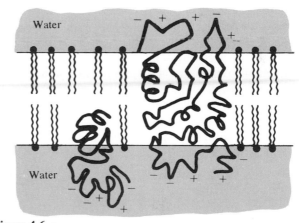

**Figure 4.5**
The fluid mosaic model of membrane structure proposed by Singer and Nicolson. The phospholipid bilayer is a cementing framework with attached peripheral proteins or integral proteins embedded within the bilayer.

**Figure 4.6**
Hydrophilic interactions occur among the charged residues of amphipathic membrane proteins, the water phase, and the polar heads of amphipathic phospholipids. Hydrophobic interactions occur between uncharged protein residues and nonpolar tails of the membrane phospholipid bilayer.

ripheral proteins so that its major thickness is due to the lipid, with a little bit of protruding integral protein added on.

Membranes of most systems are asymmetrical, having different functional proteins exposed on the outer and inner faces of the membrane. The plasmalemma of erythrocytes, and probably of other cell types as well, have saccharide residues of glycolipid and glycoprotein molecules exclusively on the outer face of the membrane and are therefore obviously asymmetrical membranes. The mitochondrial inner membrane is definitely asymmetrical, since some respiratory components are exposed only on the cytoplasm-facing side and some proteins are exposed on the inner surface side exclusively. Almost all respiratory proteins of this membrane have been shown to be situated within the membrane interior with one end of the integral protein exposed on either the outer or the inner surface of the mitochondrial inner membrane. This "sidedness" of membranes is considered by some investigators to be an important structural aspect in transport of substances across the membrane and in lipid-protein interactions required for certain enzyme activities.

The lipid matrix of the membrane has the consistency of a light oil, according to several lines of physical evidence. Lipids and proteins have been shown to move laterally within this fluid layer. The lateral, or translational, motion within the plane of the membrane provides one way in which membrane proteins can interact with one another and with lipids. This interaction would be a feature in certain metabolic interactions in which preservation of an intact membrane system is required to maintain functional activity. If the interacting components are separated during extraction, they may lose their capacity to establish physical contacts at specified times and places in order for a metabolic reaction to occur.

Photosynthetic membranes within chloroplasts show altered energy transfer activity if the tight joining between individual membranes is interrupted. Examination of the protein particles inside these membranes showed that "stacked" membranes contained a different distribution of certain particles from "un-stacked" membranes (Fig. 4.7). There was no change in the numbers of particles per unit area of membrane, only a change in their distribution within the membrane thickness. This kind of information provides support for the fluid mosaic model of the membrane as proposed by Singer and Nicolson. It also shows that membrane function is related to the *spatial arrangement* of components as well as to the *kinds* of components that are present. Relating membrane structure to membrane functions will require more than chemical analysis of the protein and lipid constituents of different cell and organelle types. The geometry of membrane organization must play an important role in underwriting function.

Membrane contacts within the chloroplast, between different organelles in the cell, and membrane junctions between adjacent cells have been proposed as facilitating exchanges. In some cases, at least, these regions of contact may form as a result of molecular redistributions within the apposed membranes, which leads to tight coupling between previously separate structures (Fig. 4.8).

## MOVEMENT OF SUBSTANCES ACROSS MEMBRANES

Substances move through membranes *selectively* in the living cell. Once the membrane has been damaged or the cell has been killed, molecules move freely across the membrane and may reach equilibrium. Equilibrium is never achieved in living systems, so that the dynamic state of the living membrane acts as a *regulatory barrier* for entry and exit of molecules and particles. There are three general routes by which substances cross the membrane barrier: (1) **free diffusion** along a gradient going from higher to lower concentration of the molecule; (2) assisted passage, or **transport,** across the membrane either by a process of **facilitated diffusion** or **passive transport,** in which the substance moves as expected in the direction of its lower concentration, or by energy-expending **active transport** against a concentration gradient; and (3) enclosure in membranous vesicles to enter the cell by the

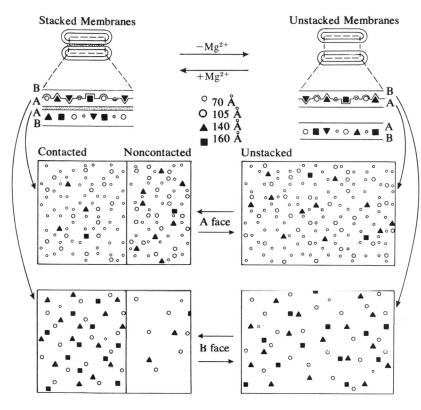

**Figure 4.7**
Summary diagram illustrating lateral particle movement during thylakoid unstacking in wild type and thylakoid stacking in mutant *Chlamydomonas* chloroplast structure. Four types of particles were identified within the membrane. Some remained with one fracture face, and others remained with the alternative fracture face in freeze-fractured membrane preparations. (From Ojakian, G. K., and P. Satir, 1974, *Proc. Natl. Acad. Sci.* **71**:2052-2056, Fig. 9.)

process called **endocytosis** or be expelled from the cell in the reverse process of **exocytosis.**

### Free Diffusion

According to a large amount of evidence, many substances move through membranes at rates of free diffusion that are directly proportional to their solubility in lipid (Fig. 4.9). Water molecules are a notable exception to this rule, since they freely diffuse through membranes regularly and rapidly. It has been suggested that membranes contain 1 nm-wide pores that are lined by hydrophilic residues (Fig. 4.10). Such openings would be large enough for water molecules but would hardly accommodate other water-soluble substances to pass through the narrow passageway. The suggestion is quite logical to explain diffusion of water through membranes, but there is little direct evidence to support the concept.

Free diffusion along a concentration gradient could be maintained for entry and exit of metabolites and waste products as long as substances were changed or washed away to keep one end of the gradient lower in concentration. If metabolites are chemically changed on entering the cell, then the concentration of the metabolite remains high outside the cell and is kept low inside the cell if it is not retained in its original form. Similarly, wastes leaving the cell could continue to move from the higher concentration inside as long as these substances were washed away or removed somehow from the immediate vicinity of the cell (Fig. 4.11).

The idea is attractive because no particular apparatus is required to maintain diffusion, and no energy expenditure is needed to keep the system operating. The major difficulty, of course, is that most biologically important molecules of cellular metabolism are insoluble in lipid and would be confronted by the lipid barrier regardless of the favorable state of

(a)
(b)

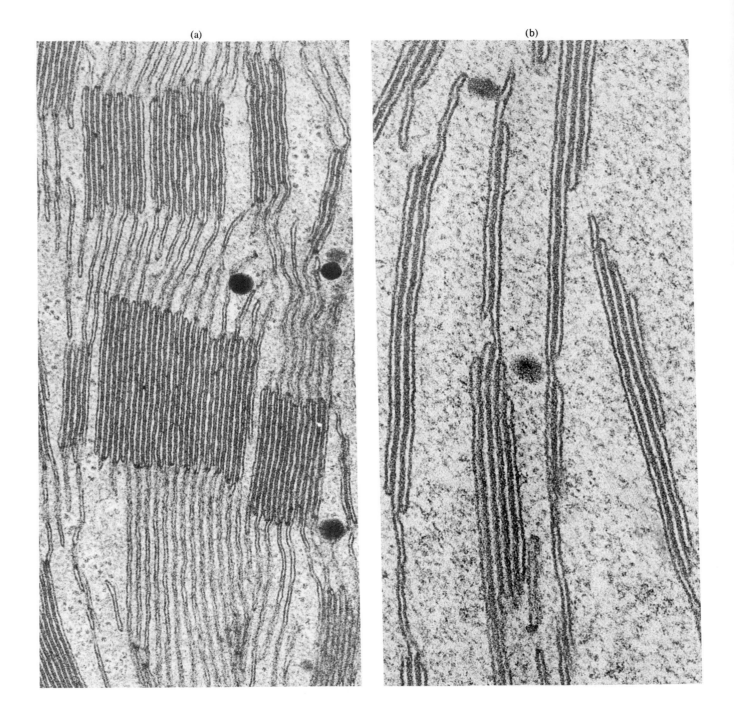

the concentration gradient. Important lipid-insoluble substances, such as sugars, move through membranes selectively and efficiently. Simple diffusion cannot explain these movements and membrane transport mechanisms have been widely accepted instead as explanations for selective movement of many biologically significant solutes across the lipid layer. Water is virtually the only liquid which penetrates by free diffusion into cells in most natural environments.

### Transport Mechanisms

Essential lipid-insoluble metabolites, such as sugars and amino acids, enter and leave the cell or its compartments through processes involving reversible combination with membrane proteins. These **carriers** are integral proteins that form part of the structure of the membrane. Carriers are highly specific. Each is assumed to have a characteristic binding site which will bind one particular kind of molecule but not another that may be almost identical. The binding between carrier and metabolite is transient. After the bound molecule has been translocated to the other side of the membrane the carrier is freed and may recycle to assist other molecules of this substance to get across the membrane. The relative solubility of the metabolite in lipid is not significant since its interaction is with protein rather than lipid constituents of the membrane.

Certain carriers, called **permeases,** accelerate transport, provide selectivity to transport, and recycle unchanged at the conclusion of the transport event. Permeases may assist molecules across the membrane in the usual direction of the concentration gradient,

**Figure 4.8**
Thin sections through a portion of a chloroplast: (a) leaf mesophyll cell of timothy grass (*Phleum pratense*). There is considerable density at the junctions between adjacent stacked membranes, but no dense material is evident at the surfaces of unstacked membranes within the common matrix of the organelle. × 80,000. (Photograph by W. P. Wergin, courtesy of E. H. Newcomb) (b) Leaf cell of *Elodea canadensis*, an aquatic flowering plant. The dense junctions between the stacked membranes are even clearer in a higher magnification view. × 135,000. (Courtesy of M. C. Ledbetter)

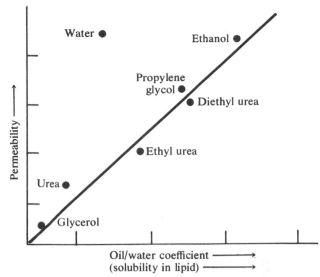

**Figure 4.9**
Graph showing that the rate of free diffusion of chemicals through an oil/water phase system (analogous to a membrane) is directly proportional to the solubility of the chemical in lipid. Water molecules are an exception to this rule.

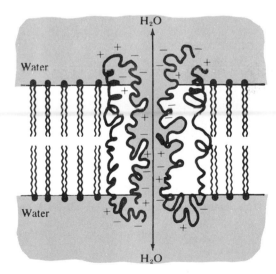

**Figure 4.10**
Illustration of the concept that narrow pores perforate the membrane allowing rapid diffusion of water molecules through their interactions with hydrophilic residues of adjacent membrane proteins.

from high to lower amounts of the metabolite, or against this usual "downhill" direction. At least two carrier-mediated transport mechanisms must exist, therefore, one that sponsors facilitated diffusion and another that aids energy-requiring active transport in the "uphill" direction of a gradient. In facilitated diffusion, the carrier helps the molecule along the direction in which that molecule ordinarily would move. The diffusion is facilitated because the molecule cannot penetrate the selectively permeable membrane on its own by free diffusion movement. In active transport, a substance can continue to be accumulated in a region where it already exists in higher concentration only if energy from metabolism is continually supplied to the carrier system. Metabolic energy drives active transport.

ACTIVE TRANSPORT AND ION PUMPS. Cells can accumulate substances in excess of expected amounts in at least three ways:

1. The substance can be precipitated from solution once it is inside the cell, effectively reducing the concentration of solute in water.
2. The molecule can be chemically changed after it has gone through the membrane, thus reducing the concentration of the specific molecule involved in the concentration gradient.
3. The transport of a metabolite can be coupled directly to a second reaction that is energetically favorable to driving the transport reaction "uphill."

The accumulation of $Ca^{2+}$ ions in channels of muscle endoplasmic reticulum operates by way of precipitation. Calcium phosphate is the precipitate in this case, and its formation maintains a favorable gradient of calcium ions. Sugars may accumulate inside the cell by a combination of facilitated diffusion and phosphorylation events, particularly in bacterial cells. If the entering sugar is phosphorylated immediately, then the inside concentration of the non-phosphorylated sugar remains low, and additional molecules will be transported into the cell along the usual "downhill" direction of the concentration gradient. Both of these situations differ from active transport in that their direction of metabolite movement is toward the region of lower concentration, whereas active transport leads to continued accumulation of a substance in the region where it occurs in its highest concentration. The end result is very similar in all these cases, namely, a situation exists by which a

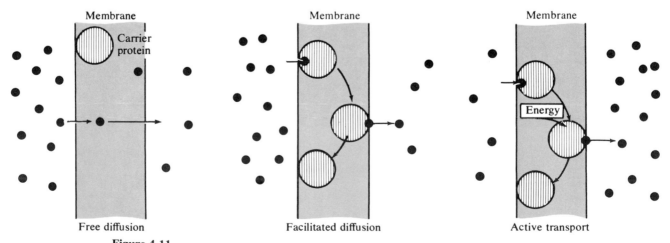

Membrane    Carrier protein     Membrane     Membrane    Energy

Free diffusion     Facilitated diffusion     Active transport

**Figure 4.11**
Three modes of movement across the membrane by metabolites. The relative concentrations of metabolite are indicated by the number of dots on either side of the membrane.

substance accumulates inside the cell or some cell compartment in great amount.

One unifying concept that has developed in recent years explains active transport on the basis of pumping actions. In particular, the active pumping of one substance out of a cell or compartment provides the driving force for active transport of various other substances inward. The pumping in of these metabolites is thus considered to be coupled to the outward transport of some one other material, but in a manner that provides the energy required to drive the needed materials inward, in the "uphill" direction of their concentration gradients. The pump is an economical as well as a simple system, since the outward movement of one kind of substance helps to drive in another kind of metabolite.

The solutes that are most actively pumped into animal cells are $K^+$ ions, sugars, and amino acids. The driving force for this inward transport is believed to be a $Na^+$ gradient across the membrane, created by active transport of $Na^+$ ions pumped out of the cell. The external $Na^+$ ion concentration remains high and the internal ion concentration remains low as $Na^+$ continues to be transported outside the cell. The energy required to pump $Na^+$ ions out of the cell is provided by ATP, which is hydrolyzed by a $Mg^+$-activated ATPase believed to be situated within the membrane. The free-energy difference between ATP and its ADP and inorganic phosphate products of hydrolysis is made available for active transport in the operation of the **primary sodium pump.** This pump acts in animal cells but apparently is not a feature of plant cells or of bacteria, neither of which requires $Na^+$ ions for its metabolism. A $H^+$ ion pump is active in bacterial cells; the pump in plant systems is not well understood.

Two distinct types of $Na^+$ pump have been described for animal cells. In one, the outward pumping of $Na^+$ is tightly linked to the inward transport of $K^+$ ions. Since $Na^+$ and $K^+$ are exchanged in a compulsory way, outward movement of $Na^+$ is always accompanied by inward movement of $K^+$. This type of sodium pump is called the **sodium/potassium exchange pump,** or the **coupled neutral pump** (Fig. 4.12). Inward transport of $K^+$ does not necessarily accompany out-

ward extrusion of $Na^+$ in the **electrogenic sodium pump** (Fig. 4.13). The electrogenic pump has been so named because a gradient of electrochemical potential may be generated when exit of $Na^+$ is not compensated by one-to-one entry of $K^+$. In most cells there is an accumulation of $K^+$ that exceeds the loss of $Na^+$ (or $H^+$ in bacteria), which is one of the indications that the electrogenic pump rather than the neutral pump is operative. Nerve and muscle cells particularly are known to have the coupled neutral pump.

A high intracellular concentration of $K^+$, regardless

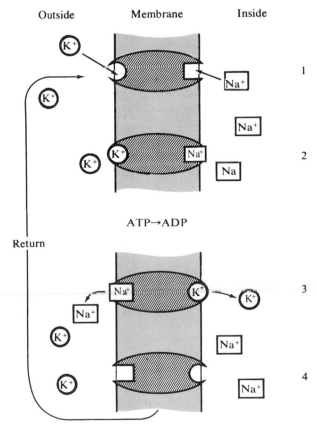

**Figure 4.12**
Diagram illustrating the action of a sodium pump of the *coupled neutral* type. Inward transport of $K^+$ ions is accompanied by compulsory outward transport of $Na^+$ ions via carrier proteins situated within the membrane. The sequence of events is shown in steps 1-4.

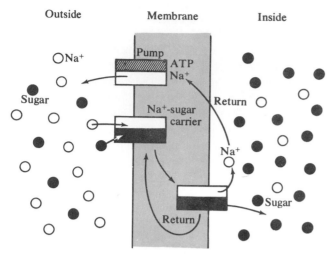

**Figure 4.13**
Diagram illustrating the action of an electrogenic sodium pump through which sugars and other metabolites enter the cell along an electrical gradient generated by the energy-requiring extrusion of $Na^+$ ions.

of the external concentrations of $Na^+$ and $K^+$, is especially needed by aerobic cells of all species. Two vital processes that require a high concentration of $K^+$ are protein synthesis at the ribosomes and one of the critical enzymatic steps in glucose processing during glycolysis. The high internal $K^+$ concentration in many kinds of cells must be balanced by loss of some cation, such as $Na^+$ or $H^+$, or there would be excessive swelling that would cause the cell to burst by creating a condition of high internal osmotic pressure.

Active transport of amino acids into cells is another consequence of the action of an electrogenic $Na^+$ pump. Extrusion of $Na^+$ from the cell generates a gradient of lower internal and higher external concentration of $Na^+$. The energy inherent in this gradient is believed to provide the driving force for transport of amino acids into the cell in the "uphill" direction, leading to accumulation of these essential compounds. The $Na^+$ gradient itself is formed at the expense of ATP. Specific carrier protein systems assist amino acids across·the cell membrane in these active transport events.

The capacity to accumulate sugars, like amino acids, is another feature of the cellular electrogenic

$Na^+$ pump activity. Both sugar and amino acid active transport are accomplished under conditions in which the outside concentration of $Na^+$ is kept high enough to create an appropriate gradient whose dissipation provides energy that drives metabolites into the cell from very dilute outside solutions of these substances. The accumulation of sugars is therefore coupled to $Na^+$ extrusion, as with amino acids, and is also assisted by specific active-transport proteins (see Fig. 4.12).

TRANSLOCATION ACROSS THE MEMBRANE. The carrier proteins assist hydrophilic molecules across a membrane thickness of 6 to 10 nm. The metabolites are considerably smaller than 6 nm, so it is important to know how these molecules are translocated across this relatively large distance by the carriers. Several alternatives have been proposed, but two have been studied more intensively than the other possibilities.

One alternative hypothesis postulates that the carrier binds with the hydrophilic molecule and that then the entire transport protein rotates across the membrane and delivers its bound metabolite to the other side (Fig. 4.14). The second alternative proposes

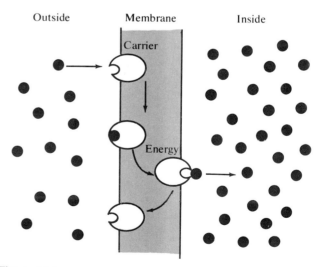

**Figure 4.14**
Diagram illustrating the carrier mechanism concept for translocation of hydrophilic molecules across the membrane barrier.

that the carrier is fixed in place within the structure of the membrane, and that the carrier molecule undergoes a conformational change that translocates the binding site across the membrane and the bound metabolite along with it at the same time (Fig. 4.15). Once the metabolite has been translocated the binding site is freed and restored to its original conformation, ready to bind another hydrophilic molecule in another transport event. This second alternative has been referred to as the **fixed-pore mechanism.** The first alternative is known as the **carrier mechanism.**

From an energetic standpoint, the carrier mechanism would require considerable energy expenditure to support rotations of integral carrier proteins across the span of the membrane. Fixed-pore mechanisms, on the other hand, have a lower energy cost and can be related more directly to membrane structure. Integral proteins with both hydrophilic and hydrophobic residues could be envisioned existing along the pore channels, with their hydrophilic portions lining the opening through the membrane while their hydrophobic regions were within the lipid matrix. The bound hydrophilic solute molecule would be

translocated across when some energy-yielding process caused a conformational change in the carrier. Peripheral proteins also could aid in translocating solutes through fixed pores in active transport processes, as suggested by S. J. Singer and others (Fig. 4.16). Although still incomplete, the weight of available evidence is more in favor of the fixed-pore than of the carrier mechanism.

**Transport by Vesicle Formation**

Most cell membranes can enclose materials in **vesicles** and bring the substances into the cell in this way, or package materials for discharge from the cell in a reverse process. The process is called **endocytosis** when materials are brought in, and **exocytosis** when vesicle discharge takes place (Fig. 4.17). Endocytosis has features that are analogous to active transport.

For example, substances enter along an "uphill" concentration gradient, and energy is required to sup-

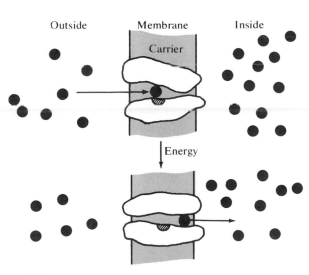

**Figure 4.15**
Diagram illustrating the conformational change of carrier protein according to the fixed-pore mechanism for translocation of hydrophilic molecules across the membrane.

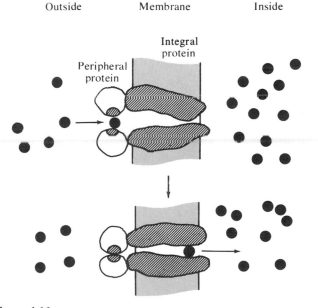

**Figure 4.16**
Participation by peripheral proteins in the translocation of solutes across the membrane according to the fixed-pore mechanism. This mechanism uses the Singer and Nicolson structural model of the fluid mosaic membrane.

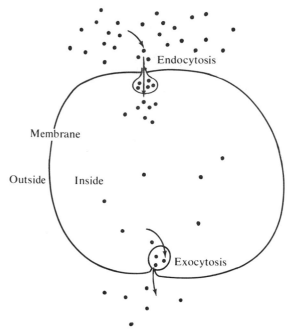

**Figure 4.17**
Entry by endocytosis and exit by exocytosis transport molecules into and out of the cell across the membrane barrier. The substances are enclosed in a membranous vesicle that forms from or fuses with the plasma membrane during entry into or exit from the cell, respectively.

port the process. Endocytosis will stop if poisons that stop energy production in the cell are added, and the process can be stimulated by the addition of ATP to cells in suspension.

Various cell secretions are discharged from the cell by exocytosis. The mucus droplets of intestinal goblet cells, digestive enzyme precursors in granule packages from pancreatic cells, and other cell secretions leave the cells in which they are produced by this general pathway (Fig. 4.18). Although more difficult to see by light microscopy, numerous endocytotic infolded regions can be seen along the plasma membrane in suitable electron micrographs (Fig. 4.19).

Endocytosis and exocytosis play essential roles in **lysosome** digestive activities within the cell. Incoming vesicles fuse with lysosomes, and the lysosomal enzymes proceed to digest the material contained in the endocytotic structures. The digested substances and occasional undigested residues then are discharged from the cell by exocytosis. These processes are discussed at greater length in Chapter 6, along with cellular events associated with the release of secretions packaged within the **Golgi apparatus** in various cell types.

At the present time relatively little information has been obtained on the way in which molecular organization of the membrane can be related to vesicle formation and substance transport. Part of this phenomenon certainly is related to membrane fluidity and repair, but membrane biogenesis must also be considered. We can anticipate more rapid progress in the next few years as experimental studies become focussed more precisely on relationships between models of membrane structure and the dynamic properties of functional membranes in living cells.

## INTERCELLULAR COMMUNICATION THROUGH JUNCTIONS

Rapid and effective exchanges of ions and metabolites is characteristic of some kinds of cells in communities of living tissues and in cell cultures. These cell-to-cell communications may operate at close range and require direct physical contact between interacting cells in a group. Both the chemical and the physical expressions of communication vary according to cell type and stage of growth and development. The chemical exchanges may include ions or organic metabolites and regulatory molecules. The physical basis for ionic and metabolic cooperation may range from the formation of actual **cytoplasmic bridges** to localized membrane **junctions** that may vary in extent from a few angstroms to micrometers of distance. Cell junctions are specialized regions of close-range contact between adjacent cells. They are associated with a differentiation of the contributing cell surface membranes and intercellular matrix, or they may only involve the matrix material. Junctional contacts range from fused membranes to areas separated by a space as wide as 20 nm.

Three major types of junctional membrane localizations have been recognized by electron microscopy: **gap junctions, tight junctions,** and **septate junctions.** In each case there is a recognizable membrane region which is physically distinguishable from nonjunctional regions elsewhere on the membrane surface (Fig. 4.20).

Gap junctions are seen as localized regions of 7-layered structure when viewed in thin sections by electron microscopy. They were clearly seen and described in 1967 as areas formed from two "unit" membranes with a space or "gap" between the adjacent plasmalemmas. The typical 8 nm thickness of each plasmalemma plus the 2–4 nm gap produces an average thickness of 17–19 nm for gap junctions in vertebrate and invertebrate animals. They are commonly observed in excitable and nonexcitable cell types, but they are absent from types such as skeletal muscle fibers and circulating blood cells. Tight junctions represent true membrane fusions. There is no space between the fused membrane width of 14–15 nm. Unlike the widely distributed gap junctions, tight junctions are characteristic of vertebrate tissues such as epithelium. Septate junctions have only been found in invertebrate tissues. They are the largest junctional types, and they sometimes extend for micrometers in length, forming a belt surrounding the apical or basal regions of a cell.

All three kinds of junctions can be described in somewhat greater detail from freeze-fracture preparations for electron microscopy (Fig. 4.21). When the membrane is laid open by fracturing, internal surface textures become visible. The membrane surface

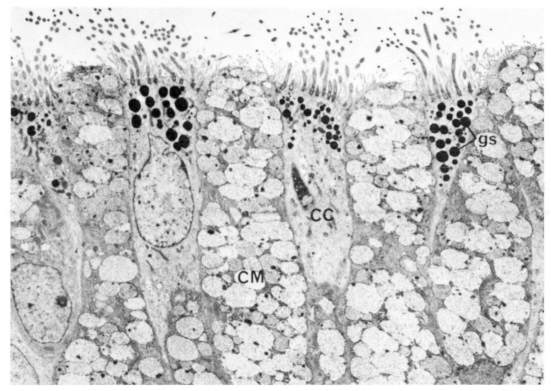

**Figure 4.18**
Release of mucus from goblet cells of epithelial tissue in the quail. × 3900. (Reproduced with permission from Sandoz, D., *et al.*, 1976, *J. Cell Biol.* **71**:460, Fig. 6.)

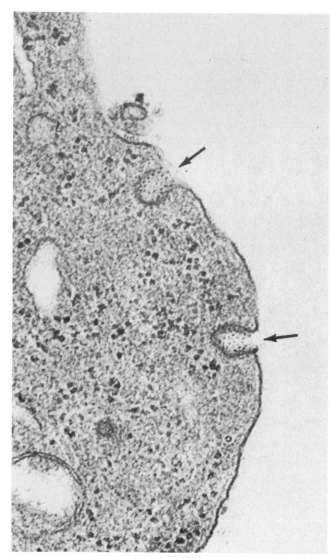

**Figure 4.19**
Electron micrograph showing invagination of endocytic vesicles at the plasma membrane of an erythroblast from guinea pig bone marrow. Dense ferritin particles adhere to these specialized areas and are carried into the cytoplasm as vesicles pinch off from the plasma membrane. Ferritin uptake by this mechanism is the normal pathway by which these cells obtain iron for hemoglobin synthesis. × 69,000. (Courtesy of D. W. Fawcett, from Fawcett, D. W., 1965, *J. Histochem. Cytochem.* **13**:75-91.)

nearer the cytoplasm-bordering part of the plasmalemma is the **A fracture face,** and the **B fracture face** is closer to the extracellular matrix around the cell. Face A in gap junction regions shows plaques of relatively small size which are distinguished by a regular latticework of homogeneous particles. Face B is the complementary surface showing pits or depressions that match the orderly arrays of face A particles. The internal pattern of bordering nonjunctional membrane consists of random arrays of different-sized particles.

The A fracture face of tight junctions has a series of ridges in a meshwork pattern, while the B fracture face is a complementary meshwork of grooves. The ridges and grooves are interwoven and represent sites of true membrane fusion. This pattern varies in different epithelial tissues, from "very tight" to "leaky" systems according to permeability criteria. In general, tight junctions exclude movements of large molecules between cells, but "leaky" epithelia are relatively more permeable to such molecules than "tight" or "very tight" tissues.

In freeze-fracture preparations of septate junctions in invertebrate tissues, there are parallel rows of particles exposed on the A face that exactly correspond to the septa between cells in the junction area. The B face contains depressions that are complementary to A face particles. The two adjacent membranes are joined by the septate or ladderlike intermembrane region, which is similar in width in various tissues and species. The spacing between septa, however, often varies from one sample to another.

The particle arrays are not artifacts of the preparative method, since the same systems of orderly subunit arrangement can be seen after negative staining of membranes (Fig. 4.22).

### Ion and Metabolite Exchange

If molecules are tagged with radioactive isotopes or a fluorescent component, they can be traced as they move from the extracellular medium into the cell and from one cell into another. Radioisotope-labeled reagents can be detected in the intercellular space of

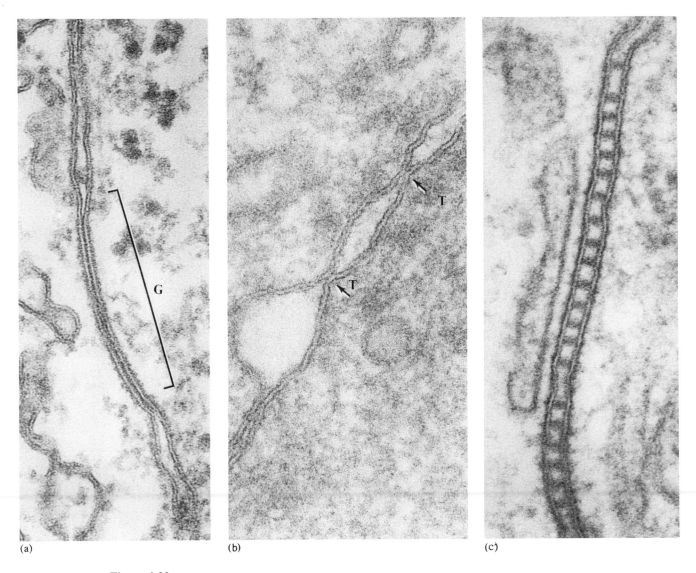

(a)  (b)  (c)

**Figure 4.20**
Thin section appearance of junctions between adjacent animal cells: (a) gap junction
between hamster fibroblasts. The two closely apposed junctional membranes are separated
by a 2–4 nm space, or "gap." × 210,000. (b) Tight junctions (T) formed by fusion of
adjacent plasma membrane areas in epithelial cells of rat small intestine. × 247,500. (c)
Invertebrate septate junction. Transverse image of the septate junction between two
molluscan ciliated epithelial cells. A periodic arrangement of electron-dense bars, or septa,
is present within the intercellular space between the two lateral plasma membranes.
× 216,000. (All photographs courtesy of N. B. Gilula. Refer to Gilula, N. B., 1974, *Cell
Communications* [ed. R. P. Cox], John Wiley & Sons, pp. 1–29.)

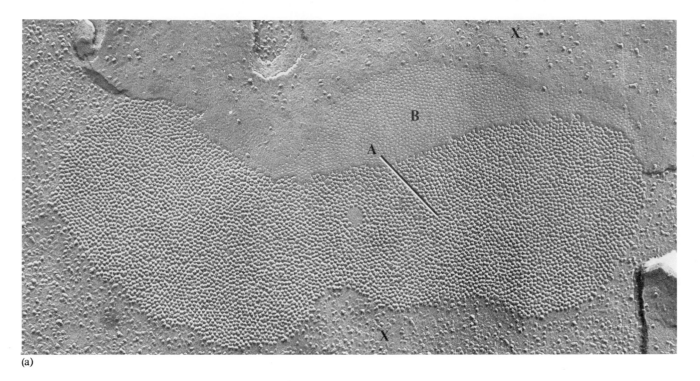

(a)

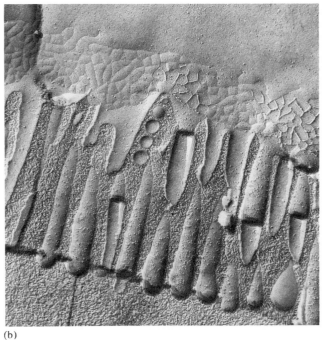

(b)

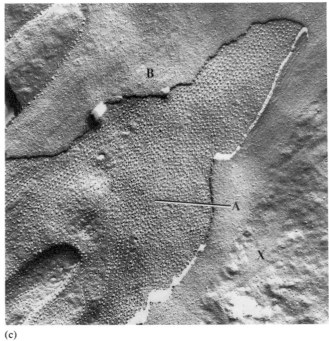

(c)

the gap junction, showing that the "gap" can serve as a channel through which substances move and penetrate the cell. Molecules of the general size of metabolites and regulatory substances can be injected into one cell and can then be located later in the adjacent cell if there is gap junction communication (Fig. 4.23). Cells which lack gap junctions or which have been dissociated into separate units do not receive the added molecules injected into the test cell in the system. Movement of ions between cells takes place, as detected by electrical measurements using microelectrodes.

Septate junctions also serve as communication links between joined cells, since tracer substances readily penetrate the differentiated region of the membranes. This kind of continuum provided by septate and gap junctions is not typical for tight junctions. The "tighter" the tight junction the more impermeable it is to movement of large molecules from cell to cell. Tight junctions are permeability barriers between cells since the intercellular space is occluded. The tighter the occlusion, the more impermeable the cell is to penetration by molecules from a neighboring cell.

Because of their wide distribution among multicellular animals and their participation in cellular continuities, more attention has been directed toward study of gap junctions. Gap junctions can be "un-

**Figure 4.21**
Freeze-fracture appearance of junctions between adjacent animal cells: (a) gap junction from mouse liver. This unique membrane differentiation is characterized by a polygonal arrangement of membrane particles on the fracture face closest to the cytoplasm (A) and complementary pits or depressions on the fracture face closest to the external border of the plasma membrane (B). Note that the gap junction occurs as a plaquelike region that is segregated from regions of nonjunctional plasma membrane (X). The nonjunctional membrane fracture faces are characterized by a random distribution of heterogeneously sized particles. × 108,000. (b) Tight junction between epithelial cells of rat small intestine. The meshwork arrangement of ridges and grooves represent sites of true membrane fusion. The arrangement of anastomosing ridges (facing the cytoplasm) and grooves (facing the extracellular surface) is responsible for the occlusion properties of the tight junction. The fracture faces are exposed when the plasma membrane is split open during freeze-fracture preparation. × 47,000. (c) Septate junction from molluscan ciliated epithelium. Two complementary fracture faces are exposed in the junctional region. The innermost fracture face (A) contains parallel rows of membrane particles that correspond to the arrangement of intercellular septa seen in thin-sections. The outermost fracture face (B) contains an arrangement of linear depressions or grooves that complement the particle rows of fracture face A. The particles in the nonjunctional membrane regions (X) are randomly arranged. × 45,000. (All photographs courtesy of N. B. Gilula. Photographs a and c from Gilula, N. B., 1974, *Cell Communications* [ed. R. P. Cox], John Wiley & Sons, pp. 1–29. Photograph b from Friend, D. S., and N. B. Gilula, 1972, *J. Cell Biol.* **52**:758.)

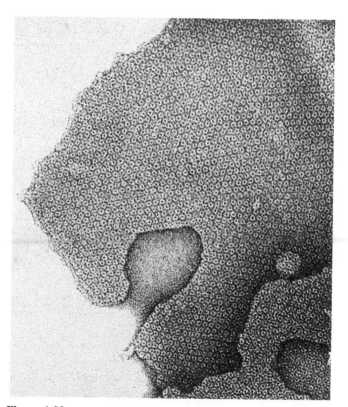

**Figure 4.22**
Negatively stained gap junction from isolated rat liver plasma membranes. An electron-dense spot is present in the center of all or most of the polygonal subunits in the lattice of the gap junction. × 240,000. (Courtesy of N. B. Gilula)

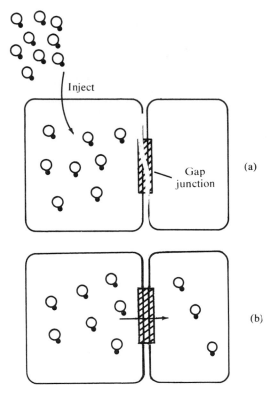

**Figure 4.23**
Diagram illustrating the movement of tagged molecules (a) into a cell and (b) from one cell to its neighbor through a communicating gap junction. Presence and movement of the molecules can be assayed during an experiment by locating and measuring the fluorescent or radioactive tag added to the molecules that are injected into the cell at the start of the experiment.

zipped'' by placing cells in hypertonic sucrose solutions. The osmotic sensitivity of gap junctions is not correlated with changes within the membrane however, since freeze-fracture faces are unchanged after the junction has been disrupted. Joined cells can be separated by treatments with protein-digesting enzymes. In such cases the entire gap junction remains part of only one of the two cells. This situation is indicated by the presence of junction areas that are 15–19 nm thick, located on the surface of a separated cell.

According to preliminary chemical studies, gap junctions contain some neutral lipid, some phospho-

lipid, and a few kinds of proteins. Since all these constituents are usually found in membranes generally, there is little known to distinguish gap junctions chemically from nonjunctional membrane regions. Some evidence from cytochemical staining tests shows that ATPase activity product is deposited in the gap junction area. If this information can be confirmed by biochemical assays, it would show that ATPase activity was a membrane-associated function of the gap junction. Since tests using inhibitors of ATP synthesis also abolish ion and molecule movements between joined cells, an energy requirement is believed to be necessary for intercellular movement of substances across the junction. ATPase activity in this area would, therefore, be expected, but clear-cut evidence has yet to be obtained.

One kind of system suited to study of metabolite exchanges in communicating cells is a culture made up of wild-type and mutant human fibroblasts (connective tissue components). Certain established lines of mutant cells cannot convert purine precursors into nucleic acids because they lack the required enzyme. If these mutants are mixed with wild-types containing the necessary enzyme, both the mutant and the wild-type cells are able to incorporate radioisotope-labeled purine precursors into nucleic acid polymers.

This phenomenon has been called **metabolic cooperation,** and it is dependent on formation of junctional contacts between the two kinds of cells. If gap junctions are disrupted, communication between cells stops and the mutant can no longer utilize purines for nucleic acid synthesis. The ability of cells to form gap junctions is inherited. Some kinds of cells do not form junctions, and are considered to be "noncommunicating" types. For example, when wild-type cells are mixed with lymphocytes from patients lacking the enzyme for purine incorporation, the two kinds of lymphocytes do not establish contact and the mutant cells remain unable to carry out the particular biosyntheses. When fibroblasts are mixed together, however, both genetic kinds of cells are able to make nucleic acids from the available precursor purines. In fibroblasts, the wild-types and mutants form gap junctions, and metabolic cooperation can take place.

There are mutant cell lines which lack gap junctions, so we may consider the capacity to be a genetic variable within a cell type as well as between different kinds of cells in the organism. When wild-type and junctional mutants are mixed together in culture, contacts are made only between wild-type cells and not between wild-type and mutant or two mutant cells. Mutants of this kind should prove useful in studying cell-to-cell communication mechanisms.

### Intercellular Adhesion

There is a class of intercellular contacts called **desmosomes** which act primarily as sites of intercellular adhesion and as anchoring sites for filamentous structures (Fig. 4.24). These differentiated regions are widely distributed among vertebrate and invertebrate tissues and may vary in form in different species or groups of cells. Desmosomes are not considered to be permeability sites, since cell-to-cell permeability remains unaffected if desmosomes are disrupted by treatment with proteolytic enzymes or by selective removal of divalent cations such as calcium and magnesium. Disruption usually is evidenced by disintegration of the condensed intercellular material that occupies the 25–35 nm-wide space between adjacent plasmalemmas of a desmosome area.

Because desmosomes are especially abundant in tissues such as cardiac muscle or outer layers of skin that are subjected to severe mechanical stresses, they are believed to act in maintaining cell-to-cell adhesion. Filaments are inserted into desmosomal plaques in several kinds of epithelia and other tissues, which indicates that desmosomes act as anchoring sites for these motility components within the cell.

### Cells and Cell Systems

One of the central postulates of the cell theory is that individual cells are separated units of function. It is quite possible, however, that interconnected cell systems rather than individual cells are functional units in the multicellular organism. At least we can say that cell-to-cell contacts provide the multicellular organism with a physical framework for metabolic cooperation and control of metabolic activities. Direct

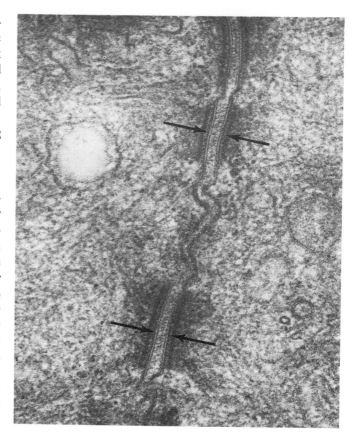

**Figure 4.24**
Desmosomes in rat intestinal epithelium. The desmosomes occur as symmetrical plaques between two adjacent cells in thin section. Their characteristics include (1) a wide intercellular space containing dense material; (2) two parallel cell membranes; (3) a dense plaque associated with the cytoplasmic surface (at arrows); and (4) cytoplasmic micro-filaments that converge on the dense plaque. × 156,000. (Courtesy of N. B. Gilula, from Gilula, N. B., 1974, *Cell Communications* [ed. R. P. Cox], John Wiley & Sons, pp. 1–29, Fig. 12.)

cell contacts are known to be involved in regulating growth and differentiation, and in development of the embryo. Cell junctions provide channels for ionic and metabolite exchanges and therefore provide ways for cells to circumvent the permeability barrier imposed by the plasma membrane. Interconnected cells can act coordinately in responses to stimuli, and order can be produced through these means for communication.

Cell junctions are present during times of active growth and differentiation in animal embryos, and are not present earlier or later in development of the embryo. Orderliness of development appears to be enhanced by cell contacts at appropriate times.

Multicellular organisms would have a considerable evolutionary advantage if intercellular communications formed between interacting cells at critical times in their activities. Cell junctions or cytoplasmic channels provide selective communication networks between particular cells at different times in their history or during different stages of activity. Direct junctional contact between plasma membranes is an obvious device in animals but cannot function in multicellular fungus or plant systems where a thick and rigid cell wall surrounds each cell in the organism. Cytoplasmic channels in plants, called **plasmodesmata,** provide communication links between cells (Fig. 4.25). Cytoplasmic channels are also a prominent feature of animal embryos during development, and they may provide a more rapid and coarser pathway for metabolite exchanges during rapid growth.

Certain kinds of cancerous cell growths are characterized by a failure of cell junctions to form. This defect may lead to uncontrolled growth if regulatory substances are transmitted slowly or erratically between cells in the group. This abnormality cannot be the only basis for malignant growths since many tumor cell populations do maintain regular junctional contacts. Indeed, it may not be the basis for any cancerous growths, and perhaps uncontrolled growth may be better explained by an altered response to regulatory substances, but the role of cell contacts in general cell communication remains a subject of intensive study. If we can learn how normal cells communicate and interact, we will then be in a better position to understand derangements in growth which take place in certain cells or cell communities.

## SUMMARY

1. Cell membranes are selectively permeable boundaries organized as a mosaic of proteins in and on a relatively fluid bimolecular layer of amphipathic phospholipids. Peripheral proteins bound to the bilayer surface and integral proteins embedded within the phospholipid phase but protruding at one or both surfaces are held together in a thermodynamically satisfactory state through maximized hydrophobic and hydrophilic interactions between protein and lipid residues. Proteins can move laterally within the lipid phase of the membrane.

2. Substances move through the membranes of a cell selectively by free diffusion, transport, or cytosis. In free diffusion and passive transport (also called facilitated diffusion), substances move along the expected concentration gradient, from their region of higher concentration to their region of lower concentration. Energy must be expended, however, in movements against the concentration gradient, as in active transport and cytosis events. Substances may be brought in by endocytosis or expelled by exocytosis, after interaction of the substances with the plasmalemma of the cell.

3. Movement of molecules across the membrane by facilitated diffusion or active transport requires the assistance of integral membrane proteins called carriers. Ion pumps located within the membrane participate in active transport of some metabolite, such as sugars or amino acids, into the cell or cell compartment. Pumping out of sodium ions or other ions against a concentration gradient, with energy provided by ATP, creates an electrical gradient across the membrane which provides energy to drive in the metabolite molecules. Carrier proteins assist molecules across the membrane barrier, but these can be effective only if an energy-generating system is also present in an active

**Figure 4.25**
Plasmodesmata in plant root cell thin sections: (a) permanganate-fixed cell of timothy grass (*Phleum pratense*) showing membranous elements extending across the cell wall and continuous between the two cells. × 31,000. (Photograph by W. Ridge) (b) High magnification view of *Potamogeton natans* cells fixed in osmium tetroxide. The plasmodesmata perforate the walls between adjacent cells. × 263,000. (Courtesy of M. C. Ledbetter)

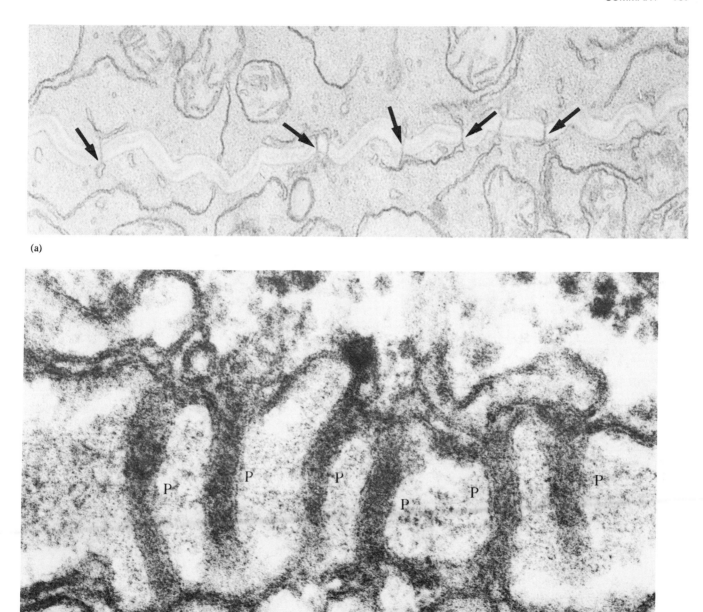

(a)

(b)

transport event. Transport by vesicle formation during cytosis is also energy-requiring, and is an essential activity in secretion and intracellular digestion phenomena associated with the Golgi apparatus and its derivatives.

4. Cell-cell communications are enhanced by cytoplasmic bridges and by some localized membrane regions called junctions. Gap, tight, and septate junctions are described and recognized by their organization as seen by electron microscope thin-section or freeze-cleave preparations of cells and cell membranes. Ion and metabolite exchanges between adjacent cells are facilitated or impeded, according to the types of junctions present in their plasma membranes.

5. Cells can establish contacts by desmosomes, which are specialized regions of adhesion between adjacent plasma membranes. Cell junctions and adhesions are important factors in communication between cells and in stabilizing cell communities such as those present in tissues and organs of multicellular animals. Plasmodesmata provide important communication links and channels for molecular exchange between plant cells. Plasmodesmata are cytoplasmic channels through pores in the rigid walls of plant cells, and are continuous between two neighbor cells.

### STUDY QUESTIONS

1. What are the major functions of cellular membranes? What is the difference between the unit membrane model proposed by Robertson and the fluid mosaic model of membranes proposed by Singer and Nicolson? What is the proposed role of the phospholipid bilayer in membrane construction according to Singer and Nicolson as compared with the role suggested by Green and Capaldi? What is the significance of there being both hydrophilic and hydrophobic interactions between molecules of the membrane? How do amphipathic molecules interact with water molecules inside and outside the cell? How does the fluid mosaic model account for functionally different cellular membranes?

2. By what major routes do substances cross the membrane barrier of cells and parts of cells? What is the basis for so few kinds of molecules crossing membranes by free diffusion? What is the major difference between passive and active transport of molecules across membranes? How are these processes similar? How do ion pumps mediate active transport of sugars and amino acids into the cell? What is an ion pump? How does the coupled neutral ion pump differ from the electrogenic sodium pump? How do bacterial

and plant cells manage ion pumps when there is little sodium present? How does an ion pump generate the required energy for active transport inward of metabolites? How do carrier proteins assist molecules across the membrane?

3. What significant contributions to the cell are provided by exocytosis and endocytosis events? How do these events relate to cell secretion and to intracellular digestion? How can we determine that cytosis is an energy-requiring process? What is the relationship of the plasma membrane to exocytosis and endocytosis events?

4. What is the distinction between cytoplasmic bridges and membrane junctions in cell-cell communication phenomena? What is a membrane junction? What types of junctions have been observed, and how are they distinguished by electron microscopy and by metabolite exchange assays? Why would we want to tag a molecule to study its movement from the environment into the cell or from one cell into another?

5. What is a desmosome? Why isn't a desmosome considered to be a cell-surface differentiation concerned with cell-cell communication? Would you expect to find desmosomes in mature plant tissues? Why not? What sorts of animal tissues are most likely to be characterized by extensive desmosome development of the cell surfaces? Why?

6. Does the concept of cell communities violate the basic principles stated in the cell theory? How can these opposing views be reconciled? What is the significance of intercellular communication in multicellular organisms, and how does such communication provide advantages to multicellular life forms? Why should we know how normal cells interact with one another if we are studying unregulated cell growth in cancerous or other abnormal cell systems?

### SUGGESTED READINGS

Bretscher, M. S. Membrane structure: Some general principles. *Science* **181:**622 (1973).

Capaldi, R. A. A dynamic model of cell membranes. *Scientific American* **230:**26 (Mar. 1974).

Cox, R. P. (ed.) *Cell Communication.* New York: John Wiley (1974).

Finean, J. B., Coleman, R., and Michell, R. H. *Membranes and Their Cellular Function.* New York: John Wiley (1974).

Fox, C. F. The structure of cell membranes. *Scientific American* **226:**30 (Feb. 1972).

Luria, S. E. Colicins and the energetics of cell membranes. *Scientific American* **233:**30 (Dec. 1975).

Quinn, P. J. *The Molecular Biology of Cell Membranes.* Baltimore: University Park Press (1976).

Rothman, J. E., and J. Lenard. Membrane asymmetry. *Science* **195:**743 (1977).

Sharon, N. Glycoproteins. *Scientific American* **230:**78 (May 1974).

Singer, S. J., and Nicolson, G. L. The fluid mosaic model of the structure of cell membranes. *Science* **175:**720 (1972).

Steck, T. L. The organization of proteins in the human red blood cell membrane. A review. *Journal of Cell Biology* **62:**1 (1974).

Tooze, J. (ed.) The external surfaces of cells in culture. In *The Molecular Biology of Tumour Viruses,* pp. 173–268. New York: Cold Spring Harbor Laboratory (1973).

Weissmann, G., and Claiborne, R. (eds.) *Cell Membranes.* New York: HP Publ. Co. (1975).

# Chapter 5

# Cellular Energy Transformations: Mitochondria and Chloroplasts

It is no exaggeration to say that the course of evolution on Earth owes some of its major features to the process of **photosynthesis.** Photosynthesis provides the foods that nourish all life, including green cells and plants themselves. Green cells in plants and in microscopic protist species are the beginning of the food chain, serving to produce food for all cells that must absorb or ingest their organic fuels from their environment. Nonphotosynthetic organisms also provide food for one another, and a cycle of carbon and nitrogen use and reuse occurs. In addition to providing foods made in reactions driven by energy originally derived from the sun, photosynthetic species constantly replenish our atmospheric oxygen.

Life originated perhaps 4 billion years ago and evolved until photosynthetic prokaryotes appeared about 3 billion years ago. It took another 2 billion years before eukaryotic photosynthesis became established. But once these eukaryotic photosynthesizers multiplied and released gaseous oxygen, beginning a little over 1 billion years ago, the Earth and its inhabitants underwent profound changes which we see all around us today. The atmosphere became about 21 percent oxygen, as it is today, and surviving life forms were predominantly of the aerobic type, as most life still is. Most of the species that could not tolerate oxygen, or whose habits were principally anaerobic, did not survive the atmospheric change. A small number of anaerobic and somewhat anaerobic species still exists, but these now represent a minute fraction of the total life on Earth.

With the appearance of eukaryotic photosynthesizing life forms, a constant source of organic fuels was assured as long as the sun kept shining and green cells captured light energy to make sugars. Our atmospheric oxygen supply is entirely due to green cells, which release oxygen during daylight as a byproduct of their light-capturing reactions. The light energy trapped by photosynthetic cells is converted to chemical energy, which is conserved in energy-carrying molecules. These energy carriers then supply the needed energy to drive reactions in which low energy $CO_2$ is reduced to higher energy sugars. These sugars, in turn, are used to make all the organic molecules of structure and function in green cells and organisms.

All organisms, whether green or not, must have energy and a variety of building blocks for growth, reproduction, and maintenance. Photosynthetic species make sugars and must then derive everything else from these sugars, so they also require metabolic systems to extract energy from sugars and other organic molecules if they are to survive. The primary fuel in virtually all life forms is carbohydrates, and the principal mechanism for extracting energy from fuel is through the processes of **aerobic respiration.** During aerobic respiration, which is the energy-yielding breakdown of sugars in the presence of oxygen, released energy is conserved in energy-carrying molecules. At the same time some of the simpler products of respiratory reactions can be channeled into pathways of biosynthesis of organic molecules.

Photosynthesis and aerobic respiration take place in particular membrane-bounded organelles in eukaryotes. Photosynthesis occurs in **chloroplasts** and aerobic respiration takes place in **mitochondria** (Fig. 5.1). In each case, organelle structure is intimately related to function. Membrane systems that carry out photosynthesis and aerobic respiration in prokaryotes are organized differently from those in eukaryotes. In prokaryotes the photosynthetic and respiratory membranes are differentiated regions of the plasmalemma. In eukaryotes, mitochondria and chloroplasts are independent of the plasma membrane of the cell. While the focus in this chapter will be on eukaryotic systems, references to prokaryotes will be given as these help us to perceive some basic concept in a broader perspective.

The first part of the discussion will deal with mitochondria and aerobic respiration, since a number of basic mechanisms are more easily explained in these than in their photosynthetic counterparts. Afterward, we will explore the more fundamental aspects of chloroplasts and photosynthesis, and also take advantage of certain explanations discussed in the first part of this chapter.

## AEROBIC RESPIRATION

Energy-yielding reactions within the mitochondrion depend on supplies of small, simple carbohydrates. Stored fats, lipids, and carbohydrates may furnish the units processed within mitochondria. The main stored foods, however, are starch and glycogen, which are degraded to soluble sugars in the **cytosol** portion of the cytoplasm. These simple sugars are processed further in the cytosol in glycolysis and other fermentative pathways. Only a small fraction of glucose free energy is released during glycolysis and conserved in ATP (see Chapter 3). The remainder of glucose free energy is released during aerobic respiration, with a substantial percentage of this energy conserved in many more ATPs per glucose than occurs in the glycolysis phase of glucose processing.

Glycolysis reactions proceed on to lactic acid if there is no oxygen available. When oxygen is present, however, pyruvic acid (pyruvate) is not processed further in the cytosol by glycolysis enzymes. Instead of the two NADH made in glycolysis being used to reduce pyruvate to lactic acid, pyruvate and NADH preferentially enter mitochondrial pathways when oxygen is present. In the mitochondrion, pyruvate is converted to 2-carbon acetate and its third carbon is

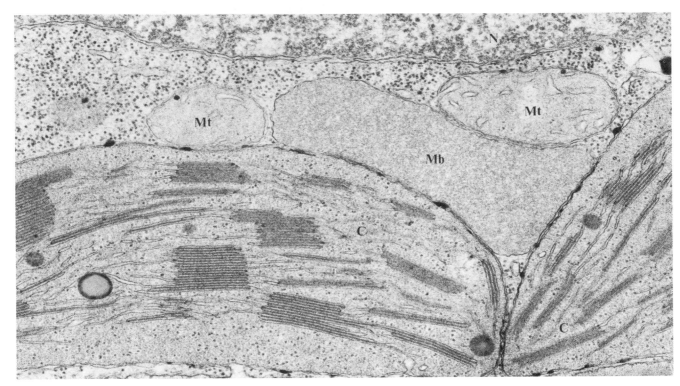

**Figure 5.1**
Thin section of leaf mesophyll cell of tobacco *(Nicotiana tabacum)*. The microbody (Mb) is wedged against the two chloroplasts (C) and one of the mitochondria (Mt). A part of the nucleus (N) is at the top of the photograph. Free flow of metabolites among the organelles is facilitated by their close physical proximity to one another. × 45,000. (Courtesy of E. H. Newcomb and S. E. Frederick)

released as $CO_2$. Acetate combines with coenzyme A to form a high-energy intermediate called **acetyl coenzyme A,** or **acetyl CoA.** It is in this form that acetate enters into oxidation-reductions and becomes fully oxidized to two molecules of $CO_2$. Coenzyme A is an energy-transferring molecule whose construction is very similar to ATP and NAD, which we encountered earlier in discussing energy transfer (Fig. 5.2). Coenzyme A engages in organic group transfer, binding through its sulfhydryl (—SH) terminus to the organic group it carries into reactions, whereas ATP transfers phosphoryl groups and NAD is an electron carrier.

Acetyl CoA (from carbohydrates, fats, and amino acids) enters the *final common pathway* in which fuel molecules are broken down in most aerobic cells. This pathway has been called the **Krebs cycle,** in honor of H. Krebs who first proposed it in 1934, and by two synonyms, the citric acid cycle (its first product) or the tricarboxylic acid cycle (some of the earlier types of intermediates in the cycle). The Krebs cycle proceeds vigorously when $O_2$ is present, since pyruvate continues to be channelled into the mitochondrion rather than being reduced to lactate in the cytosol. The Krebs cycle itself, however, does not use molecular oxygen directly, and neither ADP nor ATP participate directly in this cycle. A major function of the Krebs cycle is the oxidation of acetate, which is the remnant

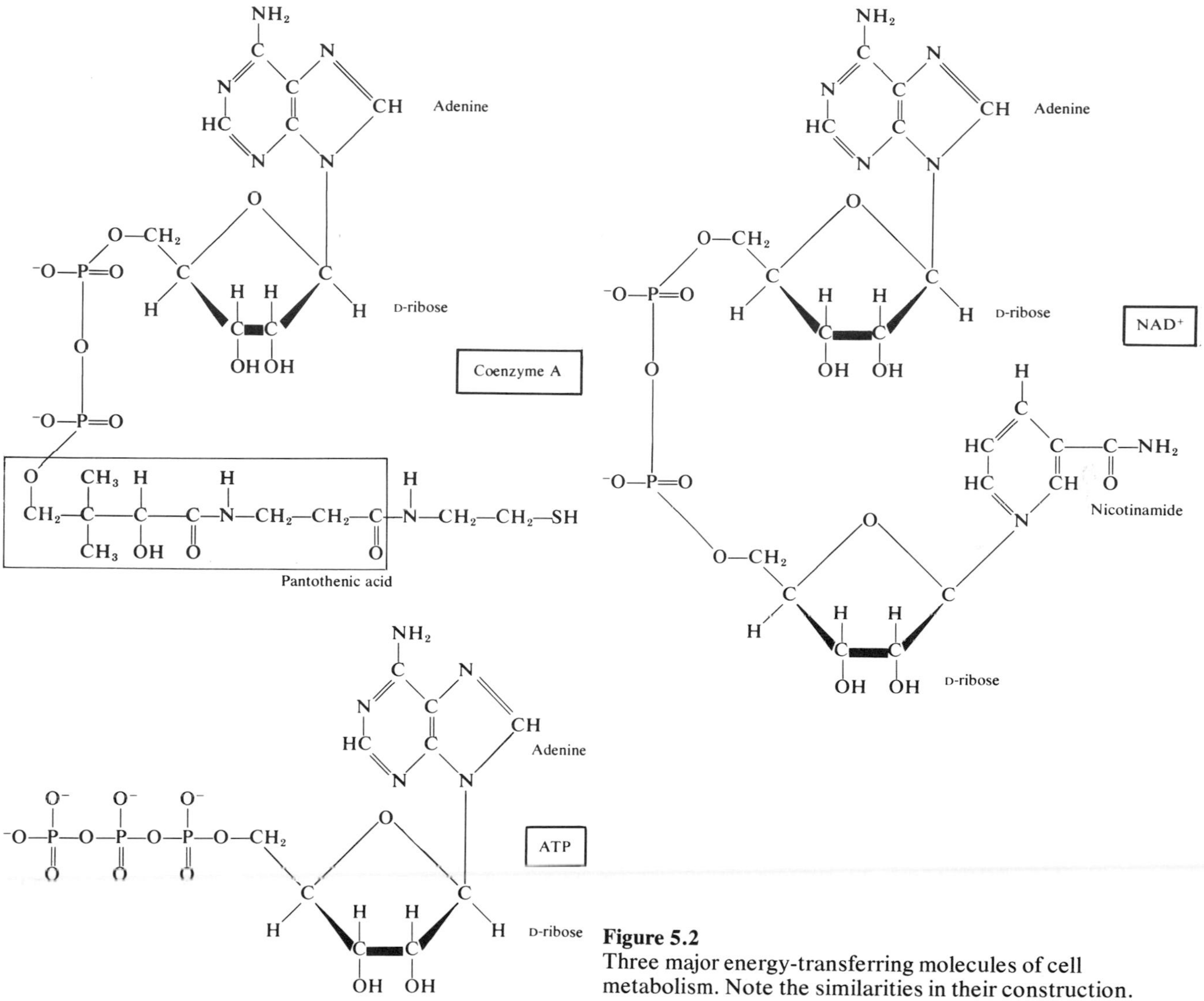

NH$_2$

Adenine

O—CH$_2$

$^-$O—P=O

D-ribose

Coenzyme A

OH OH

$^-$O—P=O

Pantothenic acid

CH$_2$—C—C—C—N—CH$_2$—CH$_2$—C—N—CH$_2$—CH$_2$—SH

CH$_3$ OH O

CH$_3$ H

H

H

O

NH$_2$

Adenine

ATP

$^-$O—P—O—P—O—P—O—CH$_2$

O$^-$ O$^-$ O$^-$

O O O

D-ribose

OH OH

NH$_2$

Adenine

O—CH$_2$

$^-$O—P=O

D-ribose

OH OH

NAD$^+$

$^-$O—P=O

O—CH$_2$

Nicotinamide

HC C—C—NH$_2$

HC CH O

N

D-ribose

OH OH

**Figure 5.2**
Three major energy-transferring molecules of cell
metabolism. Note the similarities in their construction.

of half a glucose molecule processed earlier in glycolysis, thus completing the oxidation of glucose to $CO_2$.

The free energy difference between acetate and 2 $CO_2$ is partially conserved in the reduced electron carriers NAD and FAD (flavin adenine dinucleotide). NADH and FADH$_2$ are the significant products of the Krebs cycle, each of these carriers having accepted two hydrogens (or electrons). NADH and FADH$_2$ then deliver their energy loads to a chain of electron transport enzymes. During transfers from one respiratory enzyme to another down the electron transport chain, much of the free energy differences along the way is conserved in ATP, produced in coupled synthesis reactions. The electrons are finally accepted by molecular oxygen, which can then combine with protons ($H^+$ ions), and water is formed as another end product of respiration. The significant feature of electron transport is conservation of free energy in ATP. A substantial fraction of the free energy difference between $O_2$ plus glucose and their products, $CO_2$ and $H_2O$, is extracted and conserved in ATP, and is thus made available for biosynthesis and other cellular work. We will now look at the Krebs cycle and electron transport in greater detail.

### The Krebs cycle

In each turn of the Krebs cycle, one molecule of acetate (as acetyl CoA) enters the cycle by condensing with one molecule of **oxaloacetic acid (oxaloacetate)** to form citric acid. Citric acid is oxidized ultimately to yield succinic acid, a 4-carbon compound, and two carbons are released as $CO_2$. Succinic acid is eventually oxidized to oxaloacetate, and the cycle is ready for another turn. For each turn of the cycle, one acetate enters and two $CO_2$ come out. Since oxaloacetate is regenerated at the end of each cycle, one molecule of oxaloacetate can bring about the oxidation of an infinite number of acetate molecules. There are four oxidations during the cycle, and during two of these oxidations there is removal of carboxyl ($—COO^-$) as $CO_2$. Such reactions are called oxidative decarboxylations (Fig. 5.3).

When acetyl CoA condenses with oxaloacetate, 6-carbon citric acid is formed. A series of steps occurs, finally producing the first of the 4-carbon intermediates. In this activity, two molecules of NADH are also produced and two $CO_2$ are released by oxidative decarboxylations. The first of the 4-carbon acids is coupled to acetyl CoA, the entire molecule being succinyl CoA. When succinyl CoA is hydrolyzed to form succinic acid, there is enough free energy difference to drive the phosphorylation of GDP to form GTP. GTP can be used in various energy-requiring reactions, or it can be enzymatically converted to ATP

$$GTP + ADP \rightleftharpoons GDP + ATP \qquad [5.1]$$

Succinic acid is oxidized to fumaric acid by a flavoprotein enzyme, succinic dehydrogenase. The riboflavin derivative **FAD** is covalently bound to the protein portion of the enzyme and acts as an electron carrier, producing $FADH_2$ in the coupled oxidation-reduction. Fumaric acid is hydrated next, and its double bond is changed to a single $—C—C—$ bond in malic acid (malate). Finally, malate is oxidized to oxaloacetate in an oxidation-reduction reaction during which the $NAD^+$ coenzyme portion of *malate dehydrogenase* picks up two electrons and is reduced to NADH. You may have noticed that each enzyme catalyzing an oxidation step in which NAD or FAD is reduced is a **dehydrogenase.** Dehydrogenases generally have NAD or FAD or a similar cofactor bound loosely or tightly to the protein part of the whole enzyme. FAD-linked enzymes are flavoproteins, as mentioned above.

During one turn of the Krebs cycle four pairs of hydrogen atoms were removed from substrate intermediates; three of these pairs were accepted by $NAD^+$ and one pair by FAD. Since it takes two turns of the cycle to process two acetates from one molecule of glucose, a total of six NADH and two $FADH_2$ are produced per glucose oxidized in the cycle. In addition, each acetate is oxidized to two $CO_2$ molecules. The overall equation summarizing *two* turns of the Krebs cycle is

$$2 \text{ acetate} + 6NAD^+ + 2FAD + 2GDP + 2P_i$$
$$\downarrow \qquad [5.2]$$
$$4CO_2 + 6NADH + 6H^+ +$$
$$2FADH_2 + 2GTP + 2H_2O*$$

In addition to the total of six NADH, $2FADH_2$, and 2GTP formed during Krebs cycle processing of glucose, four NADH were produced in earlier steps along

*In any phosphorylation of ADP or GDP, $H_2O$ is produced along with ATP or GTP in the reaction, as follows

$$P_i + ADP \rightleftharpoons ATP + H_2O, \text{ or, } P_i + GDP \rightleftharpoons GTP + H_2O$$

When ATP or GTP are hydrolyzed (to ADP and GDP, respectively), $H_2O$ is consumed, as an examination of the reactions indicates when going to the left.

**Figure 5.3**
The reactions of the Krebs cycle, also showing the feed-in of acetyl CoA from pyruvate after its processing from $C_3$ to $C_2$ and activation of the acetate ($C_2$) by coenzyme A. See Fig. 5.2 for molecule construction of coenzyme A, showing its sulfhydryl ($—SH$) group.

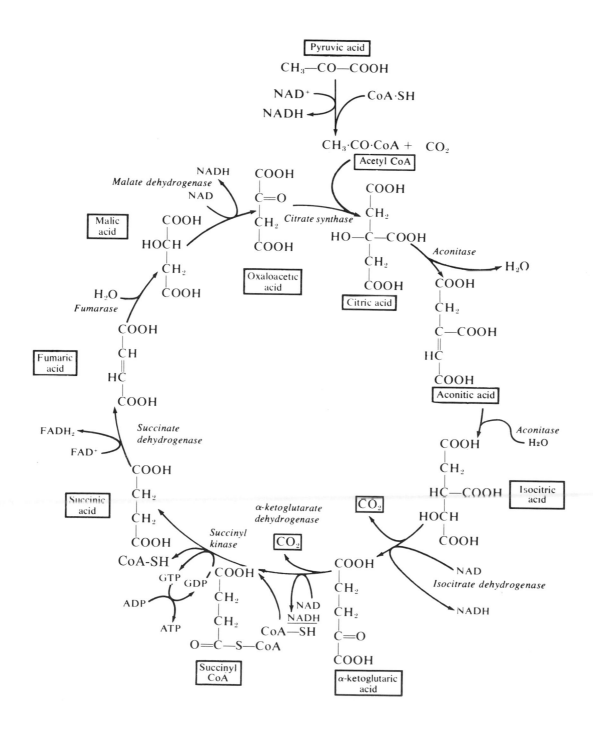

with two ATP. During glycolysis, two NADH and two ATP were formed:

$$\text{glucose} + 2\text{ADP} + 2\text{P}_i + 2\text{NAD}^+$$
$$\downarrow$$
$$2 \text{ pyruvate} + 2\text{ATP} + 2\text{NADH} + 2\text{H}^+ \qquad [5.3]$$

Each pyruvate converted to acetyl CoA also yields one NADH. Since two pyruvates per glucose are involved, we can add the equation

$$2 \text{ pyruvate} + 2\text{NAD}^+ + 2 \text{ coenzyme A}$$
$$\downarrow$$
$$2 \text{ acetyl CoA} + 2\text{NADH} + 2\text{H}^+ + 2\text{CO}_2 \qquad [5.4]$$

Taking equations [5.2], [5.3], and [5.4] together, we can summarize the oxidation of glucose through glycolysis to pyruvate, through pyruvate to acetyl CoA, and through the Krebs cycle as follows:

$$\text{glucose} + 2\text{ADP} + 2\text{GDP} + 4\text{P}_i + 10\text{NAD}^+ + 2\text{FAD}$$
$$\downarrow \qquad [5.5]$$
$$6\text{CO}_2 + 2\text{ATP} + 2\text{GTP} + 4\text{H}_2\text{O} +$$
$$10\text{NADH} + 10\text{H}^+ + 2\text{FADH}_2$$

The production of ATP, GTP, NADH, and FADH$_2$ are the significant energy-conserving achievements of glucose oxidation to CO$_2$. Release of the reservoir of energy in NADH and FADH$_2$ is the major theme of the next pathway in aerobic respiration, namely, stepwise transfers along the **electron transport chain** of respiratory enzymes. We will look at the electron transport chain in more detail, since basic mechanisms and concepts in this system are also involved in the energetics of photosynthesis.

### The Electron Transport Chain

The electron transport process involves sequential transfers of electrons from one electron carrier enzyme to another, along a pathway of decreasing free energy (reducing) potential. The electrons derived from substrate oxidations (carried by NADH and FADH$_2$) are taken up by molecular oxygen at the end of the transport chain. This reduced form of oxygen is "activated" and can combine with protons to form water. Unlike fermentative end products such as acids and alcohols, water from respiration does not pollute the environment inside or around the cell.

The most important consequence of electron transport is *release of free energy in usable-size packets*. The free-energy difference between interacting components in electron transport is substantially conserved in ATP, which is synthesized in reactions of **oxidative phosphorylation** that are *coupled* to electron transfer at several steps along the pathway. Some of the free-energy difference, of course, is lost as heat. Catalysts that drive these sets of reactions are parts of the structures of membranes, the inner membrane of the eukaryotic mitochondrion or the plasmalemma of prokaryotic cells.

Three classes of oxidation-reduction enzymes participate in transport of electrons from organic substrates to oxygen. We have already mentioned two of these, the **NAD-linked dehydrogenases** and the **FAD-linked dehydrogenases**. The **cytochromes** are representative of the third class. At least five different cytochromes are known to participate in respiratory electron transport: cytochromes $b$, $c_1$, $c$, $a$, and $a_3$, which act in the sequence given. Cytochromes act as redox couples; like NAD$^+$/NADH, a cytochrome enzyme can exist in the oxidized and the reduced states. One electron is handled at a time, and the cytochrome is oxidized on giving up an electron and reduced when it accepts an electron. Specifically, it is the iron atom in the active center of the enzyme which undergoes reversible Fe$^{2+}$–Fe$^{3+}$ valence changes during the catalytic cycle. The enzyme is alternately reduced and oxidized as electrons move from one component to another along the chain, toward oxygen (Fig. 5.4).

Each cytochrome protein is tightly complexed with a **heme** group, which is composed of a **porphyrin** ring compound that interacts with an iron atom (Fig. 5.5). The same heme is part of the active center of hemoglobin molecules, with one heme bound to each of the four polypeptide subunits of the protein (see Fig. 2.32). Another important molecule with a metal-porphyrin active center is chlorophyll, in which a $Mg^{2+}$ ion rather than Fe interacts with the same porphyrin ring structure.

The terminal member of the respiratory electron transport chain is **cytochrome oxidase,** an enzyme complex made up of cytochromes $a + a_3$. This enzyme can interact with molecular oxygen, whereas reduced forms of the other cytochromes cannot readily be reoxidized by oxygen. In addition to various lines of evidence that support the cytochrome sequence we have mentioned, the scale of standard free-energy differences is in the expected "downhill" direction (Table 5.1). If NADH is the starting material in the reaction producing water from oxygen, there is a $\Delta G°$ of $-52.6$ kcal/mole during the oxidation of NADH to $NAD^+$. If cytochrome $a$ is the starting material, the standard free-energy difference between its reduced and oxidized states is only $-25$ kcal/mole. The free-energy declines are consistent with the postulated sequence of electron carriers, which is widely accepted, although some of the components are tentatively placed at the present time.

The electron transport chain is therefore an assembly of molecules that undergo reversible oxidations and reductions and, in this way, transfer electrons from NADH and $FADH_2$ to molecular oxygen, which regenerates $NAD^+$ and FAD. Since there is a large free-energy decline when one pair of electrons from NADH is passed on to oxygen, several molecules of ATP could be formed during this multistep passage. In particular, there are three sites where there are 9–10 kcal free-energy difference (Fig. 5.6). The free-energy difference can drive ATP synthesis providing there is a mechanism that can *couple* electron transport to phosphorylation of ADP. The process of ATP synthesis during transport of electrons to oxygen is called **oxidative phosphorylation.** This term emphasizes the dependence on oxygen and distinguishes the process from other phosphorylations, such as those which occur during glycolysis (called substrate phosphorylations). First we will look at the oxidative phosphorylation process and then at the mechanisms proposed to explain coupling between oxidative phosphorylation and electron transport

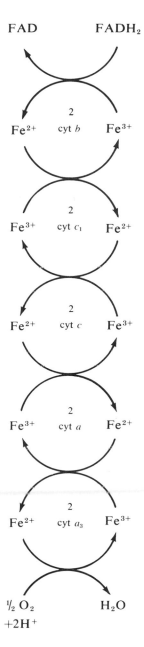

**Figure 5.4**
Diagram showing the pathway of electron transfer to $O_2$ along the cytochromes of the electron transport chain, emphasizing the valency changes between oxidized $Fe^{3+}$ and reduced $Fe^{2+}$ of the iron atom in cytochromes.

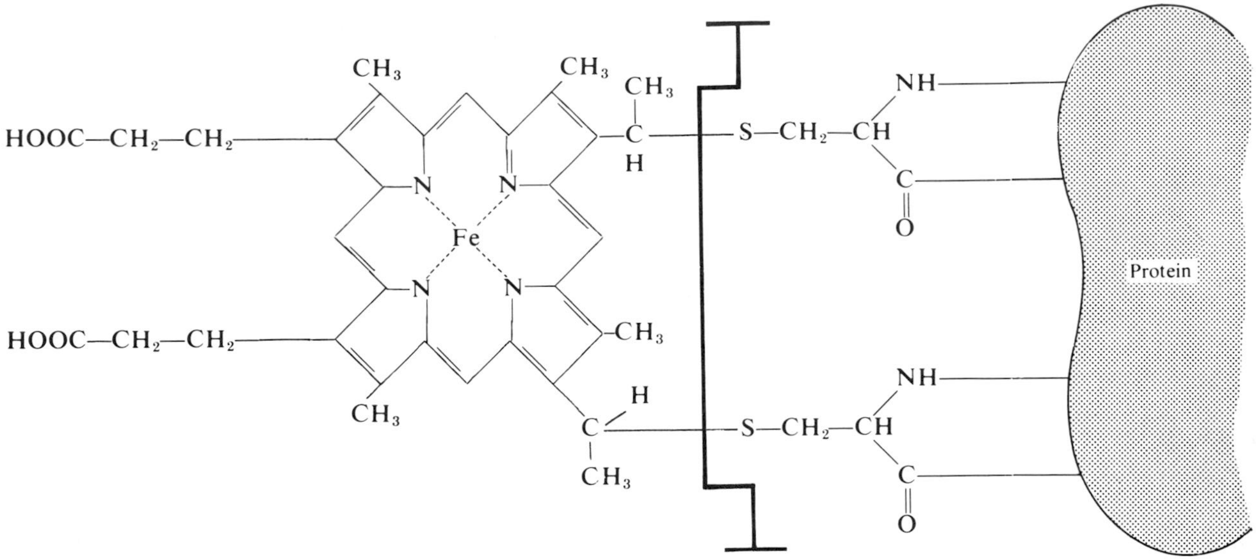

**Figure 5.5**
Construction of cytochrome $c$. The protein portion of the enzyme is bonded to a heme prosthetic group through sulfur (S) bridges.

toward oxygen. We will then be in a position to determine the energy balance sheet for oxidation of glucose to carbon dioxide and water.

### Oxidative Phosphorylation

One of the clearest indications that oxidative phosphorylation is an independent process coupled to electron transport is shown if poisons, such as dinitrophenol, are added. In the presence of the poison, electron transport continues, but energy is

**TABLE 5.1 Decline in free energy during electron flow down the respiratory chain**

| Electron Carrier (red/ox) | $\Delta G^0$ (kcal/mole) |
|---|---|
| NADH/NAD$^+$ | $-52.6$ |
| Flavoprotein red/ox | $-43.4$ |
| CoQ red/ox | $-33.2$ |
| 2 cytochrome $b$ red/ox | $-35.6$ |
| 2 cytochrome $c_1$ red/ox | $-27.8$ |
| 2 cytochrome $c$ red/ox | $-26.2$ |
| 2 cytochrome $a$-$a_3$ red/ox | $-25.0$ |
| Water/oxygen | $0$ |

released as heat, which cannot be used for cellular work. If the poison is removed, some of the free-energy difference is conserved during ATP formation. The effect is specific for oxidative phosphorylation, since the poison does not stop substrate phosphorylations in glycolysis or other metabolic pathways.

A customary expression of the ATP yield from oxidative phosphorylations is the **P/O ratio,** expressed as moles of inorganic phosphate ($P_i$) used per oxygen atom consumed (reduced). Each oxygen atom consumed represents one pair of electrons transported along the carrier chain to its terminus. When NADH electrons are taken up by oxygen the P/O ratio is 3.0, but when electrons pass from FADH$_2$ to oxygen the P/O ratio is 2.0. These figures say that 3 molecules of ATP are formed for every NADH derived from glucose oxidation, once NADH electrons are accepted by oxygen. Only two ATP are formed during transfer of each pair of electrons from FADH$_2$ made during glucose oxidation and ultimately accepted by oxygen. Using equation [5.5], the ten NADH formed per glucose lead to 30 ATP (3 per NADH $\times$ 10 = 30 ATP), and the two FADH$_2$ drive the synthesis of 4 ATP (2 per FADH$_2$ $\times$ 2 = 4 ATP), or a total of 34 ATP molecules made from NADH and $FADH_2$ derived during oxidation of a glucose molecule.

$$10NADH + 10H^+ + 2FADH_2 + 34ADP + 34P_i$$
$$\downarrow \qquad [5.6]$$
$$10NAD^+ + 2FAD + 34ATP + 34H_2O$$

### Coupling Mechanisms

There are three proposed mechanisms to explain coupling of ATP synthesis to electron transport, but we really are not sure just how the activity of the electron transport chain drives ADP phosphorylation. Each of the three mechanisms attempts to explain the means by which the free-energy differences between electron transport steps can be delivered to the system responsible for ATP formation from ADP.

The **chemical coupling mechanism** was proposed in 1953 by the Dutch biochemist, E. C. Slater. It basically postulates that several high-energy phosphorylated intermediate compounds are formed at favorable sites along the electron-transport chain. These intermediates would be "collected" by an enzyme which would then catalyze the transfer of the phosphoryl group from the intermediate to ADP, thus forming ATP (Fig. 5.7). The intermediates would thereby be regenerated for continued rounds of phosphorylation, and might therefore be expected to occur in only trace amounts. The attractive feature of the proposal is that it does not require any new mechanisms, but simply extends well-known phosphorylation reactions to include oxidative phosphorylation in a common theme. It is not surprising that the postulated high-

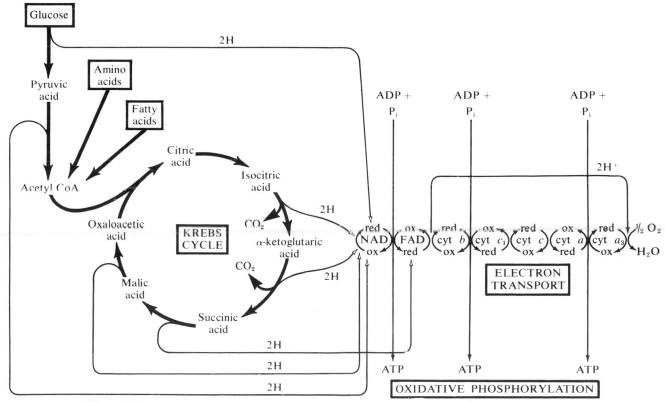

**Figure 5.6**
Summary of the steps by which glucose is oxidized to $CO_2$ and $H_2O$. A substantial part of the free-energy content of glucose is conserved in ATP formed during oxidative phosphorylation coupled to electron transport toward molecular oxygen.

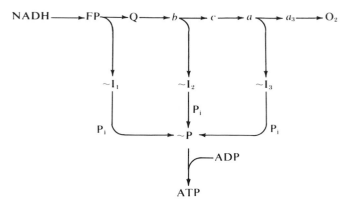

**Figure 5.7**
Summary diagram describing the postulated chemical coupling mechanism in which electrons transported from NADH to flavoprotein (FP) and cytochromes $b$, $c$, $a$, and $a_3$ via the intermediate Q are coupled to ATP formation through transfer of energy via intermediates (I). The squiggle mark ($\sim$) denotes a high free-energy level in the residue.

tively impermeable, intact mitochondrial inner membrane together could sponsor a system that would deliver the free-energy differences of electron-transfer steps to drive oxidative phosphorylation of ADP to form ATP. There are several assumptions that underlie the proposal:

1. The mitochondrial inner membrane is nearly impermeable to $H^+$ and $OH^-$ ions.
2. The electron-transport components are so situated within this membrane that $H^+$ ions are released only to the outside while $OH^-$ ions are taken up by reactants only to the inside of this membrane.
3. The ATPase, which catalyzes hydrolysis and formation of ATP, is located in the membrane in such a position that the ADP/ATP equilibrium ($ADP + P_i \rightleftarrows ATP + H_2O$) is shifted preferentially toward the ATP-forming reaction.

energy intermediates have not been detected, much less isolated, since they would be present in very small amounts. This difficulty is not considered sufficient reason, therefore, to discard the proposal. One other difficulty, however, has been emphasized more recently. The chemical coupling mechanism does not provide a rationale to explain why coupling absolutely depends on relatively intact membranes or vesicles derived from membranes. The other postulated mechanisms include membrane structure in the explanations of coupling.

The **electrochemical coupling mechanism** has been developed since 1961 by Peter Mitchell, a British biochemist, and the proposal has many enthusiastic adherents. The hypothesis states that energy released during electron transport is conserved in an ion gradient across the membrane, which then drives oxidative phosphorylation of ADP to form ATP. No high-energy intermediate compound need be involved, and intact mitochondrial membrane structure is required.

There are three pairs of $H^+$ ions produced for each pair of electrons transferred from NADH to oxygen along the electron-transport chain. It was Mitchell's conceptual achievement to develop a scheme by which $H^+$ ion release during electron transport, and a selec-

Now let's see how the mechanism has been postulated to operate in the mitochondrion, and in chloroplasts where another phosphorylation system occurs which is coupled to electron transport *in the light* (the process of *photo*phosphorylation).

If the mitochondrial membrane is impermeable to $H^+$ and $OH^-$ ions, and if $OH^-$ ions remain inside while $H^+$ ions are released to the outside, a pH gradient is created across the membrane (Fig. 5.8). The pH gradient, generated by a higher *external* $H^+$ ion concentration and a higher *internal* $OH^-$ ion concentration, also creates an electrical (charge) imbalance or voltage. This gradient is energy-rich. Mitchell suggested that the free-energy differences of electron-transport steps drive electrochemical (pH and/or electrical) gradient formation through the release of $H^+$ ions occurring at sites favorable to ATP formation. The electrochemical gradient would then transfer energy to the ATP-forming system, as the gradient assumed one steady state or another. While the gradient would "collapse" (release energy) at one time, it would be reforming at another time, and the whole gradient would remain in some energetically favorable steady state because of the selectively impermeable inner membrane and selective release of $H^+$

ions to the outside of the membrane while electron transport was in progress.

If the ATPase is appropriately situated in the membrane, as $H^+$ ions continue to be released to the outside it would lead the ADP/ATP equilibrium to shift toward ATP synthesis. This would occur because the ADP would accumulate, while $H_2O$ (H—O—H) would be removed as $H^+$ ions continue to be ejected to the other of the membrane. With removal of H—O—H, the equilibrium is tilted to ATP formation:

$$ADP + P_i \rightleftarrows ATP \ (+ \ H_2O, \ removed) \qquad [5.7]$$

In summary, it is proposed that electron transport drives oxidative phosphorylation through the coupling device of an energy-rich electrochemical gradient, which is generated during $H^+$ ion transport across the membrane as electrons are transferred along the electron-transport chain to oxygen. The free-energy differences of electron-transport steps are captured temporarily in the electrochemical gradient, which then delivers the energy to drive ATPase-catalyzed ATP synthesis during electron transport (Fig. 5.9).

The third and most recent proposal is of a **conformational coupling mechanism,** developed by David Green, an American biochemist. It is postulated that the macromolecular components of the mitochondrial inner membrane exist in at least two conformational states. The free-energy difference in going from a higher-energy conformational state to a lower-energy state is considered sufficient to drive oxidative phosphorylation. The free-energy differences of electron transport are converted to an energy-rich conformational state of inner membrane components, and when

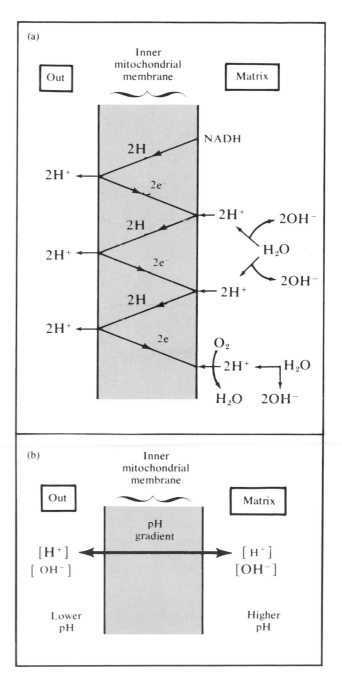

**Figure 5.8**
Summary diagram of the chemiosmotic coupling mechanism as proposed by Mitchell: (a) Transfer of energy during electron transport leads to a higher outside $H^+$ ion concentration; (b) an energy-rich pH gradient is thereby produced across the membrane.

this conformational energy is delivered to ATPase, oxidative phosphorylation occurs. Much of the evidence in favor of the idea has come from physical measurement studies, and from electron microscopy (Fig. 5.10). The usual images we photograph show mitochondria in their **orthodox state,** which is similar to mitochondrial images from cells respiring in the absence of ADP (therefore, making no ATP). When they actively respire and ATP is synthesized, mitochondria assume the **condensed state,** in which the inner membrane changes profoundly and the inner compartment is highly condensed (more opaque).

We can use an analogy to illustrate the delivery of energy through conformational change. If we put an object on a tightly coiled spring and allow the spring to relax, the object will be ejected into space. The difference in conformational energy between the relaxed and coiled states of the spring was enough to drive the energy-requiring potential movement of an object. Similarly, the conformational change of membrane components is proposed as the means for delivering free energy needed to make ATP by oxidative phosphorylation.

The three proposed coupling mechanisms differ mainly in the postulated *primary form* in which oxidation-reduction energy differences are conserved, before food energy is conserved in ATP formed by oxidative phosphorylation. Energy transformations therefore underlie all these proposals, but whether it is chemical energy, electrochemical energy, or conformational energy that is delivered for the work of oxidative phosphorylation remains uncertain.

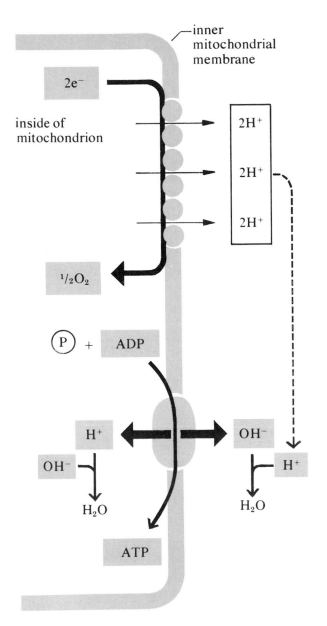

**Figure 5.9**
Simplified scheme showing the generation of ATP from ADP + inorganic phosphate, according to Mitchell's chemiosmotic theory of coupling. Two electrons pass down a chain of cytochromes situated within the membrane, and generate a $H^+$ ion gradient across the membrane. This gradient then drives synthesis of ATP and $H_2O$ in a dehydration reaction catalyzed by membrane-bound ATPase. (Reproduced with permission from Watson, J. D., *Molecular Biology of the Gene*, 3rd ed., p. 45. Copyright (c) 1976, 1970, 1965 by W. A. Benjamin, Inc., Menlo Park, California.)

## Overall Energetics of Glucose Oxidation

If glucose is burned completely to $CO_2$ and $H_2O$ in air under standard conditions, then the standard free-energy change between starting materials and end products can be shown as

$$C_6H_{12}O_6 + 6O_2 \rightarrow 6CO_2 + 6H_2O$$

$$\Delta G° = -686 \text{ kcal/mole}$$

[5.8]

The significant difference in terms of free-energy change in the cell is that some of the free energy of glucose is conserved in ATP and only a part of the potential 686 kcal is dissipated as heat. There is an approximate input of 7 kcal of energy per mole of ATP that is formed

$$ADP + P_i \rightarrow ATP + H_2O$$

$$\Delta G° = -7 \text{ kcal/mole}$$

[5.9]

Referring back to equation [5.5], we see that 2 ATP and 2 GTP were made during glycolysis and Krebs cycle activity, respectively, while 34 ATP were made in oxidative phosphorylation coupled to electron transport as shown in equation [5.6]. For simplicity we will convert GTP to ATP, and we have a total of 38 ATP formed per glucose oxidized to carbon dioxide and water. We can now calculate the efficiency of energy conservation during glucose breakdown in aerobic cells. Energy conserved as ATP equals 266 kcal/mole of glucose oxidized to completion (38 ATP × 7 kcal/mole of ATP formed = 266 kcal/mole of glucose oxidized). On this basis, the efficiency of conservation is 266/686 × 100 = 39 percent. The remaining 420 kcal is dissipated as heat. The overall equation for oxidation of one mole of glucose in which 38 moles of ATP are synthesized is shown as

$$C_6H_{12}O_6 + 6O_2 + 38ADP + 38P_i$$
$$\downarrow$$
$$6CO_2 + 44H_2O + 38ATP$$

[5.10]

$$\Delta G° = -420 \text{ kcal/mole}$$

The standard free-energy change is lower in equation [5.10] than in [5.8] because of ATP formation. The free-energy change of −420 kcal/mole in equation [5.10] is obviously derived by arithmetic (686 − 266 = 420). The free-energy change is still a large negative value, which indicates that the reactions proceed "downhill," but 39 percent of the energy stored in glucose has been extracted and retained, as ATP, for cellular work. Of the 44 $H_2O$ formed, 6$H_2O$ are made in electron transport and 38$H_2O$ are made during oxidative phosphorylations (Fig. 5.11).

Orthodox conformation

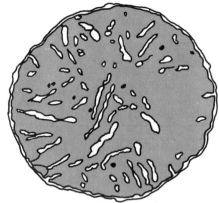

Condensed conformation

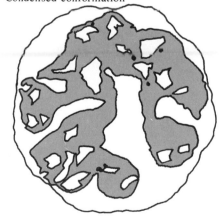

**Figure 5.10**
Drawings from electron micrographs of thin sections showing two conformational states of the mitochondrial inner membrane.

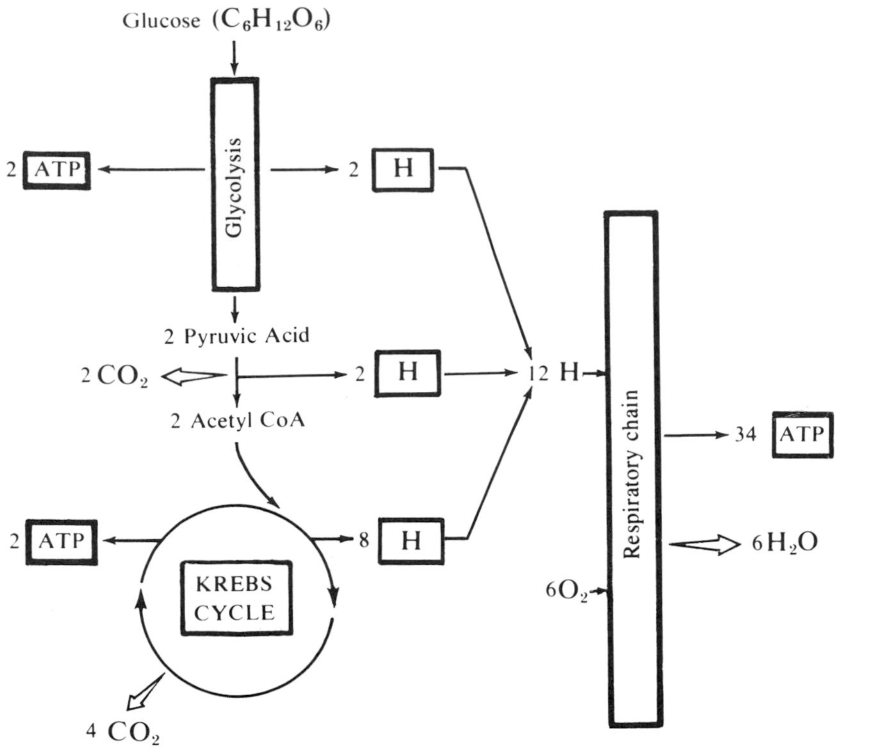

Glucose $(C_6H_{12}O_6)$

$2$ ATP $\leftarrow$ Glycolysis $\rightarrow$ $2$ H

$2$ Pyruvic Acid

$2\,CO_2$ $\Leftarrow$ $\rightarrow$ $2$ H $\rightarrow$ $12$ H $\rightarrow$

$2$ Acetyl CoA

$2$ ATP $\leftarrow$ KREBS CYCLE $\rightarrow$ $8$ H

$4\,CO_2$

Respiratory chain $\rightarrow$ $34$ ATP

$\Rightarrow$ $6\,H_2O$

$6\,O_2 \rightarrow$

Summary:

$$C_6H_{12}O_6 + 6\ O_2 \longrightarrow 6\ CO_2 + 6\ H_2O + 38\ \boxed{ATP}$$

**Figure 5.11**
Summary of the major events during the oxidation of glucose to $CO_2$ and $H_2O$.

The principle of stepwise release of small packets of energy in food breakdown is clearly illustrated in our discussion of glucose processing (Fig. 5.12). If all the energy were to be released at once or in large amounts, it would be wasteful because only about 7 kcal are needed to make one ATP. The release of 25 kcal of energy during electron transfer from cytochrome $a + a_3$ to oxygen still leads to only one ATP. The cell economizes by dealing out its energy stores in small enough packets so that explosive energy releases are prevented, and much of what is released at one step can be reasonably conserved in coupled energy-requiring reactions, such as ATP synthesis.

## FORM AND STRUCTURE OF THE MITOCHONDRION

Mitochondria were probably first described in 1886 by Robert Altmann, who was able to see granules and threadlike components after staining cells with a special dye. Nothing more could be seen using the light microscope than external shapes of tiny particles, but with the development of electron microscopy in the 1950s it was possible to study the internal structure of the organelle itself (Fig. 5.13). The typical profile of a mitochondrion seen in thin sections of cells photographed through the electron microscope shows a relatively smooth outer membrane and an infolded, inner mitochondrial membrane. These infoldings, called **cristae,** are the unique trademark of mitochondrial **ultrastructure** (structure at the electron microscope level of resolution). The infolded membrane has a great deal more surface area than a smooth outline would permit within the same volume of organelle. Infolding is an obvious device which allows more respiratory enzymes to be accommodated within the inner membrane structure, and therefore it is a device for amplifying the mitochondrial capacity for electron transport and oxidative phosphorylation, since these enzymes form part of the inner membrane construction.

The **matrix** is a semisolid system surrounded by the inner membrane. When cells are fixed in a solution of osmium tetroxide, ribosomes and DNA are evident within the mitochondrial matrix (Fig. 5.14). Cells fixed with a potassium permanganate solution rarely show preserved ribosomes and DNA, presumably because these components have been obliterated by the powerful oxidizing action of the chemical. About 50 percent of the matrix material is protein, and some of the proteins are enzymes of the Krebs cycle. In view of what we have just said and what we mentioned earlier in the chapter, you should now have a picture of mitochondrial space where pyruvate made in glycolytic reactions in the cytosol (bathing the mitochondria) passes through both membranes and enters the mitochondrial matrix, where it is processed to carbon dioxide by Krebs cycle activities. The NADH and $FADH_2$ made during glucose processing then enter into the electron transport chain situated within the mitochondrial inner membrane, where oxidative phosphorylation also takes place, catalyzed by a membrane-bound ATPase.

### Inner Membrane Subunits

Different patterns of cristae have been seen in different cell types. Cristae may be arranged transversely, parallel to the long axis of the mitochondrion, concentrically, or in the form of tubular invaginations of the inner membrane (Fig. 5.15). Cells with higher respiratory activity generally show the greatest degree of infolding, an obvious correlation with a greater amount of respiratory enzymes present in the higher amount of inner membrane surface area.

For example, liver or yeast cells which have lower activity also contain relatively few or irregular cristae, whereas highly active heart mitochondria have much of their volume filled with cristae (Fig. 5.16).

The outer mitochondrial membrane never infolds. In addition, the outer and inner membranes are distinguished by two other features: (1) different enzymatic activities, and (2) different particle displays. These two features may be related, since some of the inner membrane particles are functional enzyme proteins. The most striking particle displays of the inner membrane were first reported in 1963–1964 by D. F. Parsons and H. Fernández-Moran in independent studies using animal cells. In these studies, mitochondria were isolated from cells and then broken into small fragments of membranes. Instead of the usual method of staining membranes to contrast against a light background, negative staining was used. The stain does not interact with the biological materials, so they appear light (unstained) against the dark background of the stain itself. When viewed edge-on after

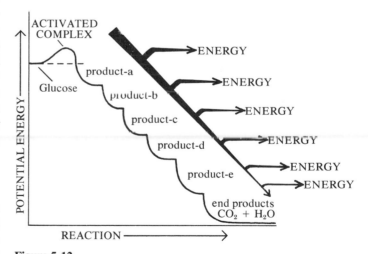

**Figure 5.12**
The energy contained in the glucose molecule is released in packets during the stepwise processing of the molecule in metabolism. The maximum energy release occurs when glucose is completely oxidized to molecules of carbon dioxide and water. Some of the energy is lost as heat, and the remainder is stored in ATP.

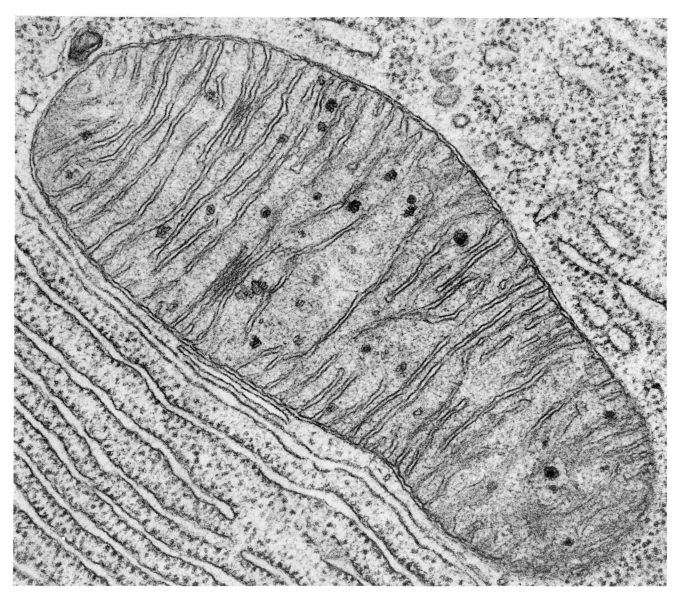

**Figure 5.13**
High magnification view of thin section through a mitochondrial profile from bat pancreas.
Cristae clearly can be seen to be infolded portions of the inner membrane of the
mitochondrion. × 85,000. (Courtesy of K. R. Porter)

extruded. Detailed measurements that were made for heart mitochondrial membrane showed that there were 2000–4000 subunits per square micrometer of inner membrane surface.

Although there were some lively controversies concerning the nature of the inner membrane subunits for a few years after they were discovered, the problem was soon resolved by careful and convincing studies conducted by Efraim Racker. They showed that the *inner membrane subunit was the ATPase of oxidative phosphorylation,* and that there was no electron

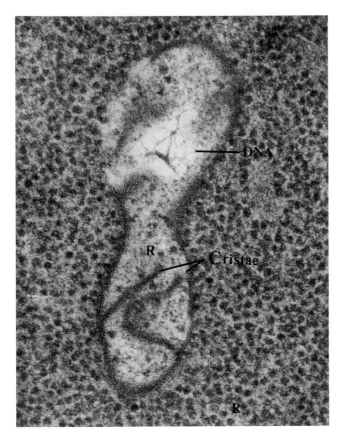

**Figure 5.14**
Thin section of mitochondrial profile from yeast, fixed with osmium tetroxide. A region of DNA fibrils is visible within the mitochondrial matrix (at arrow), as well as particles resembling ribosomes (R) that occur abundantly in the cytoplasm in which the mitochondrion is situated. × 132,600. (Photograph by H.-P. Hoffmann.)

negative staining, the mitochondrial inner membrane fragments appeared to bristle with projecting subunits (Fig. 5.17).

These **inner membrane subunits** occur only along the inner surface of the membrane, facing the matrix of the mitochondrion. Each particle consists of a spherical headpiece attached by a stalk to the inner membrane surface. It is uncertain whether or not the stalk, in turn, is attached to a basepiece that remains within the membrane even as the rest of the particle is

**Figure 5.15**
Drawings of mitochondrial profiles from thin sections showing various patterns of cristae arrangement.

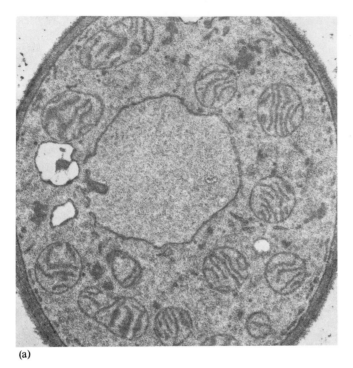

(a)

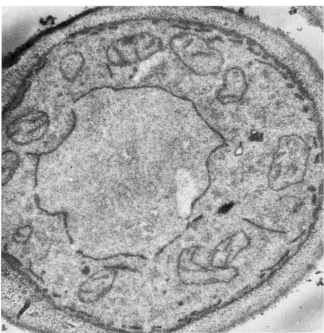

(b)

transporting activity present in the subunits. A brief look at the experiment will show how these conclusions were derived.

Racker's experiments took advantage of the fact that isolated mitochondria could be broken into fragments after exposure to high-frequency sound waves. These *sonicated* fragments were fully capable of carrying out electron transport coupled to oxidative phosphorylation, since a sufficient amount of intact structure was preserved. These fragments appeared as closed vesicles (membrane "bubbles"), formed when the edges of membrane pieces are resealed spontaneously, after membrane disruption. Using these vesicles as convenient materials for study, a reconstitution experiment was performed (Fig. 5.18). When untreated vesicles were analyzed they were found to carry out coupled electron transport and oxidative phosphorylation. When these were treated with agents that stripped proteins from the membrane surface, the vesicles were found to be able to carry out electron transport but not oxidative phosphorylation. Along with the inability to synthesize ATP, as determined biochemically, the stripped vesicles were shown to have lost their inner membrane subunits, according to electron microscopical observations. When purified preparations of inner membrane subunits were added to the stripped vesicles in the reconstitution part of the experiment, oxidative phosphorylating activity was restored. When the reconstituted vesicles were examined by electron microscopy, it was clear that the restored membranes were once again peppered with these subunits. There are many other kinds of experimental support for the association of subunits with oxidative phosphorylation exclusively, and for the identification of the subunit as the enzyme ATPase.

The physical association of oxidative phosphorylation and electron transport proteins within the inner mitochondrial membrane must certainly contribute to their close, functional interactions. The precise spa-

**Figure 5.16**
Cristae are more numerous in yeast cells that are (a) respiring at a very high rate, than in (b) comparable cells respiring at a much lower rate. × 30,000. (Photographs by H.-P. Hoffmann and M. Federman)

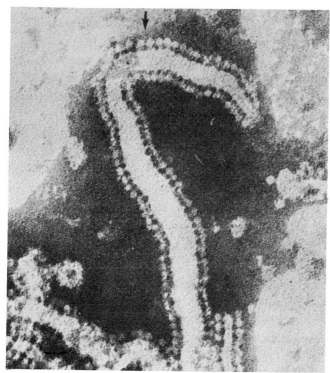

**Figure 5.17**
Negatively stained portion of cristae of isolated mitochondria from the adult bee. The inner membrane subunits (arrow) line the surface of the crista bordering the mitochondrial matrix. Subunits are not found on the opposite surface of the membrane bordering the intracristal space. × 264,500. (Reproduced with permission from Chance, B., and D. F. Parsons, *Science* **142**:1176–1180, Fig. 5, 29 November 1963. Copyright © 1963 by the American Association for the Advancement of Science.)

tial arrangements and relationships among these membrane proteins, however, have not yet been structurally determined. In a much more general perspective, we can see that mitochondrial metabolism is *compartmented*. Different sets of reactions take place in different parts of the mitochondrion for various reasons, including: (1) location of the reaction enzymes, and (2) selective permeability or impermeability to molecules that can or cannot cross the membrane barrier. Not only is the mitochondrion itself

a compartmented system, but it interacts with other cellular compartments such as the cytosol, nucleus, and a variety of cytoplasmic organelles. In very particular ways the mitochondria contribute to the relatively high efficiency of cellular metabolism, during which energy and building blocks are made available for growth, repair, and reproduction (Fig. 5.19).

### Size, Shape, and Number

The mitochondrion has been conventionally viewed as a sausagelike structure (Fig. 5.20), but other shapes have been reconstructed for mitochondria of several cell types. When they are observed in living cells, mitochondria appear to undergo rapid and dramatic changes in shape and size. More frequent studies, however, have involved measurements of the mitochondrial outlines seen in thin sections of cells photographed with the electron microscope. On the basis of an average diameter of 0.3 to 1.0 $\mu$m and a length of 1.0 to 10.0 $\mu$m, the mitochondrion has usually been compared in size and shape with a bac-

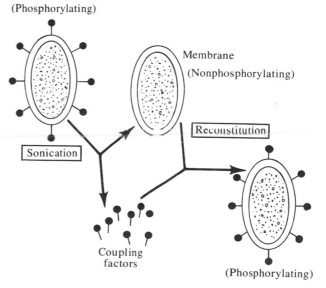

(Phosphorylating)

Membrane
(Nonphosphorylating)

Reconstitution

Sonication

Coupling factors

(Phosphorylating)

**Figure 5.18**
Diagrammatic representation of Racker's reconstruction experiment that showed oxidative phosphorylation was a function of the inner membrane subunit of the mitochondrial inner membrane. See text for details.

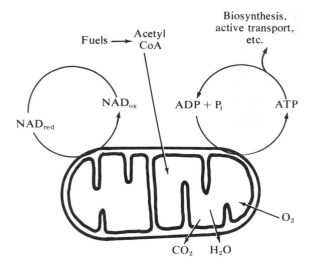

**Figure 5.19**
Compartmented metabolism involving the mitochondrion is regulated through interactions involving NADH/NAD+ and ADP/ATP. Reducing equivalents from NADH power the mitochondrial processing of fuel molecules entering in the form of acetyl CoA, and free energy is conserved in ATP, which then is available for many energy-requiring cellular activities.

terial cell. Mitochondrial numbers per cell, estimated from random samplings of thin sections of cells, vary from one in an exceptional unicellular species to more commonly hundreds or thousands in different cell types.

During the past ten years there were reports that mitochondria were large structures with branching, tubular processes (Fig. 5.21). If all the sections through a cell can be obtained, then the series of two-dimensional images in each photograph can be combined to construct a three-dimensional model of the mitochondrion as it would appear if we could see the whole cell in depth at one time.

According to three-dimensional models constructed from *serially-sectioned* cells as different as algae, fungi, insects, and rat liver, the mitochondrion may exist as a giant branching structure. There may be one or a few per cell, but hardly hundreds or thousands as the sausage model implies. Although no

model is universally accepted yet, it is important to know the size and shape of mitochondria for many reasons, including proper interpretation of their genetics, mode of formation, and metabolic interactions in the cell. In addition, the large branched mitochondrion poses quite different problems in interpreting the evolution of the mitochondrion as compared with a mitochondrion that looks like an average bacterial cell. We will discuss this further in Chapter 12.

## CHLOROPLASTS

Chloroplasts are the organelles that carry out photosynthesis in eukaryotic green cells. They are visible even at low magnification and resolution of the light microscope, and were described by Nehemiah Grew and Antonie van Leeuwenhoek in the seventeenth century. During the eighteenth and nineteenth centuries it was shown that production of gaseous oxygen by green plants was associated with light absorption by chlorophyll in chloroplasts, and

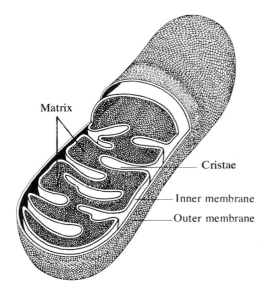

**Figure 5.20**
Conventional three-dimensional reconstruction of a mitochondrion as deduced from two-dimensional random thin sections of mitochondrial profiles in electron micrographs.

that intake of carbon dioxide led to manufacture of stored starch. Chloroplasts of land plants (mosses and other bryophytes, ferns, and seed-bearing plants) are usually discoid or lens-shaped and may number 50 or more per cell. Chloroplast size, shape, and number per cell is considerably more varied among the algae. Chloroplasts average 1–10 $\mu$m in length, but may be much larger in some of the algae. There are no chloroplasts in prokaryotic photosynthesizers, which include certain bacterial groups and all blue-green algae.

### Structure of Eukaryote Chloroplasts

Each eukaryotic chloroplast is separated from its cytoplasmic surroundings by two membranous envelopes; the space between these is sometimes obscured. Contained within the boundary of the inner membrane is a third system of membranes which are bathed in an unstructured matrix called the **stroma** (Fig. 5.22). The existence of membranous aggregates, called **grana** (sing., granum), was reported but difficult to verify by light microscopy. In 1947 Keith Porter and Samuel Granick provided substantiating evidence from electron microscopy. Grana are recognized as stacks of flattened discoid components, arranged like neat piles of coins (Fig. 5.23). Individual elements of each granum are separate, flattened, membrane-bound sacs enclosing an inner space.

The term **thylakoid** or **lamella** is generally used to refer to any flattened membranous sac in chloroplasts and related systems. The coin-shaped component in a granum is called a **grana thylakoid** or **grana lamella.** Some of the grana thylakoids in different stacks are connected by **stroma thylakoids** (or stroma lamellae), and the inner **thylakoid spaces** of connected grana and stroma thylakoids would therefore be continuous with one another. Each individual grana thylakoid is separated by its membrane from neighbors above and below it in the same stack. An opaque fusion line can be seen in the region of membrane contact.

Different cell types have from one to a hundred or more stacked thylakoids in a granum. There are some flowering plant cells that apparently lack grana, but do have long, single stroma thylakoids. On the whole, however, many of the green algae as well as most land plants contain grana or granalike assemblies of thylakoids. The stroma in which the thylakoids are suspended is granular and contains spherical, darkly-staining granules with some unknown function, as well as numerous ribosomes and aggregations of DNA fibrils.

Mitochondria and chloroplasts share many features. Both have (1) more than one membrane envelope with properties of selective permeability to ions and solutes; (2) unique genetic machinery consisting of DNA, RNA, and ribosomes; and (3) lipoprotein membranes containing tightly-bound components, including electron transport and ATP-synthesizing systems. In both mitochondria and chloroplasts it is the innermost membrane which is active in electron transport and ATP synthesis.

**Figure 5.21**
Three-dimensional model of a giant, branched mitochondrion from yeast. The model was reconstructed from photographs of a complete series of thin sections cut consecutively through one budding cell. The bulk of the mitochondrion was in the mother cell portion and a small connected mitochondrial region was in the small bud growing from the mother cell. (From Hoffmann, H.-P., and C. J. Avers, 1973, *Science* **181**:749-751.)

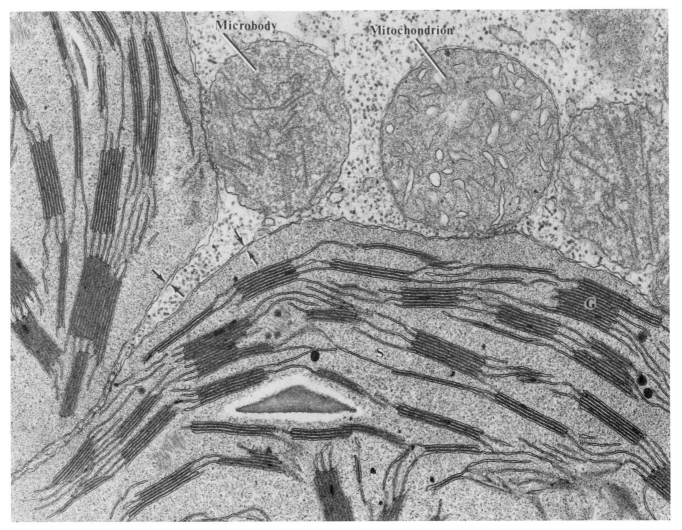

**Figure 5.22**
Thin section of part of a mesophyll cell of oat *(Avena sativa)* leaf showing thylakoids in
parts of two chloroplasts. Stroma thylakoids (S) traverse the matrix and also connect the
stacks of grana thylakoids (G). The appressed two membranes of the chloroplast envelope
are visible (at arrows). Note that the mitochondrion also has two enclosing membranes
while there is only a single limiting membrane for the microbody. × 42,000. (Courtesy of E.
H. Newcomb and P. J. Gruber)

### Thylakoids in Prokaryotes

The three important groups of photosynthetic
prokaryotes are the blue-green algae, purple bacteria,
and green bacteria. The blue-green algae are aerobic,
while the purple and green bacteria are either
anaerobic or at least carry out photosynthesis under
anaerobic conditions. This correlates with the fact that
blue-green algae produce oxygen during photosynthesis but bacteria evolve no oxygen in their activities.
These differences reflect chemical variations in their
photosynthesizing systems.

Thylakoids are distributed as single elements in the

cytoplasm of prokaryotes, and are never enclosed in organelles (Fig. 5.24). Although rarely observed in cell thin sections, some scientists believe that thylakoids are infolded regions of the plasmalemma in some species. Looking at thylakoids, DNA fibrils, and densely packed ribosomes in blue-green algae makes one immediately aware of their uncanny resemblance to chloroplasts of some eukaryotes.

There are thylakoid infoldings in blue-green algae and purple bacteria, but the green bacteria are unusual in having their photosynthetic pigments in special **chlorobium vesicles.** An unusually thin (2–3 nm) membrane encloses the vesicle, which is entirely separate from the plasmalemma. We could view these chlorobium vesicles as membranous compartments that are equivalent to eukaryote organelles, except that the very thin vesicle membrane is quite different from other known cellular membranes.

### Photosynthetic Pigments

There are three classes of photosynthetic pigments in photosynthetic cells: **chlorophylls, carotenoids,** and **phycobilins.** One or more kinds of chlorophyll and carotenoid are found in all photosynthetic species, but phycobilins are found only in the red and the blue-green algae. Except for the green bacteria whose pigments are in chlorobium vesicles, all other photosynthetic species have their pigments in tightly-bound association with (or built into) thylakoid membranes. Phycobilins are loosely associated with thylakoid membranes, and exist in granules called **phycobilisomes** (Fig. 5.25). These different kinds of pigments are therefore concentrated and localized in all cells and are not dispersed at random in the stroma of the chloroplast or the cytosol around the chloroplast. Taken in some combination, these pigments provide the cell with the capacity to absorb solar energy across the entire span of the visible spectrum from 400 to 700 nm wavelengths, and into the far-red and infra-red range in some cases (Fig. 5.26).

CHLOROPHYLLS. This predominating pigment type absorbs energy in the blue and red wavelengths, as determined from an **absorption spectrum,** which is a plot of degree of absorbance of different wavelengths of a spectrum by a test substance. By plotting the

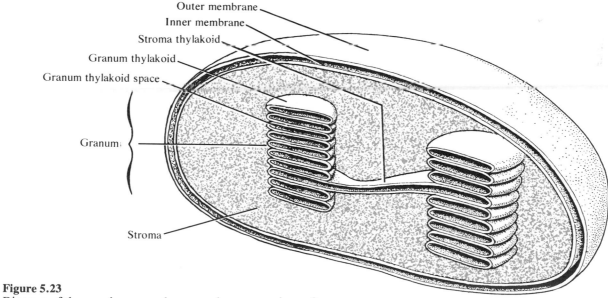

Outer membrane
Inner membrane
Stroma thylakoid
Granum thylakoid
Granum thylakoid space
Granum
Stroma

**Figure 5.23**
Diagram of the membranes and nonmembranous regions of the chloroplast.

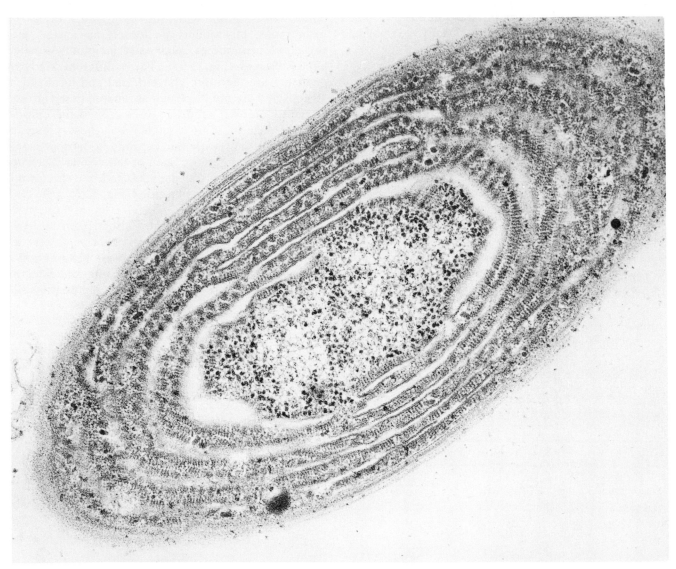

**Figure 5.24**
Thin section through the blue-green alga *Synechococcus lividus*. The photosynthetic lamellae are concentrically folded and are lined on both surfaces by granular phycobilisomes. × 62,400. (Courtesy of E. Gantt, from Edwards, M. E., and E. Gantt, 1971, *J. Cell Biol.* **50**: 896-900, Fig. 1.)

efficiency of different wavelengths of light in supporting oxygen production by some test substance, one may construct an **action spectrum** for it. Since chlorophyll's absorption spectrum and action spectrum essentially coincide, we may conclude that chlorophyll is the major light-capturing molecule in the cell and that it is involved in the oxygen-evolving process of photosynthesis (Fig. 5.27). Similar plots for other pigments have established their ranges of light-capturing abilities and contributions to photosynthesis. Since chlorophylls absorb red and blue principally, the remaining transmitted light appears green to us.

Chlorophylls are easily extracted in alcohol or ether solutions. The pigment is constructed of a **porphyrin** whose central nitrogen atoms coordinate with a $Mg^{2+}$ ion, and a long **phytol** side-chain which is added during chlorophyll biosynthesis (Fig. 5.28). Note the similarities with the porphyrin portions of hemoglobin and cytochromes, which have a central iron atom (see Fig. 5.5).

Different kinds of chlorophyll have slight differences in their porphyrin rings. Higher plants and algae, including prokaryotic blue-green algae, all contain chlorophyll *a*. Eukaryotes contain a second kind of chlorophyll, which varies according to the species group (Table 5.2). Bacteria, which do not evolve oxygen, have no chlorophyll *a*. Instead, they contain one or more of the four known **bacteriochlorophylls.** All the chlorophylls absorb the energy of visible light efficiently because of their numerous conjugated double bonds. The more double bonds, the faster energy is captured and the faster it is given up.

CAROTENOIDS AND PHYCOBILINS. The yellow, orange, or red carotenoids and the red or blue phycobilins also absorb light energy. Carotenoids absorb maximally in the violet to green range (400–550 nm), and phycobilins show maximum absorption of the green to orange region (550–630 nm) of the visible spectrum. Their colors, like chlorophylls', are due to transmission of particular wavelengths of the visible spectrum. Various carotenoids have different long-chain hydrocarbon regions with many double bonds.

Color in carrots is due to $\beta$-carotene, a commonly-occurring carotenoid pigment (see Fig. 5.28). These fatty substances are accessory pigments that can transfer absorbed light energy to chlorophyll *a* or its equivalent, although with low efficiency. Chlorophyll *a*, in turn, is directly involved in transferring energy to chemical energy-storing components in photosynthesis.

Phycobilins are accessory pigments found in eukaryotic red algae and prokaryotic blue-green algae. These pigments are different in that they are conjugated to specific proteins. The protein conjugate is a phycobiliprotein; the pigment itself is a phycobilin. The blue pigment-protein conjugate is **phycocyanin,** and its red analog is **phycoerythrin.** These substances are found together in phycobilisomes of both groups of algae, but the particular predominating pigment in the granule leads to a redder or bluer color for the cells. Phycobilins are open-chain porphyrin-type compounds, whereas chlorophylls have a ring porphyrin component (see Fig. 5.28). Like carotenoids, phycobilins transfer their absorbed light energy to chlorophyll *a*, which is the primary photosynthetic pigment in all plants and algae.

## OVERALL REACTIONS OF PHOTOSYNTHESIS

In all photosynthesizing cells, energy of solar radiation is absorbed by light-sensitive pigments and later is conserved as chemical energy for subsidizing cellular work. Water is the donor of hydrogens and electrons to reduce carbon dioxide or other electron-accepting compounds (except in the green and purple bacteria). When water gives up its hydrogens it is oxidized to oxygen. In the overall view, therefore, nonbacterial photosynthesis is a process in which light energy is used to reduce ("fix") $CO_2$ into organic compounds, with $H_2O$ as the electron donor for this reduction:

$$2H_2O + CO_2 \xrightarrow{\text{light}} [CH_2O] + O_2 + H_2O \quad \text{[5.11]}$$

As first proposed by C. B. van Niel, bacterial and plant photosynthesis could be viewed as essentially

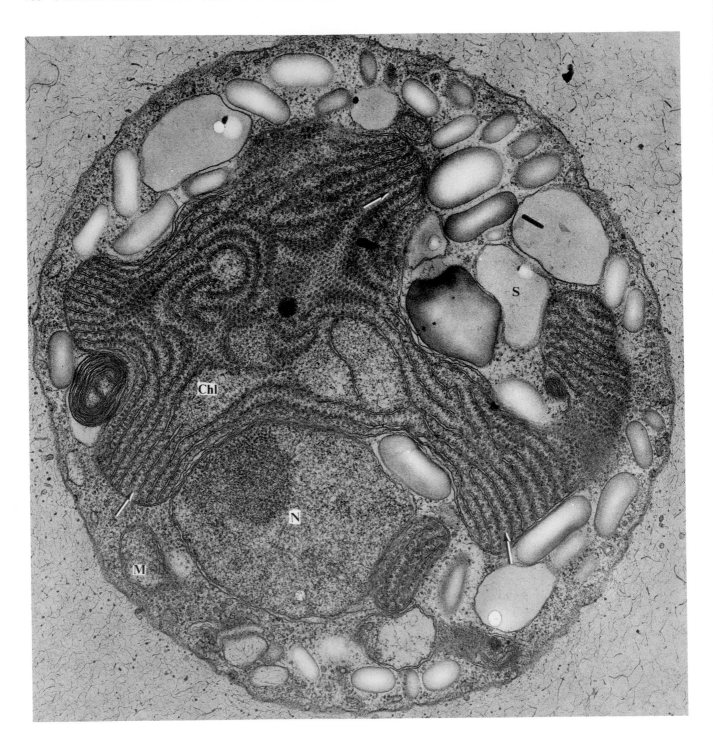

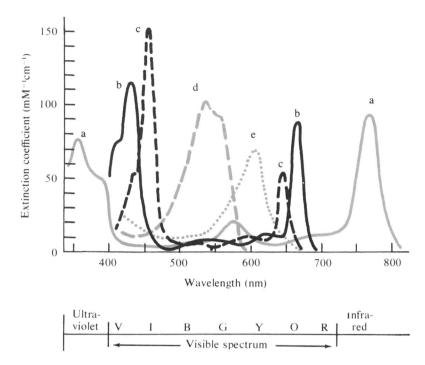

Ultra-violet | V | I | B | G | Y | O | R | Infra-red

← Visible spectrum →

**Figure 5.26**
Absorption spectra of (a) bacteriochlorophyll, (b) chlorophyll *a*, and (c) chlorophyll *b*, all in ether; and (d) phycoerythrin and (e) phycocyanin, in aqueous solution.

similar processes if the reaction reflected a more general hydrogen donor than water. For example, the hydrogen donor could be $H_2S$, or some organic compound such as lactic acid used by nonsulfur purple bacteria.

$$2H_2D + CO_2 \xrightarrow{\text{light}} [CH_2O] + D + H_2O \quad [5.12]$$

According to this formulation, various hydrogen donors ($H_2D$) may be involved. $H_2D$ could be $H_2O$, $H_2S$, or another hydrogen donor, depending on the

**Figure 5.25**
Thin section of the unicellular red alga *Porphyridium cruentum*. Granular phycobilisomes (at arrows) containing accessory pigment are displayed along both surfaces of the extensive photosynthetic thylakoid membranes, in which chlorophyll is located. In addition to the nucleus (N) near the chloroplast (Chl), there are numerous starch granules (S) and occasional mitochondrial (M) profiles visible in the cytoplasm. × 33,000. (Courtesy of E. Gantt, from Gantt, E., and S. F. Conti, 1965, *J. Cell Biol.* **26**:365-381, Fig. 7.)

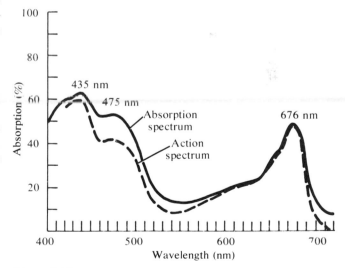

**Figure 5.27**
The absorption spectrum for a cell suspension of *Chlorella* (solid line) compared with the photosynthetic action spectrum (broken line) of these cells.

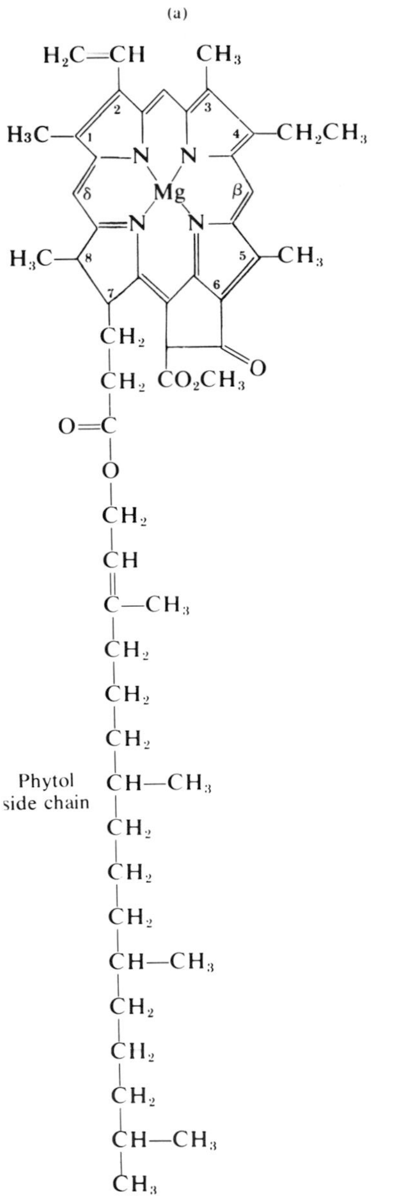

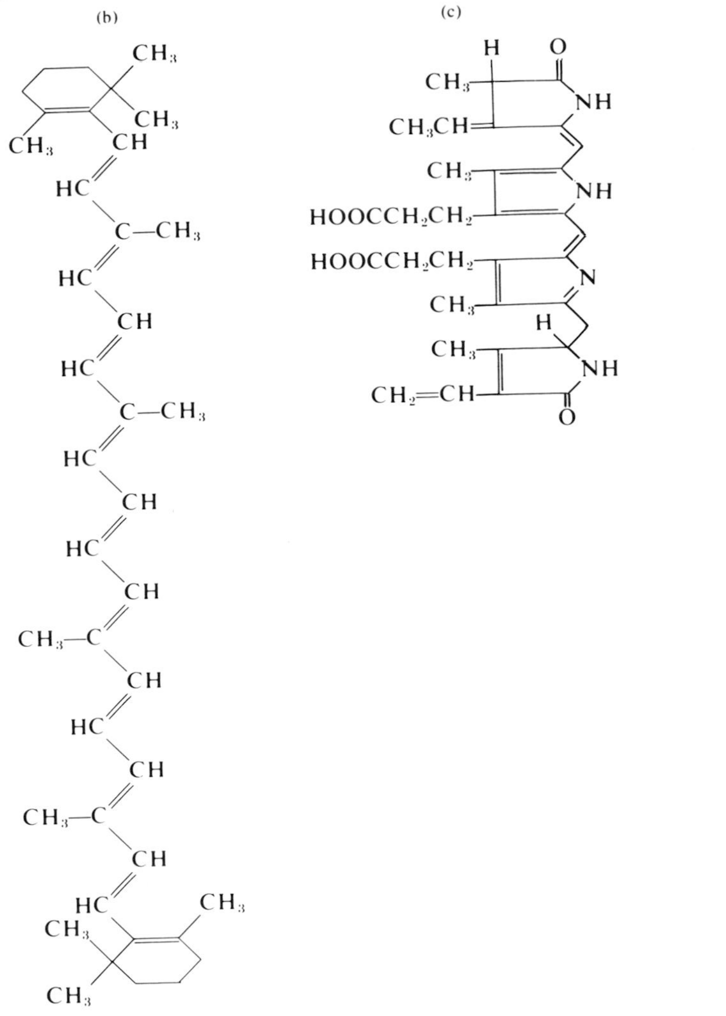

(a)

(b)

(c)

**Figure 5.28**
(a) Structural formula for chlorophyll *a*. (b) Structural formula for β-carotene. (c) Structural formula for phycoerythrobilin.

species. The donor would be oxidized to D, as its hydrogens were transferred for carbon dioxide reduction.

As van Niel noted, the molecular oxygen evolved during aerobic photosynthesis must originate from $H_2O$ and not from $CO_2$ molecules. Later studies showed this was indeed the case. Using the heavy $^{18}O$ isotope of ordinary $^{16}O$, the evolved molecular oxygen

**TABLE  5.2    Distribution of chlorophylls and other photosynthetic pigments**

| Organism | Chlorophyll a | b | c | d | Bacteriochlorophyll a | b | c | d | Carotenoids | Phycobiliproteins |
|---|---|---|---|---|---|---|---|---|---|---|
| Eukaryotes: | | | | | | | | | | |
| Mosses, ferns, seed plants | + | + | − | − | | | | | + | − |
| Green algae | + | + | − | − | | | | | + | − |
| Euglenoids | + | + | − | − | | | | | + | − |
| Diatoms | + | − | + | − | | | | | + | − |
| Dinoflagellates | + | − | + | − | | | | | + | − |
| Brown algae | + | − | + | − | | | | | + | − |
| Red algae | + | − | − | + | | | | | + | + |
| Prokaryotes: | | | | | | | | | | |
| Blue-green algae | + | − | − | − | − | − | − | − | + | + |
| Sulfur purple bacteria | | | | | + or + | − | − | | + | − |
| Nonsulfur purple bacteria | | | | | + or + | − | − | | + | − |
| Green bacteria | | | | | + | − | + or + | | + | − |

contained oxygen atoms originally present in $H_2O$, but none from $CO_2$:

$$n\,H_2{}^{18}O + n\,C^{16}O_2 \xrightarrow{\text{light}} [CH_2O]_n + n\,{}^{18}O_2 \quad [5.13]$$

In the reciprocal experiment

$$n\,H_2{}^{16}O + n\,C^{18}O_2 \xrightarrow{\text{light}} [CH_2O]_n + n\,{}^{16}O_2 \quad [5.14]$$

Furthermore, it is known that $CO_2$ is the major acceptor of hydrogens and electrons from a donor, but not the only one. Since various electron donors and acceptors may be used by different species, the more general photosynthetic equation for any organism would be

$$H_2D + A \xrightarrow{\text{light}} H_2A + D \quad [5.15]$$

in which $H_2D$ is the donor and A is the acceptor of hydrogens and electrons.

Water is a very weak reductant. It has a strongly positive standard electrode potential ($E_0 = +0.8$ volt)

and, therefore, a very high affinity for electrons. It has very little tendency, theoretically, to give up electrons and form $O_2$. Photosynthesis is the only metabolic process known to use water as an electron donor.

**Photoexcitation of Molecules**

We can see only that small portion of the electromagnetic radiation spectrum that falls between the wavelengths of 400 and 700 nm; this is the "visible" spectrum to us. These energetic wavelengths are emitted from the sun or some artificial light source as discrete packets called **photons.** The only difference between visible light and other electromagnetic radiations (x-rays, radio waves, ultraviolet light, and others) is in the frequency of vibrations of the radiation source. The energy of a photon is called a **quantum,** and the amount of energy in one quantum for a particular photon is given by the equation

$$E = h\nu \quad [5.16]$$

where $h$ is Planck's constant ($1.585 \times 10^{-34}$ cal-sec) and $\nu$ is vibrations/sec. In 1900 Max Planck developed an equation by which the energy in one quantum is related to the wavelength of the particular photon:

$$E = \frac{hc}{\lambda} \qquad [5.17]$$

where $h$ is Planck's constant; $c$ is the speed of light ($3 \times 10^{10}$ cm/sec); and $\lambda$ is the wavelength of emitted radiation. When an atom or molecule absorbs a photon of light, it is absorbing a quantum of energy. Since photon energy is inversely proportional to wavelength, photons of the shorter wavelengths have a higher energy content. This is the relationship derived from equation [5.17].

An atom or molecule may gain energy by **excitation,** an event which converts the substance to a higher-energy state. When electrons occupy orbitals of lowest accessible energy level around the nucleus of an atom, the atom is said to be in its **ground state.** When a sufficient packet of energy is absorbed by the atom so that an electron may move from one orbital to another of higher accessible energy level, the atom enters an **excited state.** In its higher-energy excited state, the atom can transfer an electron to a different atom or molecule that has a lower energy level (Fig. 5.29). The electron transfer proceeds ''downhill'' because the excited atom temporarily is at a higher energy level than the atom receiving the electron.

When photons of a particular wavelength (energy content) succeed in exciting an atom in a molecule, events occur very rapidly. It takes $10^{-15}$ second or less for excitation to occur, and the excited state only lasts about $10^{-9}$ to $10^{-8}$ second. Excited atoms are thus extremely unstable. This short period is sufficient time, however, to allow some or most of the light energy to be trapped chemically. On return to the ground state the trapped energy may be dissipated as heat, emitted as radiation (called fluorescence), or converted to chemical energy, depending on the lifetime of the excited state of each atom or molecule involved. The significant energy of photoexcitation in biological systems is the proportion of the total that is converted to chemical energy. Absorption is made even more efficient in photosynthetic cells because different pigments receive photons of different wavelengths, making more of the absorbed light energy available in the cell.

### Separability of Light and Dark Reactions

Beginning early in this century, evidence began to accumulate pointing toward two separate sets of reac-

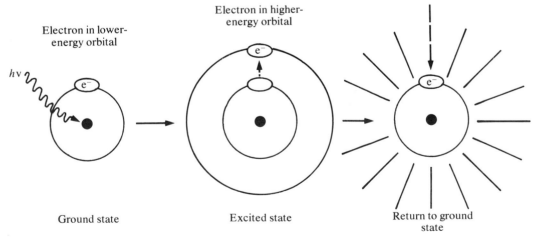

Electron in lower-energy orbital

Electron in higher-energy orbital

$h\nu$

Ground state

Excited state

Return to ground state

**Figure 5.29**
Absorption of a photon ($h\nu$) of light energy leads to excitation of an atom. Return to the ground state results in energy release (shown by radiating lines), which is equal to the amount of energy absorbed originally.

tions for photosynthesis, one set strictly dependent on light and another set that was independent of light. In 1937 Robert Hill provided the first experimental demonstration that a separate light reaction was localized in chloroplasts. When cells or isolated chloroplasts were illuminated in the presence of artificial hydrogen acceptors such as reducible dyes, molecular oxygen was evolved and the dyes were reduced simultaneously. Water served as the exclusive hydrogen donor in this system and most importantly, $CO_2$ was not required for the reaction. If $CO_2$ was reduced, its reduction products did not accumulate. These results supported the existence of separate light and dark reaction systems, since oxygen evolution could take place whether or not $CO_2$ reduction also occurred. The **Hill reaction** follows the general sequence

$$2H_2O + 2A \xrightarrow{\text{light}} 2AH_2 + O_2 \qquad [5.18]$$

where A is the hydrogen acceptor (unreduced dye) and $AH_2$ is its reduced form.

In 1950 it was shown that the coenzyme $NADP^+$ (nicotinamide adenine dinucleotide phosphate, oxidized form) could be the hydrogen acceptor in the Hill reaction. This was an important observation because NADPH was already known to be an electron donor in cellular biosynthesis, and because the observation supported the prediction that an end product of the light reaction was a reducing agent which could then act in reducing carbon dioxide during subsequent dark reactions.

Another important discovery which clearly showed that there were separate light and dark reaction systems was made in 1954 by Daniel Arnon. He showed that illuminated spinach chloroplasts could make ATP from ADP and $P_i$. $CO_2$ was neither required nor consumed in this light-dependent reaction called **photophosphorylation.**

The view thus gradually developed that $NADP^+$ reduction and ADP phosphorylation in the initial light reactions produced NADPH and ATP, which were then used in dark reaction pathways to reduce carbon dioxide and other electron acceptors. The end product

of $CO_2$ reduction in chloroplasts is sugar, often stored as starch.

During the 1950s it was also shown that oxygen evolution and photophosphorylation reactions were localized within chloroplast grana, in the same thylakoid membrane system that contains chlorophyll and accessory pigments. The dark reactions, on the other hand, took place in the stroma of the chloroplast. Once again we see a compartmentation of reaction systems within an organelle.

## THE LIGHT REACTIONS OF PHOTOSYNTHESIS

Until the late 1950s it was believed that absorption of light energy by chlorophyll was linked in one common pathway to oxygen evolution, $NADP^+$ reduction, and ATP synthesis. By the late 1950s, however, there was enough evidence to show that there were two different photosystems (or pigment systems) that somehow interacted in plant photosynthesis. With the discovery of cytochromes in chloroplasts, it seemed likely that these electron carriers acted as a link between two light-dependent systems acting in series (Fig. 5.30). A large body of substantiating evidence has now been collected to support this model of two photosystems linked in series by a chain of electron carriers, some of which are cytochromes.

### Photosystems I and II

The significant component of **photosystem I (PS I)** in plants and algae is light-absorbing chlorophyll a, with an absorption maximum of 683 nm wavelength. Oxygen evolution is *not* associated with PS I events. All oxygen-evolving photosynthetic species possess a second pigment system, **photosystem II (PS II),** which includes chlorophyll a (maximum of 672 nm) along with another kind of chlorophyll (b, c, or d) in all eukaryotes. Both photosystems also contain carotenoids, and there are phycobilins in the red and blue-green algae. The PS I primary pigment in photosynthetic bacteria is bacteriochlorophyll.

Theoretically, it takes 8 quanta of photon energy for 1 molecule of $O_2$ to be evolved (or 1 $CO_2$ to be reduced). Chlorophylls and accessory pigments

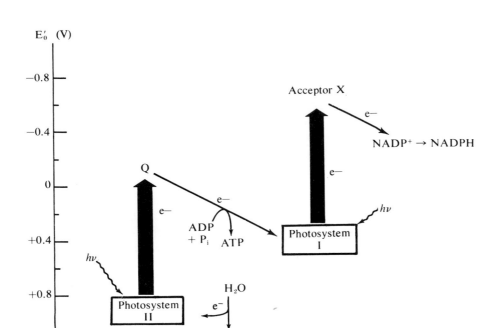

**Figure 5.30**
Summary diagram showing photosystem I and photosystem II activities coupled in series, in the light reactions of photosynthesis.

absorb light energy, which leads these molecules to become excited (raised to a higher energy state). Excitation energy is transferred from pigment to pigment in these *light-harvesting assemblies*. When special **reaction center** chlorophyll *a* molecules become excited, they alone become capable of transferring an electron to some acceptor. The reaction center is therefore the site where light energy is converted to chemical energy.

The reaction centers are believed to be special molecules of chlorophyll *a*, which have a somewhat higher-wavelength absorption maximum than the other chlorophyll *a* molecules of the pigment system. Reaction center chlorophylls for PS I (called P700) and PS II (possibly P692) absorb lower-energy photons than any of the other pigments, all of which show maximum absorptions for wavelengths that are shorter than 692 nm and 700 nm, and that are therefore more energetic. As we noted earlier in equation [5.17], photon energy is inversely proportional to wavelength. Excitation energy is transferred from one pigment to another, going toward lower-energy levels. Reaction

center chlorophylls act as traps for excitation energy funnelled from excited pigment molecules, but only an excited reaction center chlorophyll can transfer electrons to an acceptor in the photosystem. In this way, the reaction center chlorophylls act as *electron leads* from PS I and PS II pigments to electron acceptors.

### Electron Flow in Chloroplasts

There are several uncertainties in the series formulation scheme for the light reactions of photosynthesis (Fig. 5.31). We can sketch in sufficient detail, however, to follow the significant events leading to formation of ATP and NADPH, the important products of the light reactions. Water is the ultimate electron donor in eukaryotic chloroplasts and in blue-green algae. Through poorly-defined intermediates, water transfers electrons to the reaction center chlorophyll of PS II which has been excited by absorbing photons (shown as *hν*) and has lost an electron. Sufficient energy is transferred to the PS II reaction center chlorophyll that it can give up electrons to an acceptor (Q) of higher reducing potential ($E_0$). This electron transfer is thermodynamically possible because of the temporary increase in energy level of reaction center chlorophyll molecules. Electrons are replenished from water to reaction center chlorophyll of PS II. As water becomes oxidized, $O_2$ is released as a byproduct of the oxidation-reduction reaction.

Acceptor Q then transfers electrons in a conventional "downhill" sequence to a chain of electron carriers resembling the mitochondrial electron transport chain in several details. All the electron carrier molecules contain metal ions: **cytochromes, plastoquinones,** and **plastocyanin.** All these components and their precise sequence are somewhat uncertain at present. Some of the free-energy difference along this carrier chain is conserved in coupled reac-

tions of photophosphorylation. The exact number of ATP formed per electron pair transported is not certain, but two ATP per pair of electrons is a reasonable energetics estimate.

At approximately +0.4 volt, PS I reaction center chlorophyll accepts electrons from the last member of the electron transport chain. This P700 chlorophyll is raised to an excited state by absorbing photons and by energy transferred from pigments in the PS I light-harvesting assemblies. Once raised to a high enough energy level, P700 can transfer electrons to acceptor X (an unknown substance) at a much higher reducing potential of −0.6 volt. Having given up electrons, P700 is actually ionized, but is soon restored to its original state when it receives replenishing electrons from the carrier chain. Here, as in the PS II sequence, reaction

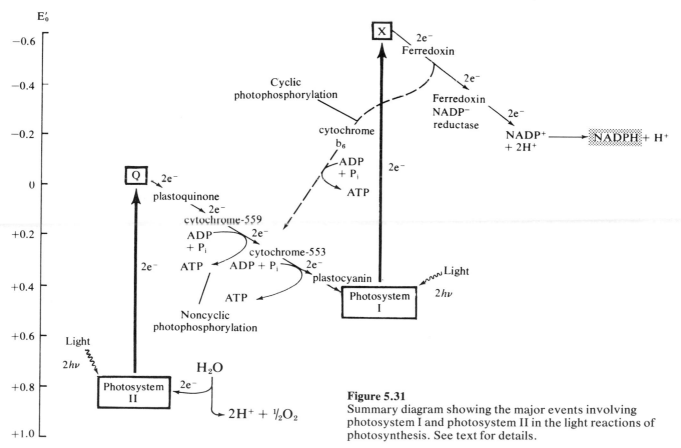

**Figure 5.31**
Summary diagram showing the major events involving photosystem I and photosystem II in the light reactions of photosynthesis. See text for details.

center chlorophyll regains its electron complement by accepting electrons from some donor, from water in PS II and from the carrier chain in PS I. By now the energy difference is the maximum between the initial electron source of water and the electron acceptor X (+0.8 to −0.6 volt, or 1.4 volt difference).

The next sequence in electron flow involves passage of electrons from X, in the expected "downhill" sequence, to the final electron acceptor, NADP⁺. On gaining electrons, NADP⁺ is reduced to the energy-carrying NADPH coenzyme. The voltage drop between compound X and NADP⁺ is sufficient to make NADPH. At this point, photon energy has been conserved in ATP and NADPH, the forms in which free energy can be extracted for sugar manufacture from $CO_2$ and used in the work of biosynthesis in general.

If we accept the assumption that two photons can cause the transfer of one pair of electrons from reaction center chlorophylls to reduce 1 NADP⁺, as implied in Figure 5.31, we can begin to look at the relative efficiency of energy conservation in the light reactions of photosynthesis. In addition, we may also assume that 2 ATP are made for each electron pair transported along the carrier chain linking the two photosystems. In the PS II reaction, water is oxidized (using $h\nu$ to represent a photon):

$$H_2O \xrightarrow{2\,h\nu} {}^1\!/_2 O_2 + 2H^+ + 2e^- \qquad [5.19]$$

and in the PS I reaction, NADP⁺ is reduced:

$$NADP^+ + H^+ + 2e^- \xrightarrow{2\,h\nu} NADPH \qquad [5.20]$$

These steps lead to the total reaction, including ATP synthesis, in which 8 photons result in release of 1 molecule of $O_2$:

$$2H_2O + 2NADP^+ + 4ADP + 4P_i \qquad [5.21]$$
$$\downarrow {\scriptstyle 8\,h\nu}$$
$$O_2 + 2H^+ + 2NADPH + 4ATP$$

For our calculations we must use moles, rather than molecules, of components so the energy values are large enough numbers to be meaningful. We can convert easily because there are $6.02 \times 10^{23}$ molecules (Avogadro's number) per mole. There also are equations to place photon energy on a "mole" basis ($6.02 \times 10^{23}$ photons per "mole"), and we can therefore calculate the energy of photons at a particular wavelength of light using the common unit of kilocalories.

There is an energy equivalence of about 41 kcal per "mole" of photons for light of 700 nm wavelength. Since it takes 8 "moles" of photons to produce 2 moles of NADPH, according to equation [5.21], 328 kcal of photon energy (8 × 41) are used to make 2 moles of NADPH and 4 moles of ATP. Each mole of NADPH represents 52 kcal of conserved energy, or 104 kcal for 2 moles, and each mole of ATP represents about 7 kcal of conserved energy, or 28 kcal for 4 moles. The overall photosynthetic light reactions therefore conserve 132 kcal (104 + 28) when 328 kcal of photon energy are put into the system. This represents an efficiency of energy conservation of about 40 percent (132/328 × 100). All things considered, the photosynthetic light reactions are very efficient processes.

**Photophosphorylation**

The particular light-dependent, ATP-forming system that is coupled to electron transport from PS II to PS I is **noncyclic photophosphorylation** (see Fig. 5.31). Since new electrons from donors are needed for each phosphorylation sequence, there is no cycle of electron transfer involved. There is also a system of **cyclic photophosphorylation,** in which electrons are fed back into the electron carrier chain. In this cyclic process, electrons from PS I are passed to acceptors and these electrons may be transferred back to the electron-transport chain, possibly by an intermediate such as ferredoxin (Fig. 5.32). There is no oxygen evolved and no NADPH made in conjunction with ATP synthesis by the cyclic process. Each cycle produces a maximum of 1 ATP for each photon of light absorbed by PS I, or for each pair of electrons passed

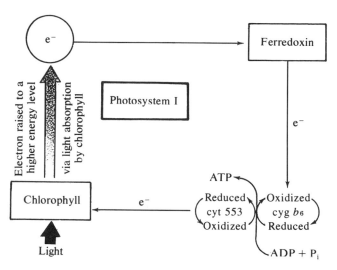

**Figure 5.32**
Cyclic photophosphorylation. When chlorophyll absorbs a
photon of light of sufficient energy, it sends an electron into a
high-energy state. The electron reduces ferredoxin and the
cytochrome system in turn. The reduction of cytochrome is
coupled to ATP formation. No "outside" electron donor is
required for the process, since the electron returns to
chlorophyll, at a lower energy level.

around the cycle. The physiological significance of
cyclic photophosphorylation is uncertain.

Noncyclic photophosphorylation is coupled to
electron transport between PS II and PS I, as shown
by use of poisons that uncouple the two processes. In
the presence of such an uncoupler, ATP synthesis
stops but electron transfers continue to occur from PS
II to PS I, so NADPH continues to be made. It is also
possible to stop electron transfer using other kinds of
poisons, in which case neither ATP nor NADPH can
be made because electrons from water are not
transported past the carrier chain.

One important experiment can serve to indicate
here that coupling mechanisms in photophosphorylation are probably similar to those in mitochondrial oxidative phosphorylation. The evidence supports
Mitchell's electrochemical coupling proposal. Briefly,
it was shown that a suspension of chloroplasts could
make ATP in response to an artificial pH gradient (Fig.
5.33). A temporary gradient of H⁺ ion concentration

was created by first exposing chloroplasts to a medium
of pH 4.0 and then transferring them rapidly to a medium of pH 8.0. In this way, the $H^+$ ion concentration
was higher inside (at pH 4.0) and lower outside (pH
8.0) the chloroplast membranes. The effect was the
same as a physiological pumping of $H^+$ ions into the
chloroplast thylakoid spaces, across the thylakoid
membranes. When ADP and $P_i$ were added to the
system at the same time the pH gradient was created,
ATP was formed.

These results are *physiologically* significant because the *amount* of ATP made in the experimental
system was approximately equal to the amount of ATP
synthesized in about 100 cycles of photophosphorylat-
ing activity in chloroplasts.

## THE DARK REACTIONS OF PHOTOSYNTHESIS

Reduction of $CO_2$ to carbohydrate can continue to
take place even when green cells are removed from the

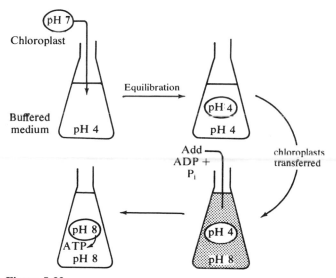

**Figure 5.33**
Spinach chloroplasts at an initial pH of 7 were transferred to
an acid medium in which they acquired a higher
concentration of $H^+$ ions after equilibrating in the acid
medium. When pH 4 chloroplasts were transferred to a
medium of pH 8 and ADP + $P_i$ were then added, ATP
synthesis was observed to occur at the expense of the
temporary $H^+$ ion concentration (pH) gradient.

light and placed in the dark. This is the basis for naming the "dark reactions", which proceed as long as ATP and NADPH continue to be available for reduction of $CO_2$. The major pathway by which $CO_2$ is reduced to simple carbohydrate end products was first made clear in the early 1950s by Melvin Calvin and associates. They showed that the first product of $CO_2$ reduction which accumulated in detectable amounts was a 3-carbon molecule that appeared before glucose or other hexoses. They followed the sequence of $CO_2$ reduction by labeling it with $^{14}C$, and looked for organic molecules with the $^{14}C$ label after very brief periods of time. The $^{14}C$ label appeared first in 3-phosphoglycerate and only later was found in glucose, or in starch made from glucose monomers.

### Fixation of Carbon Dioxide

In the initial step of the reduction pathway, $CO_2$ from the air joins with the 5-carbon sugar **ribulose 1,5-diphosphate** (abbreviated as **RuDP**) to form a transient 6-carbon compound (Fig. 5.34). The reaction is

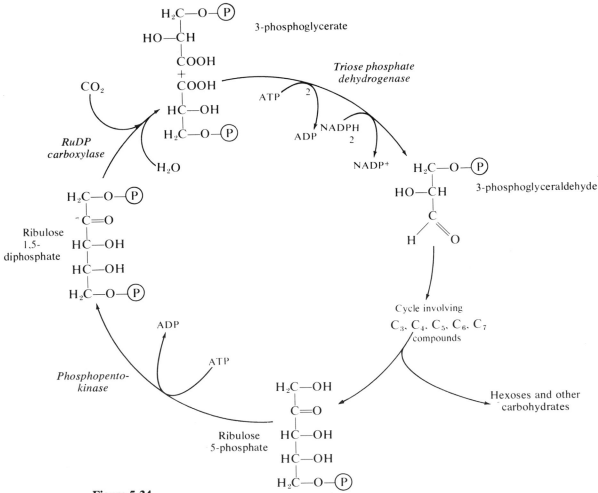

**Figure 5.34**
Reaction intermediates and enzymes of the $C_3$ cycle in which $CO_2$ is reduced to carbohydrate in the dark reactions of photosynthesis.

catalyzed by the abundant and important enzyme **ribulose 1,5-diphosphate carboxylase,** or **RuDP carboxylase,** which is found in the chloroplast stroma in loose association with thylakoid membranes. The transient $C_6$ compound breaks down to form two molecules of the 3-carbon sugar **3-phosphoglycerate** (the first [14]C-labelled product found by Calvin). Next, this triose sugar is phosphorylated at the expense of ATP to form an activated molecule that can go on to accept hydrogens and electrons from NADPH. The high energy product of these reactions is the triose sugar **3-phosphoglyceraldehyde.** Hexose sugars and more complex molecules are then formed in subsequent reactions.

Radioactive labeling studies showed that all 6 carbons of glucose become labeled during the dark reactions, but only 1 $CO_2$ is processed in each interaction with RuDP. To account for the continuous $CO_2$ fixation in a cyclic pathway that leads to all 6 carbons of hexose becoming labeled, Calvin and Andrew Benson proposed a cyclic pathway that would involve regeneration of RuDP for each $CO_2$ fixed in photosynthesis. By continuous regeneration of RuDP, continuous $CO_2$ fixation can occur. The regeneration sequence is very complex. The cycle of $CO_2$ fixation involving RuDP regeneration at each turn is called the **Calvin cycle** after its discoverer, or the $C_3$ **cycle** because of the first product in the cycle which accumulates.

During RuDP regeneration in the Calvin cycle, **fructose 6-phosphate** is formed as one of the intermediates. Diversion of this sugar into other biosynthetic pathways leads eventually to glucose formation. Many of the reactions leading to glucose, and to all other macromolecules in green plants, are the same as biosyntheses that occur in all cells. The unique feature of the dark reactions is production of a triose sugar from $CO_2$ through the mediation of RuDP. The triose sugar, in turn, enters into general biosynthetic pathways.

Each mole of hexose formed from $CO_2$ and $H_2O$ in the classic equation for green plant photosynthesis requires an input of 686 kcal. We know this because it is this amount of energy which is released when glucose is burned to $CO_2$ and $H_2O$ in air ($\Delta G° = -686$ kcal), so

it must require this amount to make glucose. We may state that

$$6CO_2 + 12H_2O \xrightarrow[\text{chlorophyll}]{\text{light}} C_6H_{12}O_6 + 6O_2 + 6H_2O \quad [5.22]$$

$$\Delta G° = 686 \text{ kcal/mole of glucose formed}$$

(*Note:* Reactions [5.13] and [5.14] state that all $O_2$ is derived from $H_2O$, so 12 $H_2O$ must be assimilated for each glucose formed in photosynthesis, to balance this equation.)

To determine the relative efficiency of photosynthetic dark reactions, we must consider the events at each turn of the Calvin cycle:

$$2H^+ + CO_2 + 2NADPH + 3ATP$$
$$\downarrow \quad [5.23]$$
$$[CH_2O] + 2NADP^+ + 3ADP + 3P_i + H_2O$$

Since it takes 6 turns of the cycle to form one molecule of glucose, a total of 12 NADPH (6 × 2NADPH) and 18 ATP (6 × 3ATP) are used to form one glucose molecule. Using the standard free-energy change for ATP hydrolysis as 7 kcal/mole, and for NADPH oxidation as 52 kcal/mole, the total free-energy input to make one mole of glucose is 750 kcal [(12NADPH × 52) + (18ATP × 7)]. The overall efficiency of photosynthetic dark reactions would then be 686/750 × 100 = 90 percent. This incredibly high efficiency is rarely achieved in biological activities.

In summary, light energy is absorbed by light-sensitive pigments and converted to chemical energy in NADPH and ATP. During the light reactions, water is oxidized to molecular oxygen, which is released into the air as a byproduct. The reduction of $CO_2$ is independent of light and takes place at the expense of NADPH and ATP made in the light. The initial product of the dark reactions is a triose sugar which is processed to fructose 6-phosphate. Fructose 6-phosphate then enters into biosynthetic pathways which lead to glucose and to other molecules required in plant growth, reproduction, and maintenance.

Despite the superficial resemblance between the overall reactions for glucose formation from $CO_2$ and

$H_2O$ in photosynthesis and the breakdown of glucose to $CO_2$ and $H_2O$ in glycolysis and aerobic respiration, these sets of processes are *not* a forward and reverse sequence of a common set of reactions. Totally different enzymes catalyze glucose formation in photosynthesis and glucose breakdown during respiration. The electron carriers are different, $NADP^+$ in photosynthesis and $NAD^+$ in respiration. Clearly, different sets of processes are involved in glucose formation and glucose breakdown.

**The Hatch-Slack Cycle**

In 1966 M. D. Hatch and C. R. Slack outlined an alternative pathway for $CO_2$ fixation in certain photosynthetic plants. They initially studied sugarcane, an important agricultural plant that was the subject of similar studies by earlier investigators. The clue to occurrence of another $CO_2$ fixation pathway was obtained from experiments using $^{14}C$-labeled $CO_2$ according to the procedures established by Calvin. Instead of finding 3-phosphoglycerate as the earliest labeled product, however, Hatch and Slack found that three particular 4-carbon acids were the first detectable labeled products: **oxaloacetate, malate,** and **aspartate.** With longer incubation times, $^{14}C$ eventually was found in 3-phosphoglycerate also, and even later the label appeared in hexoses.

Hatch and Slack proposed a scheme to explain these results (Fig. 5.35). Instead of RuDP carboxylase action as in the $C_3$ cycle, the initial enzyme in the new cycle was postulated to be **phosphoenolpyruvate carboxylase** (or, **PEP carboxylase**). The product of the initial reaction is oxaloacetate, which is formed when PEP accepts a $CO_2$ molecule and is carboxylated to the 4-carbon compound. Oxaloacetate is reduced to malate and aspartate; thus these two acids are the main components found in an equilibrium mixture during periods of active $CO_2$ fixation in photosynthesis.

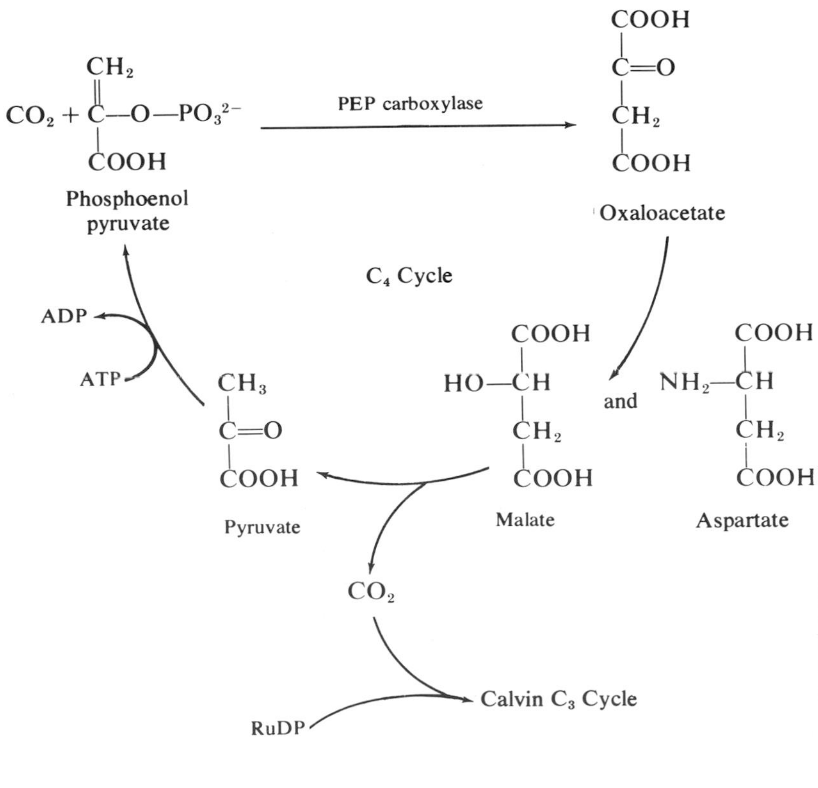

**Figure 5.35**
Reactions of the accessory $C_4$ cycle of $CO_2$ reduction as proposed by Hatch and Slack for some plants. The site of interaction of the $C_4$ and $C_3$ cycles that operate in the same plants is also shown.

Neither aspartate nor malate can continue in reactions leading to carbohydrate synthesis, however, so there must be intervening steps before hexoses can be made. It has been proposed that the $C_4$ acids are degraded enzymatically to yield free $CO_2$ and pyruvate, a 3-carbon intermediate. When pyruvate is phosphorylated at the expense of ATP to form phosphoenolpyruvate, the **$C_4$ cycle** can undergo another turn. The cycle repeats as new $CO_2$ enters the leaf from the outside air and becomes fixed into a $C_4$ acid.

The $CO_2$ which is released during the $C_4$ cycle is then picked up by RuDP carboxylase and is thus made available for $C_3$ cycle processing in exactly the same way as in plants that have only the $C_3$ cycle, as we discussed earlier. In essence, $C_4$ plants have an *additional* $CO_2$-fixing sequence which is coordinated with the usual $C_3$ cycle and RuDP carboxylase. The $CO_2$ originally fixed by PEP carboxylase action is ultimately handed over to RuDP carboxylase within the leaf, rather than being absorbed directly into the $C_3$ cycle as in $C_3$ plants (Fig. 5.36).

Hundreds of species belonging to a dozen or more plant families have so far been shown to have a $C_4$ cycle in addition to the $C_3$ cycle. All higher plants with a $C_4$ cycle appear to share a common plan of leaf anatomy, and to grow successfully under conditions of

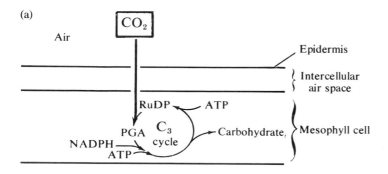

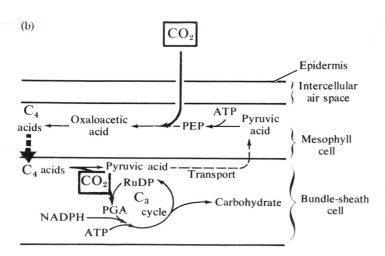

**Figure 5.36**
Comparison of the leaf anatomy and $CO_2$ reduction pathways in $C_3$ and $C_4$ plants: (a) leaves having only the $C_3$ pathway contain large mesophyll cells with chloroplasts, and air spaces saturated with water vapor. $CO_2$ enters through stomatal pores and moves into the chloroplasts of mesophyll cells, where reduction occurs via the $C_3$ cycle. (b) Leaves with the accessory $C_4$ cycle have most of their chloroplast-containing cells in a unique arrangement of mesophyll cells in a layer around a layer of bundle-sheath cells, which surrounds a leaf vein. $CO_2$ enters the mesophyll cell chloroplasts first and undergoes a reaction with the 3-carbon compound phosphoenolpyruvate (PEP) to form $C_4$ oxaloacetic acid. Further processing of the $C_4$ acids occurs in the bundle-sheath chloroplasts, where $C_4$ acids are decarboxylated to yield free $CO_2$. $CO_2$ enters the major $C_3$ reduction cycle. The remaining 3-carbon pyruvate is returned to the mesophyll cells where it is phosphorylated to PEP, and the $C_4$ cycle is completed and ready for another $CO_2$ fixation event.

low $CO_2$ concentration in the surrounding air. In fact, some $C_4$ plants can survive laboratory conditions that incapacitate or even kill $C_3$ plants.

$C_4$ plants can thrive when there are only about 5 parts per million (ppm) of $CO_2$ in the surrounding air, whereas $C_3$ plants stop photosynthesizing near 50 ppm of $CO_2$. The usual concentration of $CO_2$ in our atmosphere is about 300 ppm, or 0.03 percent. Gas exchanges between leaves and outside air take place mainly through **pores** (called **stomates**) in the leaf epidermis and not through the thick, waxy epidermal areas all around the pores. Stomates open and close, and gas exchange requires stomates to be open or at least partially open.

In regions where there is high light intensity, high temperature, and little water, leaf pores are almost closed. In this way the leaves lose the least amount of water by diffusion toward the drier air. When these pores are almost closed, however, very little $CO_2$ can enter the leaf. In $C_4$ plants, PEP carboxylase efficiently sweeps up what little $CO_2$ may enter under hostile outside conditions. By removing $CO_2$ almost as fast as it enters, through carboxylating pyruvate to form PEP, the enzyme helps to maintain a favorable $CO_2$ *diffusion gradient,* and $CO_2$ continues to enter the leaf even when outside concentrations are very low and pores are barely open. In plants without a $C_4$ pathway, RuDP carboxylase catalyzes $CO_2$ removal, but it is much less efficient than PEP carboxylase in $C_4$ plants. When $CO_2$ in the air around the leaf drops below 50 ppm, RuDP carboxylase is not efficient enough to remove $CO_2$ to internal levels that would maintain a favorable diffusion gradient. $CO_2$ no longer enters the $C_3$ leaf, and photosynthesis (sugar manufacture) essentially stops.

Two kinds of programs of plant improvement are currently under way, each directed toward increasing the food supplies in hot, dry climates. One program involves cross-breeding favorable $C_3$ crop plants with $C_4$ plants, hoping thereby to incorporate suitable genes from $C_4$ plants into the $C_3$ crops usually grown in hostile climates. Another program is aimed at finding higher-yielding strains of crops that already have a $C_4$ pathway. Larger harvests would be produced under such conditions if suitable strains could be made available. Both kinds of programs would have the greatest impact in underdeveloped parts of the world in which crops must be grown in scrubby or semidesert zones. Three high-yield $C_4$ crop plants already form important foundations for world agriculture, namely, sugarcane, sorghum, and corn.

## BIOGENESIS OF CHLOROPLASTS AND MITOCHONDRIA

Mitochondria and chloroplasts are unique cytoplasmic organelles because they are formed only from *preexisting* mitochondria or chloroplasts, respectively. Every other known eukaryotic organelle (except for chromosomes. of course) can be formed anew *(de novo)* from one or more of the cellular membrane systems. We will deal with some of these other organelle types in the next two chapters, but we will look at special features of mitochondrial and chloroplast biogenesis now.

### Chloroplasts and Proplastids

New chloroplasts either *develop* from primordia (undeveloped structures) called **proplastids,** in higher plants and certain lower plants and green protists, or arise by *division* of mature chloroplasts in plants such as ferns. There is no substantiated case for chloroplasts appearing in cells that lack these organelles or their undeveloped primordia.

Proplastids are usually about 0.5–1.0 by 1.0–1.5 $\mu$m in size. Their unstructured matrices are surrounded by two enveloping membranes, and the matrix itself contains ribosomes, DNA fibrils, and perhaps some starch or other carbohydrate. Occasionally there may be some bits of thylakoid membrane visible in the undifferentiated proplastid (Fig. 5.37). In the light, blisters and invaginations of the inner envelope may pinch off and gradually become dispersed within the proplastid stroma. These membranous elements gradually and ultimately become organized into long, flattened stroma thylakoids and stacks of grana thylakoids (Fig. 5.38). Synthesis of photosynthetic pigments takes place simultaneously so that, at maturity,

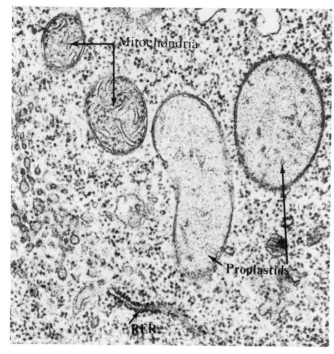

**Figure 5.37**
Undifferentiated proplastids in a thin section of *Arabidopsis thaliana* root cell. × 37,000. (Courtesy of M. C. Ledbetter.)

typical chloroplast ultrastructure and chemistry have differentiated from the proplastid primordium. There seems to be no fusion of proplastids to produce the larger chloroplasts; instead, it is believed that the mature organelle enlarges because new membrane material is synthesized and incorporated into the existing membrane framework.

Proplastids increase in number by a cleavagelike process in which there is a pinching-in and a later separation of the two products. In species where mature chloroplasts divide to produce new chloroplasts, there is also a type of cleavage process. Eukaryotic chloroplasts are therefore formed by some kind of constriction process that involves either proplastids or mature chloroplasts, depending on the species.

There is another sequence of ultrastructural changes that has been described for mature chloroplasts moved from light to dark and back again. Chlo-

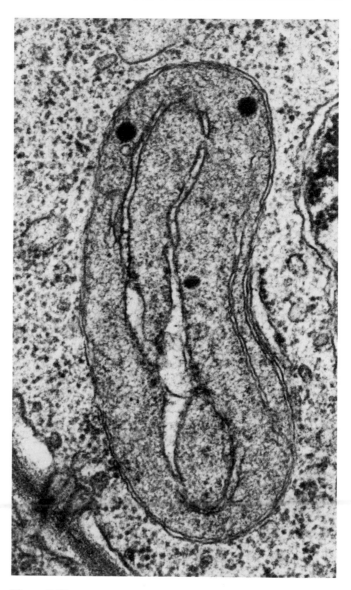

**Figure 5.38**
The beginnings of thylakoid development in a proplastid from a cell of plantain (*Plantago*). × 62,000. (Courtesy of M. C. Ledbetter)

roplasts are transformed into **etioplasts** when green plants are placed in the dark. These modified chloroplasts have no thylakoids; instead, the internal membranes may become disorganized or they may enter a latticework arrangement. In either case the membrane aggregate is called a **prolamellar body.** If plants are returned to the light, etioplasts are converted back into functional and recognizable chloroplasts (Fig. 5.39). If plant cells are grown in the dark, proplastids will differentiate into etioplasts rather than into typical chloroplasts. On exposure to light, however, these etioplasts also revert to typical chloroplasts. Etioplast ultrastructure, therefore, arises in dark-grown cells from either proplastids or mature chloroplasts, and reverts to mature chloroplast ultrastructure in the light regardless of the developmental stage in the dark phase of growth. The organizational conversion of prolamellar body membranes into flat sheets of thylakoids usually begins within minutes after etioplasts have been illuminated. It must therefore be membrane reorganization and not synthesis of new membranes that underlies the first stage of the phenomenon. Chlorophyll synthesis, however, takes several hours after dark-grown cells are put in the light.

We have very good evidence showing that the *organization* of thylakoids into particular groupings typical for a species is an important aspect of photosynthetic abilities in these species. In the unicellular green protist *Chlamydomonas reinhardi,* a favorite system for chloroplast studies, thylakoids are usually fused into stacks of three. In various mutant strains that have been examined ultrastructurally, some deficiency in an electron carrier or a photosystem component has been correlated with a change in thylakoid fusion patterns within the stack. Some mutants have no stacks, and their thylakoids exist as single elements within the stroma. Other mutants have larger stack sizes than the typical number of three thylakoids. The differences in "stacking" and "unstacking" of thylakoids in mutants and in wild type strains (Fig. 5.40) underscore the importance of membrane organization in relation to photosynthetic activities. This adds another dimension to the relationship between structure and function in an organelle system. It is quite possible that normal function occurs in normally-stacked thylakoids because the particles in the membrane are arranged in particular organizational groups. In the study illustrated in Fig. 4.7, particle distributions were specific and distinct in stacked and in unstacked thylakoid groups. Much remains to be done, but we now have a better appreciation of organelle biogenesis in relation to organization of components as well as to the more generalized summary view of how new chloroplasts arise from preexisting ones.

**Mitochondrial Biogenesis**

We know that mitochondria do not arise *de novo,* based on different kinds of experimental evidence. It is widely believed that new mitochondria arise through **growth and division** of existing mitochondria. The manner of *growth,* that is, new membrane formation and its insertion into an existing framework, is undetermined. The manner of *division* of enlarged mitochondria (after growth) has been examined more extensively, since it is somewhat easier to do than analyzing the membranes themselves.

From occasional electron micrographs and a relatively few biochemical experiments there has emerged a belief that mitochondria divide by some *active* fissionlike process, resembling bacterial cell fission in some respects. Two-dimensional electron micrographs are subject to various interpretations, and should not be considered to be very easily translated into three dimensions on the basis of a random sample of thin sections through large cells. According to one particular study of yeast cells that were analyzed by reconstruction of entire series of thin sections into three-dimensional models of the mitochondrion, organelle division seemed to be a *passive* rather than an active process. Yeast cells divide by budding, that is, growth of the new cell occurs until it is about as large as the mother cell and then the mother and daughter-bud cell are separated after new cell membrane and cell wall have formed. During budding to produce new cells, the mitochondrion in the mother cell enlarges to about twice its original size. About half of this enlarged mitochondrion is within the mother cell and the other half is contained within the bud portion of the division complex (Fig. 5.41). When the new cell wall separates the mother and daughter-bud into two cells, the large

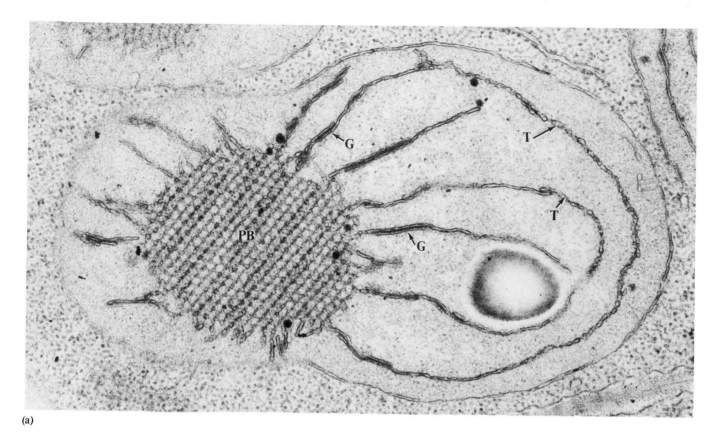

(a)

(b)

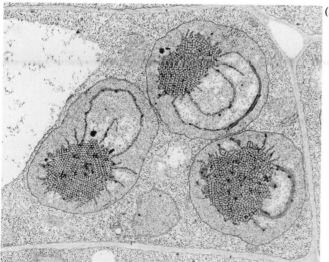

**Figure 5.39**
Thylakoid differentiation in etioplasts of bean (*Phaseolus vulgaris*) leaf grown in the light. (a) Note the crystalline display of the prolamellar body (PB) from which the thylakoids develop, and that grana thylakoids (G) have already begun to stack (at arrows). Thylakoids form from fused vesicles (T, at arrows). × 39,000. (Courtesy of M. C. Ledbetter) (b) Etioplast differentiation is at approximately the same stage throughout the leaf cell population in 11-day-old seedlings. Three organelles in part of one cell are shown here. × 12,500. (Courtesy of E. H. Newcomb and P. J. Gruber)

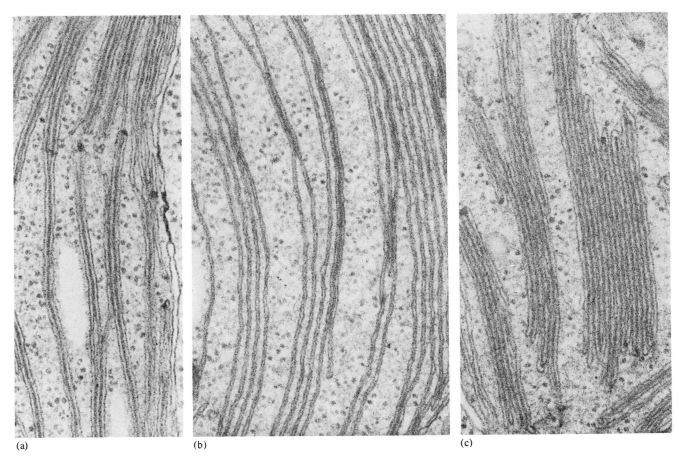

(a)     (b)     (c)

**Figure 5.40**
Variations in thylakoid "stacking" pattern in *Chlamydomonas reinhardi*: (a) wild type; (b) mutant strain *ac-115*, deficient in cytochrome-559, with single, unstacked thylakoids predominating; (c) mutant strain *F-1*, deficient in P700 chlorophyll, with "superstacked" thylakoids. × 65,000. (Courtesy of U. W. Goodenough, from Goodenough, U. W., and R. P. Levine, 1969, *Plant Physiol.* **44**:990-1000.)

mitochondrion is simply cut in two by the new wall. Each cell then has a mitochondrion, but not as a result of active mitochondrial division; the separation into two mitochondria occurred passively as the consequence of cell division. If this information can be verified in other systems, we will have to view mitochondrial division as a different process from bacterial cell division, and also as different from chloroplast or proplastid active-division processes. Once again, we would have to think differently about

mitochondrial origin and evolution depending on the extent of similarities among mitochondria, chloroplasts, and bacteria, a topic to be discussed in the last chapter.

## SUMMARY

1. The major systems for energy transformation in cells are aerobic respiration, which takes place in eukaryotic mitochondria, and photosynthesis, which takes place in

**Figure 5.41**
Three-dimensional reconstruction of a single, giant mitochondrion from a yeast cell with a
large bud. The mitochondrion is continuous in the mother cell and bud (at the left). The
model was reconstructed from a complete series of consecutive thin sections through the
entire budding cell complex. (Courtesy of H.-P. Hoffmann.)

eukaryotic chloroplasts. Equivalent membrane systems in
prokaryotes exist, but these are not organized as separated
subcellular compartments. During respiration, energy is extracted from fuel molecules and some of the free-energy difference between the fuel and its oxidation products is
conserved in ATP. During photosynthesis, sugars are manufactured using chemical energy transformed from light
energy absorbed by chlorophyll and other light-sensitive pigments.

2. During oxidation of glucose, which begins with fermentation reactions in the cytosol, its derivative molecules
of acetyl CoA are oxidized to carbon dioxide in Krebs Cycle
reactions within the mitochondrial matrix, and much of the
released energy is transferred to electron-carrying NAD and
FAD coenzymes. Reduced NAD and FAD then transfer
electrons to a chain of cytochrome enzymes and other
electron-carriers situated within the inner membrane of the
mitochondrion. Some of the free-energy differences of oxidation-reductions during electron transport are conserved in
ATP, which is made in coupled reactions of oxidative phosphorylation. Of the 686 kcal of free energy released per mole

of glucose oxidized to $CO_2$ and $H_2O$, about 39 percent is conserved in ATP and the remaining 420 kcal are dissipated as
heat.

3. Molecular oxygen participates directly in glucose oxidation as the final electron acceptor from cytochrome oxidase, the terminal member of the electron transport chain of
respiratory enzymes. Oxygen which has been activated on
accepting electrons can then accept protons and be reduced
to water. The end products of glucose oxidation are
therefore made in different pathways: $CO_2$ during processing
of pyruvate to acetate (or acetyl CoA) and in Krebs Cycle
reactions, and $H_2O$ at the completion of transport of
electrons from reduced NAD and FAD to molecular oxygen
along the chain of cytochrome enzymes. Most of the conserved fuel energy is found in ATP synthesized during oxidative phosphorylation coupled to electron transport (34 out of
38 ATP per mole of glucose oxidized to $CO_2$ + $H_2O$).

4. Stepwise release of energy in small, usable packets is
mediated by all the major systems for energy transfer: phosphoryl group transfer in the ADP-ATP cycle, organic group
transfer by coenzyme A, and electron transfers by the

$NAD^+/NADH$ and $FAD/FADH_2$ systems of coenzymes and by cytochrome electron carriers.

5. Mitochondria have two enveloping membranes, of which the innermost is folded into tubular invaginations called cristae. The shape of a mitochondrion has been proposed to be rodlike or sausagelike by many people, but some studies have led to a model of a mitochondrion showing a large, tubular, branching organelle. On the basis of the former model there would be hundreds or thousands of mitochondria in a cell, and on the basis of the branching model there might be only one or a few mitochondria per cell.

6. Chloroplasts in eukaryotic algae and in tissue-forming green plants are disc-shaped organelles in which an internal system of thylakoid membranes is bathed in a granular stroma all of which is bounded by two membrane envelopes. Thylakoids or lamellae are flattened membranous sacs which are organized into stacks, called grana, or which may traverse the stroma between grana. The stroma contains many substances in solution and suspension including DNA, ribosomes, and the enzymes which catalyze reduction of $CO_2$ to sugars in photosynthesis. Thylakoid membranes are the sites of the light-capturing reactions of photosynthesis, and these membranes house chlorophylls and other light-sensitive pigments, enzymes, and factors which participate in energy transfer from chlorophyll to $NADP^+$. They also house the ATPase which catalyzes the coupled synthesis of ATP during electron transport in the light—a process called photophosphorylation.

7. There are two photosystems or pigment systems linked together in series in eukaryotic chloroplasts. Each system contains chlorophylls and carotenoids which absorb different wavelengths of the visible light spectrum, and which transfer the energy of excitation that results from absorbing light energy to other excited pigment molecules until energy is transferred to reaction-center chlorophyll $a$ molecules. This special form of chlorophyll $a$ alone is capable of transferring electrons to some acceptor, and is therefore the site of transformation of light energy to chemical energy. The ultimate source of electrons is water, and the ultimate acceptor of these electrons is $NADP^+$ which is reduced to NADPH, an energy-transferring coenzyme that participates in sugar manufacture from $CO_2$ in the subsequent dark reactions of photosynthesis. In addition to NADPH, reduction of $CO_2$ is subsidized by ATP made in the coupled photophosphorylation reactions during electron transport along a carrier chain between photosystem II and photosystem I. The evolution of molecular oxygen is a byproduct of the light-dependent reactions in which water donates electrons to reaction-center chlorophyll $a$ of photosystem II.

8. During the reduction ("fixation") of $CO_2$ to hexose sugars in the dark reactions of photosynthesis, subsidized by chemical energy in NADPH and ATP made at the expense of light energy, 6 moles of $CO_2$ are processed in endergonic reactions to form one mole of glucose. The usual pathway of $CO_2$ reduction is the $C_3$ pathway in which a triose sugar is made as the initial product of $CO_2$ reduction, catalyzed by ribulose 1,5-diphosphate carboxylase (RuDP carboxylase). RuDP itself is regenerated in a cyclic pathway, so that the cycle accounts for all the $CO_2$ reduced per mole of hexose formed. It takes 750 kcal [(12NADPH × 52 kcal/mole) + (18ATP × 7 kcal/mole)] to make one mole of glucose, with a $\Delta G°$ of 686 kcal/mole. The overall efficiency of the photosynthetic dark reactions is therefore 686/750 × 100 = 90 percent.

9. More than 100 species of green plants have an additional pathway for $CO_2$ reduction. This $C_4$ pathway feeds into the conventional $C_3$ pathway, but it provides a means for efficient extraction of $CO_2$ from the air around the leaf. Plants with the $C_3$ + $C_4$ cycles may carry on photosynthesis, while plants with only the $C_3$ cycle will stop photosynthesizing under certain atmospheric conditions. $C_4$ plants continue to photosynthesize under limiting $CO_2$ conditions because their phosphoenolpyruvate carboxylase (PEP carboxylase) in the $C_4$ cycle is much more efficient than RuDP carboxylase of the $C_3$ cycle in binding $CO_2$, so that at certain times a favorable $CO_2$ diffusion gradient is maintained in the vicinity of the $C_4$ leaves but not in that of the $C_3$ leaves.

10. Mitochondria and chloroplasts form only from pre-existing mitochondria or chloroplasts, respectively. Chloroplasts often mature from proplastid initials, which can grow and divide, or mature chloroplasts may grow and divide directly to produce new plastids. Mitochondria grow in volume as new membrane is added to the existing framework. Division of the enlarged organelle may take place passively when a new cell membrane or wall cuts across the organelle, giving roughly equal parts to each daughter cell. Whether or not there is an active division process in mitochondria as there is for chloroplasts or proplastids is uncertain at the present time.

### STUDY QUESTIONS

1. What is the importance of green plant photosynthesis to life on Earth? What would happen on Earth if all green plants were wiped out by catastrophes? What is the relation-

ship between photosynthesis and aerobic respiration in organisms? Where do these processes take place in eukaryotic cells?

2. Where are stored foods metabolized in cells? What is the advantage of processing glucose by aerobic respiration rather than by glycolysis alone? What is the tie-in between glycolysis and oxygen-dependent respiration reactions in the cell? Why do we consider the Krebs cycle to be the final common pathway for fuel breakdown in most cells? Since molecular oxygen is not used directly in the Krebs cycle, why do we consider the cycle to be a part of aerobic respiration?

3. What are the major functions of the Krebs cycle? What happens to the acetate derivatives of glucose during Krebs cycle reactions? What is the source of $CO_2$ released in Krebs cycle reactions? What happens to the hydrogens originally present in the acetate derivatives of glucose during oxidations in the Krebs cycle? in electron transport? in oxidative phosphorylation?

4. What oxidation-reduction enzymes participate in electron transport from organic substrates to $O_2$? How do cytochromes contribute to electron transport? What is a cytochrome? How is oxidative phosphorylation related to electron transport in mitochondrial membranes? What happens when oxidative phosphorylation is uncoupled from electron transport? Why? How do we think these processes are coupled? What is Mitchell's proposal for a coupling mechanism? How does it work?

5. When glucose is burned in air in a test tube to $CO_2$ and $H_2O$, $\Delta G° = -686$ kcal/mole, but $\Delta G° = -420$ kcal/mole for the same reactions in living cells. Why? What is the importance of the differences in $\Delta G°$ in a cell and a test tube? What is the overall equation for glucose oxidation to $CO_2 + H_2O$ in living cells? Where does $H_2O$ come from during glucose oxidation? Are the $CO_2$ and $H_2O$ end products formed in the same or in different reactions during glucose oxidation? What are these reactions? What is the role of molecular oxygen in aerobic respiration? What happens if $O_2$ is absent?

6. What are the relationships between mitochondrial structural components and the reaction systems of aerobic respiration? What kinds of proteins are included in the mitochondrial inner membrane construction? How do we know that the ATPase of oxidative phosphorylation is not an integral protein of the mitochondrial inner membrane? What is an inner membrane subunit? What are the opposing ideas concerning mitochondrial size, shape, and number per cell? How can the controversy be resolved?

7. How are chloroplasts constructed? Why are we more certain about size, shape, and numbers per cell of chloroplasts than of mitochondria? What is a thylakoid? What is a granum? What are some basic similarities between mitochondria and chloroplasts? What are the differences between thylakoids in prokaryotes and eukaryotes?

8. What kinds of photosynthetic pigments participate in absorption of light? How can we determine what parts of the visible spectrum are absorbed by each pigment? How can we be sure that absorption of light by pigments is a part of photosynthesis? What are the differences and similarities among chlorophylls, carotenoids, and phycobilins?

9. What is the difference between green plant photosynthesis and photosynthesis in bacteria? How can we generalize the equation for photosynthesis to show that there are fundamental similarities between eukaryotic and prokaryotic photosynthesis? Where does the $O_2$ come from during green plant photosynthesis activities? Is $O_2$ important to photosynthesis reactions?

10. What happens during photoexcitation of a substance? What is a photon? a quantum? What is the significance of photoexcitation in biological systems? What is the relationship between photoexcitation energy and chemical energy in biological systems? How did we discover that photosynthesis consists of separate light-dependent and light-independent pathways?

11. What are the activities of photosystems I and II? What is the function of reaction center chlorophyll *a* in photosynthesis? How are photosystems I and II linked? What is the relationship between photophosphorylation and electron transport in the light reactions of photosynthesis? What are the significant results of the light reactions of photosynthesis? Why is there no $O_2$ evolved during bacterial photosynthesis? What is the overall equation for the light reactions of green plant photosynthesis?

12. What is the pathway for reduction of $CO_2$ to hexose sugars in most green plants? What is the role of ribulose 1,5-diphosphate in $CO_2$ reduction ("fixation") during the dark reactions of photosynthesis? What is the role of RuDP carboxylase in these reactions? How did we discover that glucose is not the first $CO_2$ reduction product to accumulate during the dark reactions? How is glucose formed in these reactions?

13. What is an overall statement for green plant photosynthesis? What is the contribution made by NADPH and ATP to sugar synthesis in chloroplasts? In what reactions is $O_2$ made and in what reactions is glucose made in chloroplasts? Where within the chloroplast does each of

these reaction systems occur? Why isn't photosynthesis considered to be a reverse reaction sequence of aerobic respiration?

14. How does the $C_4$ pathway interact with the $C_3$ pathway for $CO_2$ reduction in some plants? What is the advantage of the $C_4$ pathway in such plants? What is the role of PEP carboxylase and of RuDP carboxylase in the maintenance of a favorable $CO_2$ diffusion gradient in $C_3$ and $C_4$ plants? How can we incorporate favorable features of the $C_4$ pathway into available or potential crop plants by breeding programs? What would be the advantage of such breeding programs to world agriculture?

15. What is the relationship of proplastids to mature chloroplasts? What is an etioplast? What is the significance of thylakoid "stacking" to photosynthetic activities in green cell chloroplasts? What is the evidence that both chloroplasts and mitochondria arise by some process of growth and division? How are these processes similar and how are they different for the two kinds of organelles?

### SUGGESTED READINGS

Arnon, D. I. The role of light in photosynthesis. *Scientific American* **203:**104 (Nov. 1960).

Atkinson, A. W., Jr., John, P. C. L., and Gunning, B. E. S. The growth and division of the single mitochondrion and other organelles during the cell cycle of *Chlorella,* studied by quantitative stereology and three dimensional reconstruction. *Protoplasma* **81:**77 (1974).

Bassham, J. A. The path of carbon in photosynthesis. *Scientific American* **206:**88 (June 1962).

Björkman, O., and Berry, J. High-efficiency photosynthesis. *Scientific American* **229:**80 (Oct. 1973).

Bolin, B. The carbon cycle. *Scientific American* **223:**124 (Sept. 1970).

Clayton, R. K. *Light and Living Matter,* vol. 2. New York: McGraw-Hill (1971).

Cloud, P., and Gibor, A. The oxygen cycle. *Scientific American* **223:**110 (Sept. 1970).

Fernández-Morán, H., Oda, T., Blair, P. V., and Green, D. E. A macromolecular repeating unit of mitochondrial structure and function. *Journal of Cell Biology* **22:**63 (1964).

Goodenough, U. W., and Levine, R. P. Chloroplast ultrastructure in mutant strains of *Chlamydomonas reinhardi* lacking components of the photosynthetic apparatus. *Plant Physiology* **44:**990 (1969).

Govindjee, and Govindjee, R. The absorption of light in photosynthesis. *Scientific American* **231:**68 (Dec. 1974).

Hatch, M. D., and Slack, C. R. Photosynthesis by sugarcane leaves: A new carboxylation reaction and the pathway of sugar formation. *Biochemical Journal* **101:**103 (1966).

Hill, R. Oxygen evolved by isolated chloroplasts. *Nature* **139:**881 (1937).

Hoffmann, H.-P., and Avers, C. J. Mitochondrion of yeast: Ultrastructural evidence for one giant, branched organelle per cell. *Science* **181:**749 (1973).

Janick, J., Noller, C. H., and Rhykerd, C. L. The cycles of plant and animal nutrition. *Scientific American* **235:**74 (Sept. 1976).

Levine, R. P. The mechanism of photosynthesis. *Scientific American* **221:**58 (Dec. 1969).

Lloyd, D. *The Mitochondria of Microorganisms.* New York: Academic Press (1974).

Munn, E. A. *The Structure of Mitochondria.* New York: Academic Press (1974).

Racker, E. The membrane of the mitochondrion. *Scientific American* **218:**32 (Feb. 1968).

Stryer, L. *Biochemistry.* San Francisco: W. H. Freeman (1975).

Whittingham, C. P. *The Mechanism of Photosynthesis.* London: Edward Arnold (1974).

# Chapter 6

# Cellular Packaging: The ER and Its Derivatives

Eukaryotic cells are highly compartmented, with further subdivisions within compartments. These systems are highly coordinated, as we saw in the case of mitochondria and of chloroplasts. The system we will examine next is the **endoplasmic reticulum,** a folded membrane sheet with various regions of differentiated function, and some of its derivative structures.

Proteins are synthesized at the ribosomes, but these proteins must be *delivered* to various parts of the same cell or to different cells where they function. In some cases the proteins are simply released into the surrounding cytosol from their ribosomal sites of synthesis, and enter pools of metabolites, monomers, and reaction systems in the cytosol. For example, enzymes of glycolysis are retained in dilute solution in the cytosol after newly-synthesized enzymes have been released from the ribosomes. Other kinds of proteins are packaged in some membrane-bound container, and are sequestered from external churnings and metabolic activities of the cytosol. Some membranous vesicles are transported to the cell surface where they release their contents into extracellular fluids or spaces. Other vesicles are a convenient means for transport of coordinated sets of molecules to other parts of the same cell or to different cells. Still other membranous containers are cellular organelles which persist for some time while they carry out their share of cellular metabolism. Proteins are not passed around in helter-skelter fashion. They are channelled by some means of delivery system to those places in cells where they carry out functions which are related to the specific nature of the molecules, whether catalytic, structural, or regulatory.

After we have looked at the structure and activities of the endoplasmic reticulum and the **Golgi apparatus,** a smooth membrane derivative of the reticulum, we will discuss selected packaged products of these membrane systems, particularly in relation to cell secretion and intracellular metabolism. At best we can only skim the surface of such a vast array of topics.

## THE ENDOPLASMIC RETICULUM

According to electron micrographs of cell thin sections, the endoplasmic reticulum (**ER**) appears to be a tortuously folded sheet of membrane(s). Different cells have more or less ER within the cytoplasm, but actively-synthesizing vertebrate cells display considerably more ER than less active cells in the same species. On the whole, there is more ER in animal cells than in protists, algae, fungi, or higher plants.

It is very difficult to see ER with the light microscope, even if cells are stained to increase contrast. Most of our information has therefore come through electron microscopy, particularly since the 1950s.

The two basic differentiations of ER are the **rough ER** and **smooth ER.** The rough ER is studded with ribosomes on its outer surfaces and the smooth ER lacks attached ribosomes (Fig. 6.1). This difference immediately implies that protein synthesis must be associated with rough ER, but not with smooth ER regions. Ribosomes also occur free in the cytoplasm, but their proportions relative to amounts of *bound* ribosomes of rough ER vary from one cell type to another.

Because of extensive folding of ER membrane, two kinds of spaces between membrane surfaces are created: (1) **ER channels,** or **ER lumen,** are created by the facing *inner* surfaces of the folded membrane sheet, and (2) cytosol spaces are distributed between folds of the ER, bathing the *outer* surfaces of the folded sheet (Fig. 6.2). In rough ER it is easy to distinguish these two kinds of spaces because facing ribosome-studded outer membrane surfaces encase

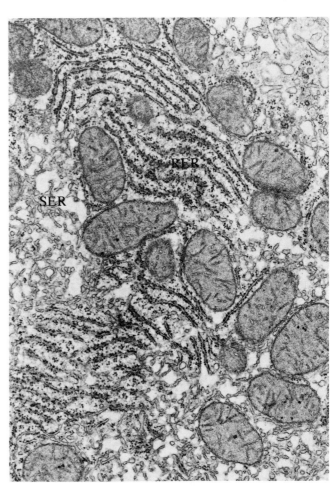

**Figure 6.1**
Portion of rat liver cell showing rough endoplasmic reticulum with attached ribosomes (RER), and smooth endoplasmic reticulum (SER). × 5,000. (Courtesy of H. H. Mollenhauer)

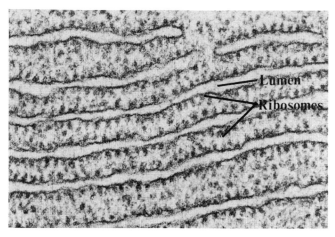 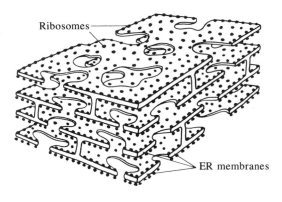

**Figure 6.2**
The two-dimensional perspective of an electron micrograph is contrasted with a three-dimensional representation of the system of membranes and channels that form the rough endoplasmic reticulum of eukaryotic cells. × 85,000. (Courtesy of K. R. Porter)

cytosol, while facing nonribosome inner surfaces define the ER channels. In a view showing the cut edges of folded ER membrane, these two kinds of spaces alternate with one another. This leads to the interpretation that ER consists of a flat membrane sheet that is repeatedly folded back on itself.

Proteins are synthesized at ribosomes attached to the outer surface of rough ER. Each ribosome is attached by its larger subunit to the membrane surface. A convenient system for studying polypeptide synthesis at the rough ER is the **microsome.** When cells are broken up and sedimented by centrifugation, particles settle out and can be collected in separate fractions of cell-free material. One of these subcellular fractions contains microsomes, which are membranous vesicles filled with fluid and studded on the outside with ribosomes. Microsomes as such do not actually occur in living cells. They are produced when the extensive ER system is fragmented during preparations for centrifugation (Fig. 6.3). Microsomes are, however, representative *pieces* of rough ER (the minor amounts of smooth ER vesicles can be ignored, for practical purposes) and therefore provide a useful system for biochemical and microscopical studies.

When radioactively labeled amino acids are presented to microsome preparations *in vitro* (outside the cell, that is, a test tube system), labeled polypeptides can be collected. This shows that protein synthesis occurred in the microsome system just as it does at the rough ER in intact cells. If the synthesizing microsomes are treated with the antibiotic drug *puromycin,* polypeptide synthesis stops and the unfinished polypeptide chains are released prematurely from the ribosomes. Such released polypeptides can be traced by radioactivity of their incorporated amino acids, and they are found mostly *inside* the microsome cavity (Fig. 6.4). This observation indicates that most polypeptides are elongated *into* the ER lumen (the equivalent space in intact cells). We may further infer that finished polypeptides in intact cells are then transported through ER channels to other parts of the cell. Many of these proteins are then packaged at other places in the ER system, or at the Golgi apparatus, which is a region of smooth membranes. Whether such packaged proteins remain in the same cell or are exported out of the cell, or are modified in some way, depends on the cell type and the nature of the proteins themselves. Some polypeptides do not elongate into the ER channels. Perhaps such proteins are destined to remain in the

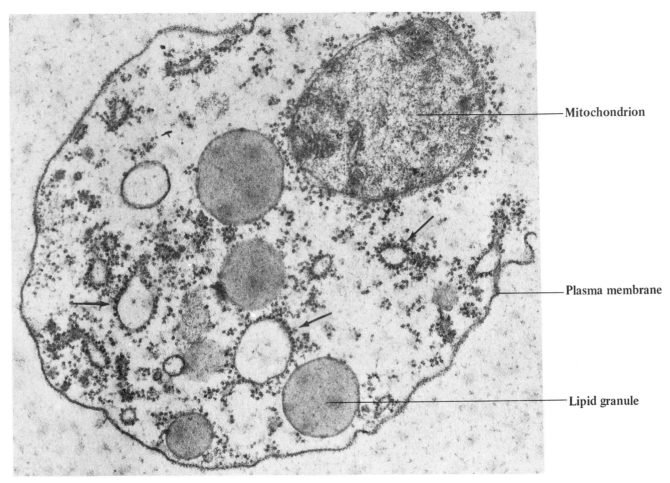

Mitochondrion

Plasma membrane

Lipid granule

**Figure 6.3**
Part of a broken yeast cell showing membranous vesicles with ribosomes attached to their outer surface (at arrows). Such components constitute a substantial part of the microsome fraction of centrifuged cell-free lysates but are rarely observed in intact or undamaged yeast cells. × 43,300. (From Szabo. A., and C. J. Avers, 1969, *Ann. N.Y. Acad. Sci.* **168**:302-312, Fig. 2.)

cytosol, where they undergo synthesis. Until specific proteins can be identified in relation to site of synthesis and site of eventual activity in the cell, we can only speculate on the details. Although there are such variations, the main functions of the ER are *synthesis and transport of proteins within the cell.*

## THE GOLGI APPARATUS

Although difficult to resolve in most cells, a particular system of threads and granules was first reported in nerve cells by the Italian biologist Camillo Golgi in 1898. Until electron microscopy was developed and improved, however, the components we now call the Golgi apparatus were quite elusive and little studied.

### Ultrastructural Organization

The **Golgi apparatus** is composed of smooth membranes organized as flattened membranous sacs, called **cisternae,** and assorted **vesicles** and **vacuoles**

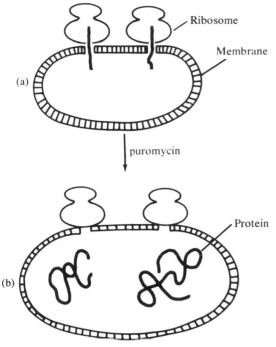

**Figure 6.4**
Diagrammatic illustration of polypeptide chain growth at 80S ribosomes attached to the outer surface of the *ER* or to a microsomal representative of the rough *ER*: (a) The growing polypeptide chain extends through a groove in the 60S subunit into the microsomal vesicle (or *ER* lumen). This process is demonstrated after puromycin treatment in b, showing released nascent polypeptide chains within the microsomal vesicle.

(Fig. 6.5). The cisternae, vesicles, and vacuoles (larger vesicles) are fluid-filled cavities enclosed by membrane when seen in cell thin sections. When Golgi fractions are carefully isolated from cells and contrasted by negative staining, the whole apparatus can be seen as a network of tubules and flattened plates (Fig. 6.6). Only an edge-on view of this intricate network can be seen in thin sections, but the vesicles and vacuoles are actually swollen ends of tubules that anastomose throughout the network and interconnect with each other and with the flattened cisternal regions.

When seen in thin sections, which is the usual way to study the Golgi apparatus in cells, cisternae are usually organized into stacks, which are called **dictyosomes** (Fig. 6.7). There are usually 5 to 8 cisternae

per dictyosome, but 30 or more are not uncommon in dictyosomes of simpler organisms. A single Golgi apparatus may consist of one or more dictyosomes, depending on whether or not they are linked by membrane continuities of the anastomosing network. This network is difficult to detect in random thin sections of cells.

Dictyosomes show a definite *polarity*. The proximal region, or **forming face,** in a dictyosome is oriented toward the nucleus or the ER. The distal region, or **maturing face,** is oriented toward the cell surface. There are differences in membrane thickness, size of associated vesicles and vacuoles at each face, and density of the contents in cavities of cisternae and

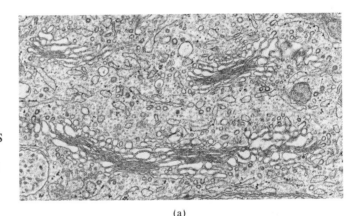

(a)

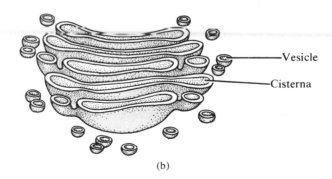

Vesicle

Cisterna

(b)

**Figure 6.5**
Part of a Golgi apparatus from epididymis of rat testis: (a) Electron micrograph showing the pattern of parallel cisternae, some of which have terminal dilations. Numerous vesicles are usually associated with the flattened membranous sacs, or cisternae. × 12,600. (Courtesy of H. H. Mollenhauer) (b) Three-dimensional representation of a stack of cisternae and associated vesicles.

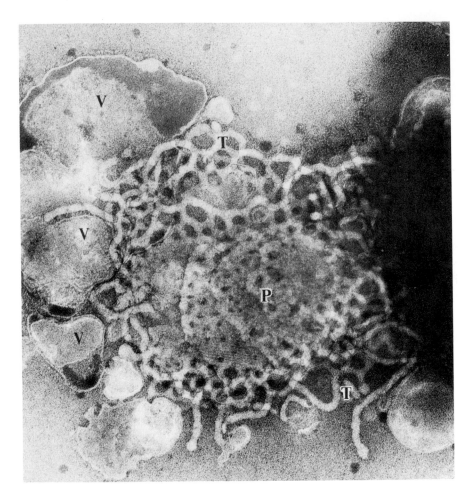

**Figure 6.6**
Negatively stained, isolated dictyosome from radish (*Raphanus sativus*) root. The dictyosome is partially unstacked and shows the central platelike region (P), peripheral tubules (T), and secretion vesicles (V). × 75,000. (Courtesy of H. H. Mollenhauer)

vesicles (Fig. 6.8). Dictyosome polarity reflects the developmental sequence of packaging of cell products into membranous containers; the packages begin to develop deep within the cell and become mature as they progress toward the cell surace.

Because of difficulties in interpreting cell thin sections, and because the extent and form of the Golgi apparatus vary according to cell type and metabolic state, there is some question about organizational similarities in different species groups. A single region of Golgi smooth membranes is usually present near the nucleus in mammalian cells, so it appears to be a localized, unified compartment (Fig. 6.9). Cell thin sections look quite different in protists, fungi, in-

vertebrates, and higher plants. In these organisms there generally seem to be single dictyosomes dispersed throughout the cytoplasm (see Fig. 6.7). Whether these are parts of one or more larger Golgi apparatus systems, or many individual dictyosomes not interconnected into larger systems, cannot always be determined. According to studies of isolated Golgi systems (see Fig. 6.6), there is not much difference between mammalian and plant Golgi apparatus organization. There are cell types that seem truly distinctive, however. Some fungi have single cisternae rather than stacks, or systems of vesicle-producing tubules that function as the Golgi apparatus. Whatever their external appearance, the cisternae and tubules of the

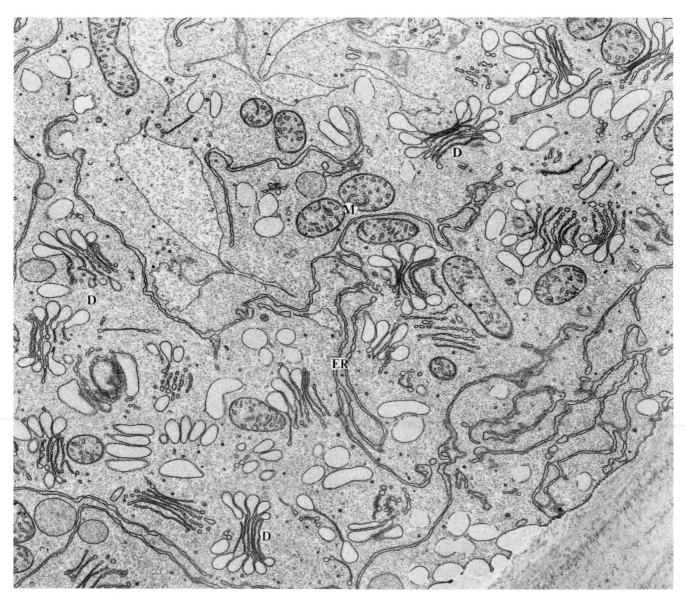

**Figure 6.7**
Dictyosomes (D) in permanganate-fixed thin section of root cap cell of maize (*Zea mavs*).
Mitochondria (M) and endoplasmic reticulum (ER) are also present. × 19,500. (Courtesy of
H. H. Mollenhauer)

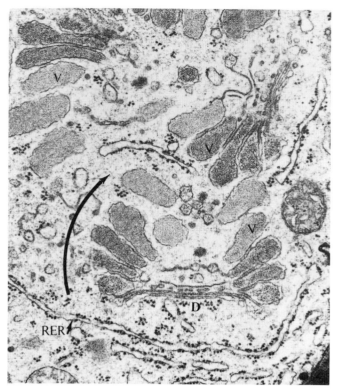

**Figure 6.8**
Dictyosomes (D) in osmium-fixed thin section of root cap cell of maize. The secretion vesicles (V) formed from dictyosomal components of the Golgi apparatus are filled with a dense material secreted elsewhere in the cell and packaged at the Golgi region. The arrow shows the direction of maturation, toward the maturing face of the dictyosome. × 35,000. (Courtesy of H. H. Mollenhauer)

Golgi apparatus function in very similar ways in almost all eukaryotic cells.

### Functions of the Golgi Apparatus

The Golgi apparatus acts as a way station for processing and packaging cell products, many of which are **cell secretions** (proteins to be exported from the cell interior to the cell surface or outside the cell). We know that all or most of the molecules that enter the Golgi region are processed in some fashion before they are wrapped up in a membranous package. Using radioactive tracer methods, it has been possible to see

where the molecules are made, how they are changed during processing, where the processing occurs, and what happens to the containerized products. The study of mucus formation in goblet cells of rat intestine will illustrate these points.

Goblet cells are squeezed between other cells of the intestinal lining and are therefore long and narrow in form. The Golgi apparatus in these cells is cup-shaped and includes dictyosomes which are considerably compacted into a region of smooth membranes. Each dictyosome is made up of 8 to 10 cisternae, with the forming face near the nucleus and the maturing face directed toward the apex of the cell. The cisternae at the forming face appear empty and greatly flattened; those at the maturing face are swollen with mucus contents. Globules filled with mucus detach from the topmost cisterna and migrate to the cell membrane where these containers discharge their contents outside the cell into the intestinal lumen (Fig. 6.10). The cycle occurs repeatedly in active goblet cells.

In a group of experiments, labeled amino acids were injected into rat intestine and the time course of migration of incorporated amino acids was followed by their radioactivity. Using a photographic emulsion over the cell sections obtained at particular intervals during the experiment, radioactivity could be located in specific parts of the cell by the silver grains that form in the emulsion wherever a radioactive event has occurred in the cell under the emulsion (Fig. 6.11). Labeled amino acids in proteins were first seen at the rough ER, then in the Golgi cisternae, still later in the mucus droplets, and finally in the mucus which had been discharged at the cell apex. The series of **autoradiographs** of cell sections therefore indicated that proteins made at the rough ER pass through the Golgi apparatus and exit the cell via mucus droplet formation. But mucus is a glycoprotein; that is, a protein with covalently-linked sugar groups, as are many cell secretions. At what point was the sugar added on to the protein? The next series of autoradiographic studies provided the answer, and showed that the Golgi apparatus was more than a packaging station.

Radioactive glucose was injected into rats, since it was known that glucose was the source of covalently-linked sugars in glycoprotein molecules. Within 15 minutes after injection of glucose, all the radioactivity was concentrated in the Golgi cisternae and vesicles; after 20 minutes radioactivity began to appear in mucus droplets; after 4 hours the radioactively-labeled mucus was discharged into the intestinal lumen. From these observations it was found that a stack of 8 to 10 cisternae was converted to mucus droplets in about 40 minutes, and a replacement stack had formed in this time interval.

From these and similar experiments we can say that proteins made at the rough ER pass on to the Golgi apparatus where carbohydrates, manufactured from simple sugars, are then added to the protein (Fig. 6.12). Various secretory cells have a similar system of packaging their processed proteins for export from the cell. Before we consider the very important question of membrane transformations, which must take place as new dictyosomes are formed and as vesicle membranes are added to the plasmalemma, there are a few interesting secretion systems that bear examination. These systems help to solidify the case for Golgi apparatus mediation in cell secretion phenomena.

### Cell Secretion

**Secretion** is defined as the process by which cells synthesize products that will be used elsewhere than their site of origin. Secretions which are expelled into

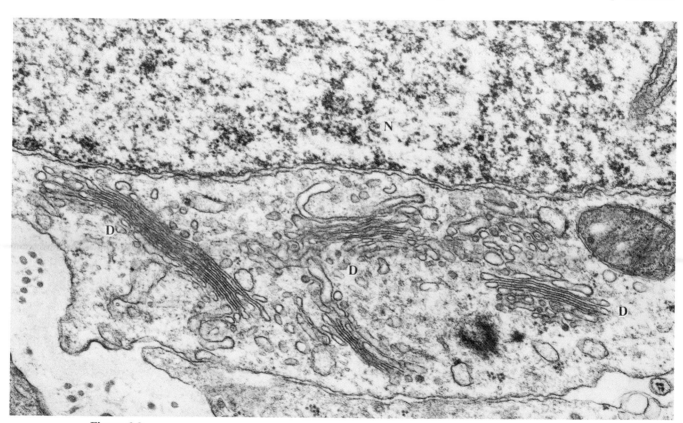

**Figure 6.9**
Portion of the Golgi apparatus of spermatogonium of rat testis. The closely spaced dictyosomal (D) components are localized adjacent to the nucleus (N). × 50,000. (Courtesy of H. H. Mollenhauer)

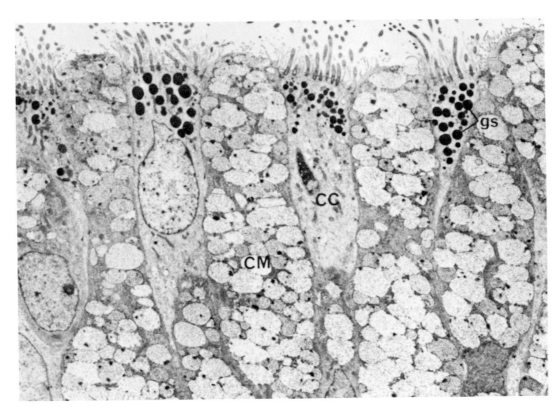

**Figure 6.10**
Release of mucus from goblet cells of epithelial tissue in the quail. × 3,900. (Reproduced with permission from Sandoz, D., *et al.*, 1976, *J. Cell Biol.* **71**:460, Fig. 6.)

the outer environment or, more frequently, into natural cavities such as the digestive tract and others, are called exocrine or **external secretions.** We may include in this group those secretions that are deposited at the cell surface, outside the plasma membrane boundary. Endocrine or **internal secretions** enter directly into the circulation or other internal pathways, and act on other tissues of the organism. Internal secretions are typical of endocrine glands, such as the adrenal glands, the islets of the pancreas, and others.

Many secretions are packaged in membranous granules or vesicles, but some are not. For example, antibodies made and released by plasma cells are not packaged. Their synthesis and release is more or less simultaneous with their transport elsewhere. Whether packaged or not, secretions may be made continuously

or discontinuously, depending on the cell type, the nature of the secretion product, and the cellular environment. Packaged secretions generally are transported more rapidly through the cell, probably because they are helped along their directional pathway by fibrous components in cells. Free molecules tend to move more slowly, and with less direction, undoubtedly because their movement is mostly random and unaided by any cellular apparatus.

**Figure 6.11**
Protocol for autoradiography and a photograph showing the localization of silver grains over those chromosome regions that contain $^3$H-thymidine-labeled DNA. (Photograph courtesy of P. B. Moens, from Moens, P. B., 1966, *Chromosoma* **19**:277-285.)

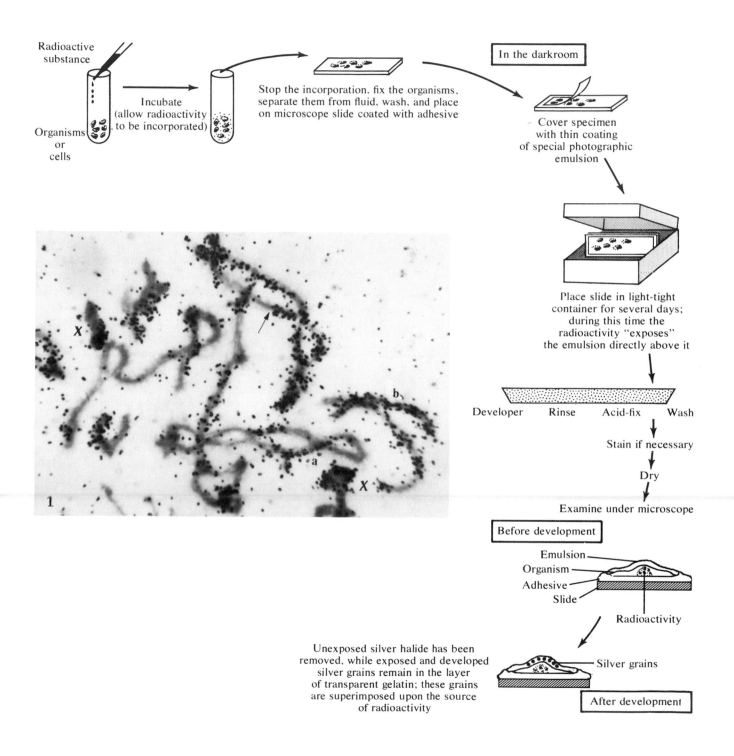

Radioactive substance

Organisms or cells

Incubate (allow radioactivity to be incorporated)

Stop the incorporation, fix the organisms, separate them from fluid, wash, and place on microscope slide coated with adhesive

In the darkroom

Cover specimen with thin coating of special photographic emulsion

Place slide in light-tight container for several days; during this time the radioactivity "exposes" the emulsion directly above it

Developer    Rinse    Acid-fix    Wash

Stain if necessary

Dry

Examine under microscope

Before development

Emulsion
Organism
Adhesive
Slide
Radioactivity

Unexposed silver halide has been removed, while exposed and developed silver grains remain in the layer of transparent gelatin; these grains are superimposed upon the source of radioactivity

Silver grains

After development

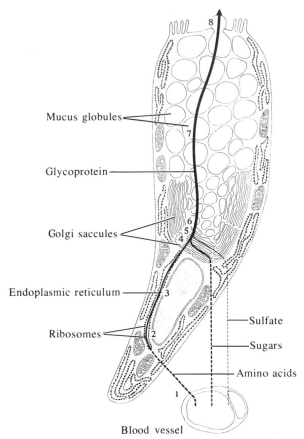

Mucus globules

Glycoprotein

Golgi saccules

Endoplasmic reticulum

Ribosomes

Sulfate

Sugars

Amino acids

Blood vessel

**Figure 6.12**
Summary diagram showing the flow of precursors and glycoprotein products through the intestinal goblet cell. This process has been deduced from autoradiographic experiments described in the text. (From ''The Golgi Apparatus'' by M. Neutra and C. P. Leblond. Copyright © 1969 by Scientific American, Inc. All rights reserved.)

One well-studied system is **zymogen** secretion in pancreatic cells. Zymogens are digestive enzymes of the intestinal tract, which are secreted and transported out of the cell in membrane-bound granules. The enzymes are in an inactive, or precursor, form in zymogen granules and are modified to their active form once they arrive at their place of activity. The precursors *pepsinogen* and *trypsinogen,* among others, are converted to active *pepsin* and *trypsin* once they reach the intestinal cavity. Zymogens are made at the

rough ER, packaged and processed within the Golgi apparatus, expelled from the cell, and function in the digestive tract. The processes are basically the same as we found for mucus in goblet cells. In the case of zymogens, however, the proteins move slowly through the pancreatic system and then are diluted by other secretions before entering the intestinal cavity. Mucus which is expelled from goblet cells soon spreads over the surface of the intestinal lining, and is not transported very far from its site of origin.

The process of expulsion of vesicle contents involves an **exocytosis** event. By exocytosis we mean elimination of cell contents through the plasma membrane, by a process that involves fusion of vesicle and plasma membranes (see Fig. 4.17). A mobile cell surface must be available, as in animal cells; therefore exocytosis is probably less common in plants and fungi, which have rigid, thick cell walls. The reverse process of **endocytosis** involves intake of materials through infolding of the plasma membrane and subsequent packaging of the incoming material in membrane derived from the plasmalemma itself. We will refer to these processes often during later portions of this chapter.

Cell secretions contribute to cell surface construction and its unique features in plant and animal systems. Plant cell walls must be constructed at each cell division, leading to separation of the daughter cells. Carbohydrates, such as cellulose and pectins, are packaged in the Golgi apparatus and are directed toward the newly-forming cell plate, which eventually thickens into new cell wall. The vesicles at the cell plate and in the Golgi region are very similar in appearance according to electron micrographs, and similar chemically as well (Fig. 6.13).

The cell coat which covers the surface of animal cells is a composite of secretion products made and packaged in the cell and transported via the Golgi apparatus to the cell envelope. These materials are released by exocytosis events and form a coating of immense importance to the cell. In particular, recognition systems between cells involve specific molecules in the coating. Such molecules regulate cell-to-cell communications and interactions; for example, cells

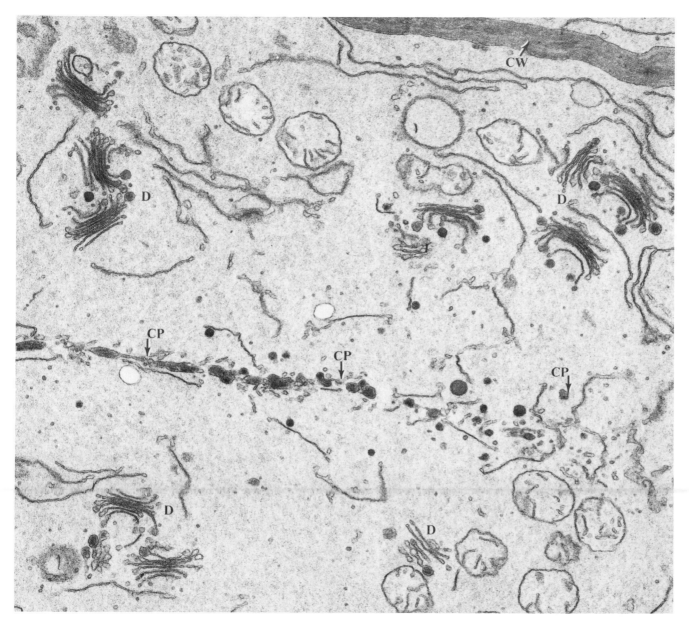

**Figure 6.13**
Thin section of permanganate-fixed maize epidermal cell. The dense vesicles within the region of the developing cell plate (CP) resemble vesicles formed at the dictyosomes (D) of the Golgi apparatus. This resemblance leads to the suggestion that some of the cell-plate material may originate in the Golgi apparatus. Part of the cell wall (CW) is at the upper right of the photograph. × 14,000. (Courtesy of H. H. Mollenhauer)

often stop moving when they make contact with other cells (a phenomenon called *contact inhibition* of cell movement). In other cell communities mitosis comes to a stop once cells touch each other after spreading over an area, and they remain one layer thick rather than piling up into mounds of cells (a phenomenon called *density-dependent inhibition* of mitosis). The surface of red blood cells includes the antigens of the A-B-O major blood groups, and whether or not cells will clump in a clotting reaction depends on interaction between antibodies made in plasma cells and antigens covering the cell surface. Another important group of surface antigens are those which interact in cell fusions, and which underlie problems of incompatibility and graft rejection between dissimilar cell and tissue systems.

## LYSOSOMES

Membranous vesicles called **lysosomes** contain a number of hydrolytic enzymes whose optimum pH is around 5.0. At least 40 different acid hydrolases have been localized in these organelles, which give the cell the capability of digesting all the biologically important groups of large molecules (Fig. 6.14). The general hydrolysis reaction catalyzed by these enzymes is

$$R_1—R_2 + H_2O \rightarrow R_1—H + R_2—OH \qquad [6.1]$$

Lysosomes were not seen until 1955 when Christian de Duve published electron micrographs showing an ovoid organelle measuring about 0.5 $\mu$m, with an unstructured matrix surrounded by a single membrane. These organelles had been sought for six years by de Duve, who had predicted their existence on the basis of biochemical information. The lysosome is more specifically identified by light or electron microscopy after a **cytochemical** treatment, which causes an enzymatic hydrolysis reaction to precipitate an opaque reaction product. The usual tested lysosome activity is of an **acid phosphatase** which leads to deposits of a lead phosphate compound that is very dense and easily seen (Fig. 6.15).

Lysosomes have been found in virtually all animal

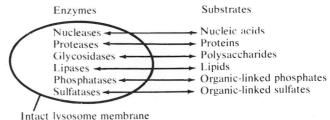

**Figure 6.14**
Summary of the general groups of digestive enzymes contained within the lysosome and the types of substrates upon which they act under appropriate cellular conditions.

cells and protozoa, and also in yeasts and other fungi, some green protists such as *Euglena,* and in certain types of plant cells. There was some early problem in interpreting lysosomal variation in size and contents, but this was later resolved when it was learned that lysosomes underwent various changes during a sequence of digestion events. Lysosomes function as *intracellular digestive systems;* in fact, they are sometimes considered to be the "garbage disposals" of the cell.

### Formation and Function

Lysosomal enzymes are synthesized at the rough ER, as are other proteins. They are packaged within a single enveloping membrane, but there is some difference of opinion concerning the site of lysosome formation. According to one view the lysosome membrane is derived from the Golgi apparatus; that is, the enzymes are packaged in Golgi cisternae and then

**Figure 6.15**
Rat liver tissue fixed 4 hours after partial hepatectomy and then incubated for acid phosphatase activity in a cytochemical test system. (a) Light microscope photograph showing dark precipitates of the enzyme activity product in structures presumed to be developmental stages in the lysosome cycle. × 600. (b) Electron micrograph reveals precipitated product of acid phosphatase activity in a secondary lysosome (L) and two autophagic vacuoles (A). Note the single limiting membrane of these structures. × 31, 000. (Courtesy of A. B. Novikoff and M. Mori)

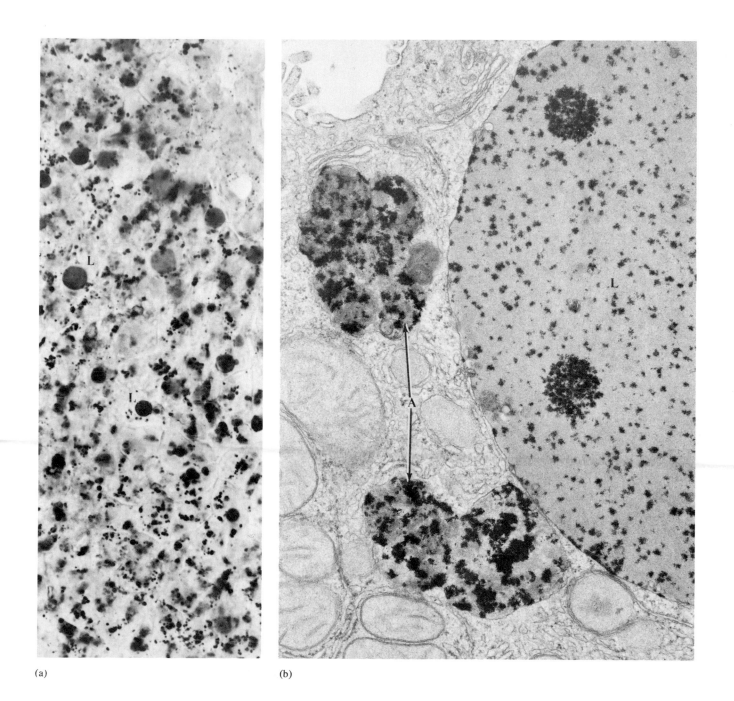

(a)

(b)

enter the cytoplasm as membrane-bounded systems. Another viewpoint, developed by Alex Novikoff, is that lysosomes actually form in a special ER region that borders the Golgi apparatus. This membrane system of *Golgi-associated ER* which produces *Lysosomes* has been called *GERL*. At the moment, the prevailing view favors a Golgi origin for lysosomes.

Lysosomal enzymes are inactive in newly formed structures, and they remain latent until the organelle membrane is damaged or until some substrate enters the lysosome itself. When a substance enters the cell by endocytosis, the newly-formed endocytic vesicle which encloses the substance may fuse with a **primary lysosome** (containing latent enzymes). The fusion between endocytic vesicle and primary lysosome leads to activation of lysosomal enzymes, which can then digest the substance in the common container of the structure we can now call a **secondary lysosome,** or **digestive vacuole.** If digestion is complete or nearly complete, the soluble remainder of digestion can move into the cytosol or can be expelled from the cell by an exocytosis event (Fig. 6.16). If some undigested

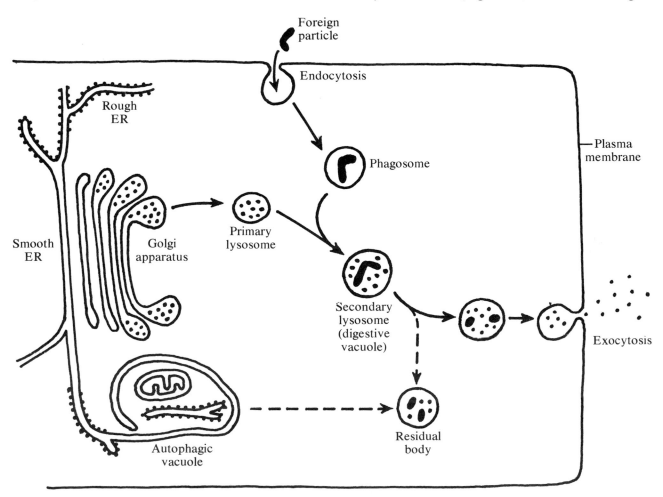

**Figure 6.16**
Summary diagram of the lysosome cycle. See text for details.

residue remains, the membranous organelle is retained within the cell as a **residual body.** These activities constitute the so-called **lysosome cycle.**

There are occasional incidents in which some part of the cell itself undergoes digestion, in an **autophagic vacuole.** It is not entirely clear how these structures form, but it is believed that fusion of cell structure with primary lysosome must occur at some point in the procedure. Starved protozoan and animal cells often develop autophagic vacuoles and can survive these incidents. In some cases during normal development or because of some pathological condition, the membrane of the primary lysosome may dissolve or leak its contents into the cell. Self-dissolution and death of the cell then occurs. In each mammalian ovarian cycle in which there is no fertilization, lysosomal enzymes cause the corpus luteum to degenerate. Lysosomal enzymes are also responsible in large measure for degeneration of larval tissues during metamorphosis. For example, just before tadpole metamorphosis there is an increase in the concentration of lysosomal enzymes, and continued increase as the tail is resorbed (Fig. 6.17).

The head of a spermatozoan is covered by an **acrosome,** a structure derived from the Golgi apparatus of the spermatocyte cell (Fig. 6.18). From detailed study of its formation, the acrosome appears to be a giant, specialized lysosome uniquely associated with most animal spermatozoa. Within seconds after a sperm has become attached to the outer coat of the egg, the acrosome membrane and plasma membrane of the sperm fuse together, thus creating perforations through which acrosomal contents are dispersed into the surrounding medium. The outer coat and adhering cells around the egg are digested quickly and the sperm reaches the plasma membrane of the egg cell. The egg and sperm plasma membranes fuse as the sperm nucleus enters and becomes surrounded by cytoplasm of the egg.

There are a number of almost instantaneous reactions by the egg to sperm penetration, one of which is disruption of the egg's cortical granules, which themselves have enzymes and staining properties common to lysosomes. The outer layers of the egg are

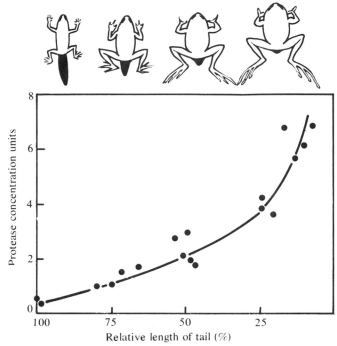

**Figure 6.17**
Increase in proteolytic enzyme activity during tadpole morphogenesis is correlated with resorption and dissolution of the tail as the tadpole develops to its adult form. The digestive enzymes are located in lysosomes in these cells.

broken down after the cortical granule contents have dispersed, and a new ("fertilization") membrane that is resistant to enzyme degradation is formed. Cleavages are initiated afterward and the embryo is on its way to development. In addition to lysosomal participation in these fertilization events, there are also factors related to recognition between the egg and sperm. Each is an exclusive target of the other, so there are no fusions between sperm or between eggs, only between egg and sperm. The surface properties of these sex cells are crucial factors in cell-cell recognition and in subsequent membrane fusions. Cell secretions must be involved at each step.

### Lysosomes and Disease

Except for protozoa and mammalian white blood cells, very few other kinds of cells regularly engulf materials by **phagocytosis** (a type of endocytosis in which the ingested material is an insoluble particle or structure); many cell types do so on occasion. White blood cells, or leucocytes, are an important part of the body's defense against infection and disease. Lysosomes develop quite extensively in leucocytes, even as membranous systems such as mitochondria and ER diminish, and these intracellular digestive organelles eventually fill a large part of the cell volume. When a leucocyte engulfs some foreign material, such as bacteria, lysosomes rapidly fuse with and digest this ma-

terial. The white blood cells usually die shortly afterward. Phagocytic cells (phagocytes) in tissues such as liver, lung, and spleen also contain large lysosomes that are important in digesting foreign materials.

Nonliving materials, such as silica particles and asbestos fibers, are not destroyed by lysosomal enzymes. Similarly, the thick waxy coat of living tuberculosis bacteria makes them immune to attack by lysosomal enzymes. These organisms can therefore initiate infections, since they are not stopped by leucocyte or other phagocyte encounters.

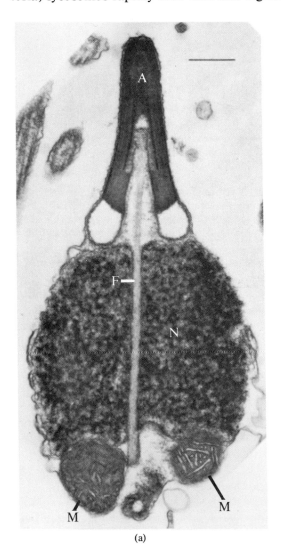

(a)

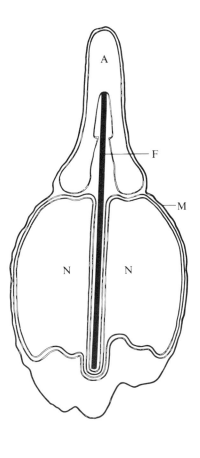

(b)

**Figure 6.18**
Longitudinal section through the head of a spermatozoan from the mussel *Mytilus*: (a) electron micrograph showing the acrosome (A), nucleus (N), and bundle of filaments (F) passing through the nucleus to a point of attachment at the acrosome membrane; (b) drawing illustrating these features and also showing the plasma membrane (M) around the headpiece of the spermatozoan. Two mitochondrial profiles, with cristae, are visible at the bottom of the photograph in a. (Reproduced with permission from Tilney, L. G., and M. S. Mooseker, 1976, *J. Cell Biol*. **71**:402, Fig. 1.)

Studies by Anthony Allison in London and by others have provided important insights into the relationship of lysosomes to disease processes and other aspects of cellular development. When silicon dioxide particles (for example, glass, quartz, sand) are inhaled into the lungs, they are taken up by phagocytes in the tissues. These cells die, releasing the silica which can then be ingested by other phagocytes with the same result. Repeated phagocytic deaths ultimately stimulate the deposition of nodules of collagen fibers by cells called fibroblasts. The fibrous collagen deposits decrease lung elasticity and thus impair lung function. The development of fibrous tissues in lungs of people who have been exposed to silica particles is one symptom of the disease called silicosis. Similarly, inhalation of asbestos fibers leads to fibrous tissues that are a symptom of asbestosis, another debilitating and often lethal disease.

Recent medical genetics studies have provided new information concerning certain kinds of "storage" diseases, in which a macromolecule remains incompletely processed and ultimately accumulates leading to some pathological condition. Many of these inherited disorders affect the central nervous system, leading to its deterioration and death of the individual.

One well-studied example of a metabolic defect which leads to accumulations of fatty materials in neurons and some other cells is **Tay-Sachs disease,** a condition inherited as a simple recessive trait. The disease occurs predominantly among Eastern European Jews and their descendants, but some 5–10 percent of babies with this trait come from other Jewish groups and from non-Jewish families. Afflicted children begin to show symptoms by about 6–8 months of age and undergo rapid deterioration from then on. Death occurs at some time between 2 and 4 years of age.

In Tay-Sachs disease, a 100- to 300-fold excess of a particular glycolipid, called a ganglioside, accumulates in the brain tissues of afflicted children. The accumulation is due to greatly reduced degradation of the glycolipid, and not because the compound is produced at an increased rate. There is one enzyme which is defective or missing in tissues of Tay-Sachs children, an enzyme normally found in lysosomes. Lysosomes in children with the disease are greatly enlarged and filled with masses of glycolipid-containing membranes.

In other storage diseases there is a similar accumulation of some macromolecule because of a missing or defective enzyme. In these cases, too, there is considerable enlargement of lysosomes which are filled with undigested molecules (Fig. 6.19). According to many studies, a defect in lysosomal hydrolase activities leads to an accumulation of undigested materials in bloated lysosomes, which indicates that the enzyme defect is the primary cause of the "storage" disease process.

## MICROBODIES

Microbodies are ovoid organelles that look very much like primary lysosomes. The two organelle types, however, are very different in function and distribution in different cells and species. Microbodies were first described in electron micrographs published in the mid-1950s. In some kinds of cells there is a crystalline inclusion that makes the microbody very distinctive, but such inclusions are not found in many species or cell types within a species (Fig. 6.20). Microbodies are packages of cell products, particularly enzymes, that are produced at the smooth ER. The protein contents are made at the rough ER and are transported through ER channels to smooth ER regions where packaging takes place (Fig. 6.21).

Unlike all the other subcellular components we know, microbodies have sets of functions that vary from one tissue to another and from one species to another. All mitochondria carry out aerobic respiration and all chloroplasts carry out photosynthesis and all lysosomes engage in intracellular digestion, but all microbodies do not participate in a common set of activities. This feature makes microbodies unique. A great deal of information has come to light in a very short time, but we will look only at some of the known microbody activities. In particular, it should be instructive to see how microbody activities coordinate with reaction systems in other organelles, together producing a level of metabolic flexibility that can only

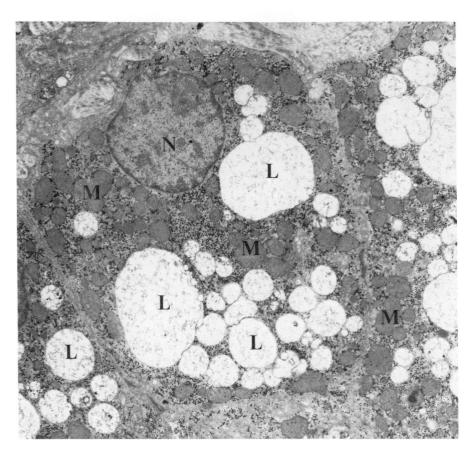

**Figure 6.19**
Thin section of liver biopsy from a patient with Hurler disease. The largest of the bloated lysosomes (L) are about the same size as the nucleus (N) and considerably larger than the mitochondrial (M) profiles in the group of cells included in the section. × 5,500. (Courtesy of F. Van Hoof)

be advantageous to cells growing in constantly changing environments (internal and external).

### Occurrence and Identification

Microbodies or microbodylike structures have been seen in species belonging to all kingdoms of eukaryotes. While they have been found in protists, fungi, plants, and animals, they are not necessarily present in every cell or every developmental stage. Since microbody ultrastructure is utterly simple, identification can be a problem. They can be identified confidently by biochemical criteria, however, usually in assays for particular enzymes that are found most of the time in most microbodies in the majority of cell types. The most commonly occurring microbody enzyme is **catalase,** a powerful catalyst that degrades hydrogen peroxide to water and oxygen. When catalase is present, as is usually the case, there

generally is another kind of enzyme in the same structure, a **flavin oxidase.** There are different flavin oxidases, but we usually identify and name such enzymes according to the substrate that is oxidized. One of the first flavin oxidases found was urate oxidase, which acts on uric acid (urate). In fact, it is this enzyme that makes up the crystalline core of certain microbodies (see Fig. 6.20). A much more common flavin oxidase is **glycolate oxidase,** which catalyzes the oxidation of glycolate to form glyoxylate and hydrogen peroxide. Considering catalase and flavin oxidases as a coordinated system, we find that hydrogen peroxidase produced by the flavin oxidase reaction is subsequently degraded by catalase action:

$$RH_2 + O_2 \xrightarrow{\text{flavin oxidase}} R + H_2O_2 \qquad [6.2]$$
$$H_2O_2 \xrightarrow{\text{catalase}} H_2O + \tfrac{1}{2}O_2$$

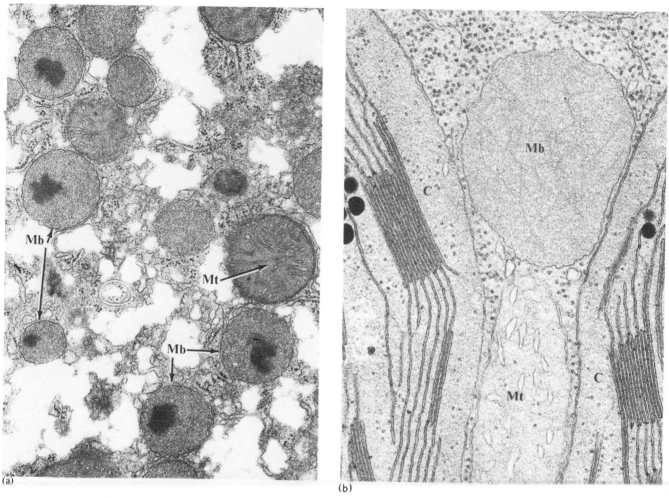

**Figure 6.20**
Microbodies. (a) A core inclusion occurs typically in microbodies (Mb) from rat liver.
× 25,000. (Courtesy of M. Federman) (b) Cores do not occur regularly in microbodies (Mb)
from leaf mesophyll cells, although fine fibrils are often present. The cells are from timothy
grass *(Phleum pratense).* Note the membrane organization in the chloroplasts (C) and
mitochondrion (Mt). × 66,000. (Courtesy of E. H. Newcomb and S. E. Frederick)

In the specific sequence involving glycolate oxidase and catalase, we have

$$\underset{\text{glycolate}}{\overset{\text{COOH}}{\underset{\text{CH}_2\text{OH}}{|}}} + \text{O}_2 \xrightarrow{\text{glycolate oxidase}} \underset{\text{glyoxylate}}{\overset{\text{COOH}}{\underset{\text{CHO}}{|}}} + \text{H}_2\text{O}_2 \qquad [6.3]$$

followed by the catalase-guided reaction as shown in [6.2].

Hydrogen peroxide is a powerful substance that is potentially harmful to living cells. Its disposal by catalase would therefore be advantageous to aerobic life, and catalase does occur in all aerobic species.

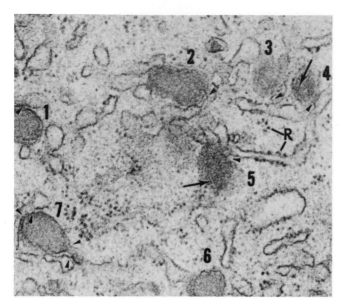

**Figure 6.21**
Thin section of guinea pig duodenum. All seven microbodies are associated with smooth endoplasmic reticulum and not with rough *ER*. Continuities of microbody membrane and smooth *ER* are shown at arrowheads; arrows indicate wavy, tubulelike structures in No.4 and 5 microbodies. Ribosomes (R) are present both on the *ER* and apparently free in the cytoplasm. × 45,000. (Courtesy of A. B. Novikoff, from Novikoff, P. M., and A. B. Novikoff, 1972, *J. Cell Biol.* **53**:532-560, Fig. 14.)

This reaction is wasteful, however, because energy is not conserved during the hydrogen transfer.

It is now possible to identify microbodies in cell thin sections seen by electron microscopy. Based on a **cytochemical reaction,** catalase activity leads to the formation of an opaque reaction product that can be seen very easily in electron micrographs (Fig. 6.22). Since this is a specific catalase test, it serves as a means for recognizing microbodies and determining their distribution even in those cells that cannot be analyzed easily by biochemical methods. It was through this and other cytochemical tests that some cells were shown to have microbodies with glycolate oxidase activity but not to have catalase within the microbodies themselves. In certain algae, catalase is in the cytosol exclusively. These same algae also excrete large amounts of glycolate into the surrounding waters. Because most microbodies have a system for hydrogen peroxide disposal, they are often referred to as **peroxisomes.**

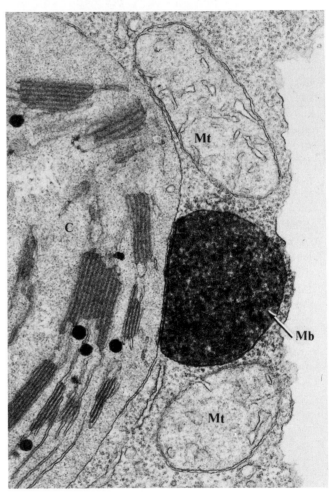

**Figure 6.22**
Thin section of leaf mesophyll cell from tobacco (*Nicotiana tabacum*), incubated for catalase activity in a cytochemical test. Enzyme activity product is localized in the microbody (Mb) but is absent from the closely appressed mitochondrion (Mt) and chloroplast (C) in the same cell. × 31,000. (Courtesy of E. H. Newcomb, from Frederick, S. E. *et al.*, 1975, *Protoplasma* **84**:1-29, Fig. 5.)

## Coordination Among Organelles

Some cells in fatty seeds of plants such as castor bean contain stored food reserves in fat deposits. During seed germination, fats are converted to sugars and these sugars in turn provide the energy and building blocks required to sustain the seedling until it emerges from the ground and can carry out photosynthesis. Conversion of fats to sugars is a remarkably efficient process; 1 gram of sugar is made for every gram of fat that is processed, and involves coordinated activities of enzymes in the cytosol, microbodies, and mitochondria of the food storage cells (endosperm), all of which leads to development of the embryo into a seedling. It takes about 11 days for the seedling to become independent of its stored foods, at which time it is able to photosynthesize.

Fats are converted to **fatty acids** by lipases in the cytosol, and the fatty acids then enter the microbodies (here called glyoxysomes; Fig. 6.23). Once in the microbody, fatty acids are dismantled to 2-carbon fragments in a cycle that produces **acetyl CoA.** Acetyl CoA next enters the **glyoxylate cycle** at either of two places: (1) by condensing with 2-carbon glyoxylate to form 4-carbon malate, or (2) by condensing with 4-carbon oxaloacetate to form 6-carbon citrate. One of the products of the glyoxylate cycle is **succinate,** an intermediate in the Krebs cycle. Succinate leaves the microbodies and enters the mitochondria where it undergoes Krebs cycle reactions. The added succinate is converted to oxaloacetate during these reactions, and oxaloacetate leaves the mitochondrion to enter the cytosol where it is converted to sugars.

There are several important features of the coordinated system. Fatty acid oxidation usually takes place in mitochondria, but the fact that fatty acids are oxidized in microbodies leads to retention in microbodies of the acetyl CoA produced by fatty acid oxidation. The acetyl CoA feeds into the glyoxylate cycle, which provides intermediates that can be channeled through the mitochondrion and out into the cytosol. In this way, sugars enter the many pathways required to construct all the macromolecules needed for seedling growth. If these seeds processed fatty acids in mitochondria, the acetyl CoA would be used almost exclusively in Krebs cycle reactions leading to ATP synthesis (see Chapter 5). Growth certainly requires ATP, but it also requires molecular building blocks to make proteins, nucleic acids, and all the rest of a cell's equipment. In bypassing the mitochondrial system, fatty seeds manage to get building blocks as well as energy for many days of growth during germination, and all of it is derived from stored fats.

Fatty seed microbodies are very unusual in having so many coordinated enzyme systems. Interestingly, once the seedling grows above ground and makes food by photosynthesis in its leaf cells, there is an entirely different microbody system in the leaf cells of the very same plants. Leaf microbodies are also unusual, but in ways that are related to their unique metabolism, just as seed microbody reaction systems provide certain advantages in a different set of growing conditions. Hydrogen peroxide disposal by catalase action occurs in seeds and leaves, but appears to be a relatively minor aspect of the whole spectrum of microbody metabolism.

Although mammalian microbodies were the first to be studied, we still know very little about them or about their contributions to cellular activity. We do know that none of the enzymes of the glyoxylate cycle occur in mammals (or in green leaves), and that a few other kinds of enzymes are present in addition to catalase and glycolate oxidase (Table 6.1). In particular we are anxious to know whether microbodies make some contribution to **gluconeogenesis** pathways, that is, pathways for production of sugars from noncarbohydrate, organic precursors. Microbodies are especially prominent in mammalian liver and kidney, two organs known to engage in gluconeogenesis. But it is uncertain at present whether or not these microbodies are involved in the gluconeogenic reaction sequences in these organs.

There are two major pathways for synthesis of sugars from noncarbohydrate precursors. One pathway is not gluconeogenic; it is formation of sugars from carbon dioxide in photosynthetic organisms (see Chapter 5). The second major pathway proceeds from fats and fatty acids via acetyl CoA and the glyoxylate cycle in fatty seeds and certain microorganisms, and

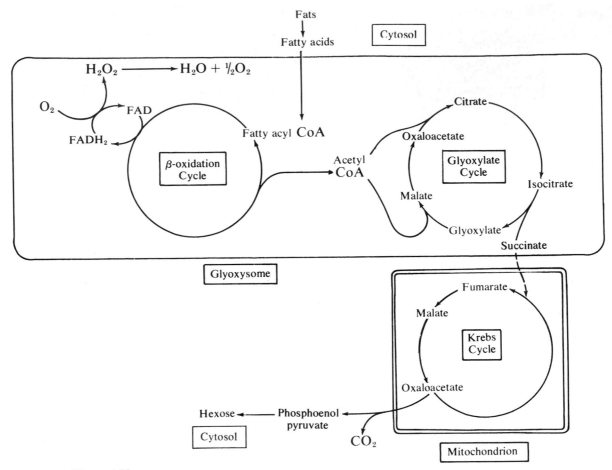

**Figure 6.23**
Interactions among compartmented metabolic pathways in castor bean endosperm cells.
Fats are digested to fatty acids which then enter the glyoxysome and undergo β-oxidation
to acetyl CoA units. These units in turn are shunted to the glyoxylate cycle within the same
organelles. Succinate produced during glyoxylate cycle reactions is transferred to the
mitochondrion where it is converted to oxaloacetate during Krebs cycle oxidations.
Oxaloacetate leaves the mitochondrion and is converted in the cytosol to sugars in a
reverse glycolysis sequence. This pattern of gluconeogenic coversion of fats to sugars is a
major feature of glyoxysome metabolism during castor bean seedling germination and
growth. Note that reactions associated with the β-oxidation cycle lead to the production of
$H_2O_2$, which is then disposed of by catalase activity. These latter reactions are typcial of
microbodies.

through pyruvate or Krebs cycle intermediates in
other kinds of cells (Fig. 6.24).

## MEMBRANE TRANSFORMATIONS

Various cell membranes are made and broken
down in continuous cycles in living cells. We have already mentioned several ways in which membranes
are assembled, for example during secretion granule
formation from the Golgi apparatus and during
endocytosis when plasma membrane pieces are taken
for endocytic vesicle formation. Almost all cellular
structures have a limited life span; for example, mi-

**TABLE 6.1  Enzyme activities found in microbodies of selected cell types and species***

| Enzyme | Rat Liver | Rat Kidney | Frog Liver | Frog Kidney | Tetra-hymena | Yeast | Spinach Leaves | Castor Bean Endosperm |
|---|---|---|---|---|---|---|---|---|
| Glyoxylate cycle: | | | | | | | | |
| Isocitrate lyase | − | − | − | − | + | + | − | + |
| Malate synthase | − | − | − | − | + | + | − | + |
| Malate dehydrogenase | − | − | − | − | − | − | + | + |
| Citrate synthase | − | − | − | − | − | − | − | + |
| Aconitase | − | − | − | − | − | − | − | + |
| Catalase | + | + | + | + | + | + | + | + |
| Glycolate oxidase | + | + | | | + | + | + | + |
| Urate oxidase | + | − | + | + | − | − | − | + |
| Allantoinase | | | + | | + | | | + |
| Transaminases (various) | | | | | | | + | + |
| Fatty acid oxidation system | | | | | | | − | + |

*Enzyme present, +; enzyme absent, −; enzyme not tested, left blank.

crobodies exist for about 4 days, on the average. Microbodies and other cell components are in a constant state of **turnover,** old systems degenerating and being replaced by new ones. In fact, except for DNA there is regular turnover for all the molecules and structured components of living cells. DNA, of course, provides the continuity from one generation of cells or organisms to the next.

The plasma membrane and ER apparently can assemble spontaneously from phospholipids and other lipids, carbohydrates, and proteins. Each membrane constituent lends unique properties of construction, recognition, catalysis, regulation, and other characteristics of membranes. In addition to membrane assembly, we know that membrane transformations also take place. These transformations particularly involve the ER, Golgi apparatus, and plasmalemma. In fact, there seems to be a direction of membrane "flow" toward the outer surface in most cells (Fig. 6.25).

Various kinds of observations and experimental evidence support the notion that ER membranes are transformed into Golgi-specific membranes, which in turn are transformed into plasma membrane via the mediation of exocytic vesicle fusions. For example, radioactively labeled membrane precursors can be found first in the ER and then in Golgi membranes and finally in the plasmalemma. The gradations of membrane thickness and staining patterns flow from thinner and barely three-layered staining in ER membranes to thicker and definitely dark-light-dark three-layered patterns of the plasmalemma, with Golgi membranes being intermediate in these respects. Specific membrane molecules differ in ER and plasmalemma, with Golgi membrane molecules once again being intermediate between those two systems. Because of these and other kinds of information, it has been proposed by some investigators that most of the membranes in a cell may be morphologically and functionally diversified components of a single **endomembrane system.** Such a postulate is an attempt to explain the dynamic interactions among cellular membranes and their coordinated activities in eukaryotic cells. Even though membrane transformations do not take place in every case, many membranes are in direct contact or interconnected at one or more places.

It is energetically more effective for membranes with different functions to be transformed from existing membranes than to produce new membranes *de novo.* A considerably larger amount of energy would be required to sponsor many new chemical bonds in totally new membrane than is needed to modify some components of existing membrane. The energy stores of the cell can be diverted to other activities which otherwise might be competing for the same limited

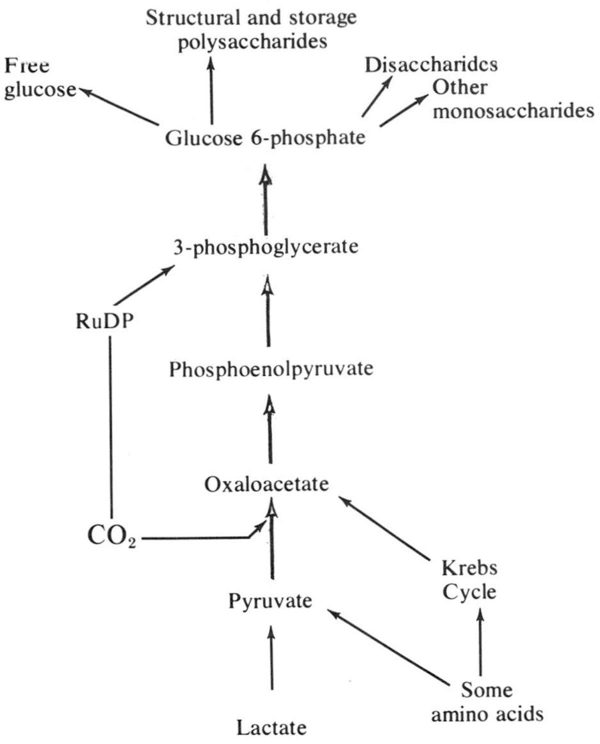

Structural and storage
polysaccharides

Free
glucose

Disaccharides
Other
monosaccharides

Glucose 6-phosphate

3-phosphoglycerate

RuDP

Phosphoenolpyruvate

Oxaloacetate

$CO_2$

Krebs
Cycle

Pyruvate

Lactate

Some
amino acids

**Figure 6.24**
A general summary of some gluconeogenic sequences leading to carbohydrate production. Amino acids may serve as precursors by mediation of pyruvate or by Krebs cycle intermediates to form oxaloacetate. Or, pyruvate may be formed from lactic acid and other compounds and then be carboxylated to form oxaloacetate. Sugars may also be formed from RuDP and $CO_2$ in the dark reactions of photosynthesis. Other pathways exist, such as conversion of fats to sugars through acetyl CoA formation and subsequent reactions in the glyoxylate cycle or the Krebs cycle (see Fig. 6.23).

energy supplies if all new membranes were assembled from scratch.

So far we have discussed two of the three general types of eukaryotic organelles, categorized according to their membranes or lack of surrounding membrane:

1. Double-membrane mitochondria and chloroplasts, with their own portion of DNA and ribosomal machinery.
2. Single-membrane organelles and systems that are containers of protein products of gene action, or which channel these products into containers.
3. Organelles lacking their own surrounding membranes, such as chromosomes, centrioles, and ribosomes (to be discussed in the following chapters).

Each of these categories includes systems with quite different capabilities in the spectrum of cellular activity. The double-membrane mitochondria and chloroplasts are very complex and are uniquely able to specify and synthesize some of their own protein constituents. Single-membrane-bound structures may modify some molecules, but they are quite limited in their biosynthetic activities. Nonmembranous organelles can direct macromolecular biosyntheses: protein synthesis at ribosomes, microtubule synthesis at centrioles, and synthesis of DNA and RNA at chromosomes. By such varied compartmentation, eukaryotic cells can manage enormously complicated but flexible and coordinated sets of activities. Among other features, eukaryotic cell organization clearly underlies the rapidly established dominance and continued successful evolutionary progress of eukaryotes as compared with their prokaryotic ancestors and modern descendants among bacteria and blue-green algae.

**SUMMARY**

1. The folded membrane(s) of the endoplasmic reticulum (ER) provide a system for distribution of proteins made at ribosomes attached to one surface of the rough ER membrane. ER without attached ribosomes, generally called smooth ER, is another differentiated region of the system, as are the smooth membranes that comprise the Golgi apparatus of a cell. The Golgi apparatus, made up of one or more dictyosomes, acts as a packaging and processing station for various cell secretions and for other kinds of proteins which remain within the cell rather than being distributed to other cells or to extracellular spaces. Various proteins which are made at the rough ER are channeled through the ER lumen to the Golgi apparatus, where the molecules are chemically processed before being packaged in vesicle containers made from Golgi membrane.

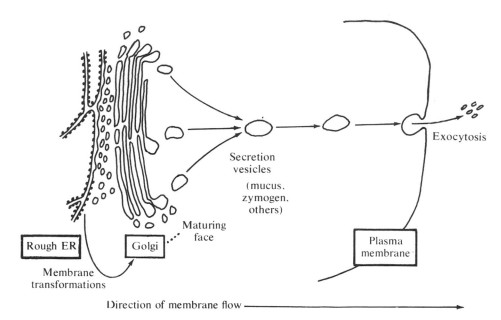

Exocytosis

Secretion
vesicles

(mucus,
zymogen,
others)

Maturing
face

Rough ER

Golgi

Plasma
membrane

Membrane
transformations

Direction of membrane flow

**Figure 6.25**
Summary diagram showing a
postulated sequence of cellular
membrane transformation and
travel from sites of synthesis at
the rough *ER* through the Golgi
apparatus and the final fusion
with the plasma membrane
during exocytosis.

2. Secretion vesicles move toward the cell apex, where the secretions are released after an exocytosis event involving the cell membrane. Lysosomes are also made at the Golgi apparatus, according to many investigators, but these membranous organelles remain within the cell in which they are produced. Lysosomes contain powerful digestive enzymes, all with a pH optimum of about 5. These enzymes can dismantle virtually every organic substance of biological importance, whether native or foreign to the cell. Lysosomes serve on the front line of defense against disease, since they fuse with incoming endocytic vesicles containing particles or molecules that can be harmful to the cell, and digest such substances during a sequence of events called the lysosome cycle. Digested residues are expelled from the cell by exocytosis, but vesicles containing undigested residues (structures now called residual bodies) may remain within the cell. Consumption of parts of the cell itself may take place within autophagic vacuoles, another type of lysosome system. Malfunction of lysosomal enzymes has been associated with the development of inherited "storage" diseases, such as Tay-Sachs disease and others.

3. Microbodies are organelles bounded by a single membrane, and they are believed to arise as vesicles pinched off from parts of the smooth ER. Microbodies are unique among cellular organelles because they have different enzyme repertories and therefore perform different func-

tions in cells of various tissues and species. Catalase is one enzyme found in virtually all microbodies, usually in conjunction with some kind of oxidase which produces hydrogen peroxide in an oxidation-reduction reaction in which a substrate is oxidized while $O_2$ is reduced to $H_2O_2$. Catalase disposes of the $H_2O_2$ in a subsequent reaction yielding $H_2O + \frac{1}{2}O_2$.

4. Except for vertebrate animal and green leaf microbodies, all or most other systems have microbodies containing enzymes which are part of the glyoxylate cycle. The only microbody system known so far to have all five enzymes (rather than two) of the glyoxylate cycle are so-called glyoxysomes found in germinating fatty seeds of plants like castor bean, peanut, and others. Glyoxysomes are the most metabolically complex of the known microbody systems, and are directly responsible for conversion of stored fats to carbohydrates during the short interval between the beginning of seed germination and the emergence of a photosynthesizing green seedling plant. By these processes of carbohydrate formation from noncarbohydrate precursors, called by the general term gluconeogenesis, the seedling obtains all its energy and building blocks for growth before it begins to manufacture organic molecules by photosynthesis. The leaf microbody system in seedlings and mature plants lacks every enzyme of the glyoxylate cycle and most other enzymes found in seed

glyoxysomes, although an oxidase—catalase system is present in both cases.

5. Membranes and other cellular structures and molecules, except for DNA, are in a constant state of turnover during which new components replace existing ones. One economical means of replacing membrane is by transformation from one membrane type to another. The transformation of ER to Golgi membranes, and Golgi membranes to plasma membrane is one such pathway, in addition to self-assembly of new membranes of these and other types in cells.

### STUDY QUESTIONS

1. How is the endoplasmic reticulum organized? What kinds of ER have been found, and how are they recognized in electron micrographs? What is the relationship of ER organization to its functions of intracellular protein synthesis and protein transport? How are microsomes formed, and what is their relationship to rough ER?

2. How is the Golgi apparatus organized, and what are the relationships between cisternae and dictyosomes? What information can we obtain about Golgi apparatus organization in different cell types using thin sections and whole-mount preparations of these membrane systems? What are the main functions of the Golgi apparatus? How do we know these are its functions? What is autoradiography, and how can the method be used to study synthesis and processing of proteins in parts of the cell?

3. What is secretion? How do cells export their secretion products? What kinds of secretions have been studied, and how do they differ from one another? What is the contribution made by the cell surface to secretion, cell-cell recognition, and inhibition of cell growth and movement?

4. How can we recognize and identify lysosomes? What enzyme usually is localized in lysosomes by a cytochemical test, and why is such a test helpful in lysosome studies? What is the major function of lysosomes? How is this function carried out during the lysosome cycle? Where do lysosomes form in the cell? Why do we consider the acrosome of a spermatozoan to be a type of lysosome?

5. What is the relationship between phagocytosis and intracellular digestion? What kinds of cells regularly carry on phagocytosis? How is the process related to the body's defenses against disease? Why does long-term inhalation of silica particles or asbestos fibers contribute to impairment of lung function in diseases such as silicosis and asbestosis? What is a "storage" disease? What happens to lysosomes in people who develop Tay-Sachs disease and similar inherited metabolic defects?

6. What are the similarities and differences between microbodies and lysosomes in relation to structure, composition, mode of formation, and function? What is a characteristic enzyme found in microbodies and routinely detectable by a cytochemical test? What does catalase do in microbody-related reactions? How is $H_2O_2$ usually formed in microbodies?

7. How does the germinating fatty seed obtain energy and building blocks for growth? What is the importance of acetyl CoA entering the glyoxylate cycle pathway in fatty-seed glyoxysomes rather than the Krebs cycle pathway in mitochondria of the same cells? What is gluconeogenesis and how does a green plant make sugars during seed germination and afterward when the plant can carry out photosynthesis?

8. What do we mean by "turnover" in cells? How do new membranes form in cells? What is the advantage of transforming one kind of membrane into another kind of membrane as an alternative means for producing specific kinds of cell membranes? What do we mean by the "endomembrane system" of the cell?

9. What are the different abilities of organelles bounded by two membranes, bounded by one membrane envelope, and having no surrounding membrane envelope? What is an example of an organelle of each of these three classes, and what function characterizes each example you have selected?

### SUGGESTED READINGS

Allison, A. Lysosomes and disease. *Scientific American* **217**:62 (May 1967).

Brady, R. O. Hereditary fat-metabolism diseases. *Scientific American* **229**:88 (Aug. 1973).

Breidenbach, R. W., Kahn, A., and Beevers, H. Characterization of glyoxysomes from castor bean endosperm. *Plant Physiology* **43**:705 (1968).

Claude, A. The coming of age of the cell. *Science* **189**:433 (1975).

deDuve, C. Exploring cells with a centrifuge. *Science* **189**:186 (1975).

Frederick, S. E., and Newcomb, E. H. Cytochemical localization of catalase in leaf microbodies (peroxisomes). *Journal of Cell Biology* **43**:343 (1969).

Hogg, J. F. (ed.) *The Nature and Function of Peroxisomes (Microbodies, Glyoxysomes).* Annals of the New York Academy of Sciences **168**:209–381 (1969).

Kappas, A., and Alvares, A. P. How the liver metabolizes foreign substances. *Scientific American* **232:**22 (June 1975).

Kolodny, E. H. Current concepts in genetics. Lysosomal storage diseases. *New England Journal of Medicine* **294:**1217 (1976).

Morré, D. J. Membrane biogenesis. *Annual Reviews of Plant Physiology* **26:**441 (1975).

Neutra, M., and Leblond, C. P. The Golgi apparatus. *Scientific American* **220:**100 (Feb. 1969).

Palade, G. Intracellular aspects of the process of protein synthesis. *Science* **189:**347 (1975).

Satir, B. The final steps in secretion. *Scientific American* **233:**28 (Oct. 1975).

# Chapter 7

# Cellular Movements: Microtubules and Microfilaments

Movement is a basic property of living systems. The work of motion requires an input of energy which may be provided in various forms. The most obvious and familiar motions of single cells are swimming and creeping. Many kinds of cells are propelled through liquids by means of **cilia** and **flagella.** These motility organelles act like oars that carry the cell from one place to another. Many simple protozoa and algae have cilia and flagella that are basically the same as structures that permit mammalian sperm to swim. The familiar amebae creep along solid surfaces by **ameboid movement;** cells in our own blood system creep along by this type of locomotion; some eukaryotic cells grown in culture also move across surfaces by inching forward.

Movement of cell parts occurs commonly and may even be continuous in some cases. Muscles contract, expand, and relax; chromosomes move from one place to another during nuclear divisions; protoplasm streams or churns in most cells. When particles or organelles are carried along with the protoplasmic flow, the streaming movement is called **cyclosis.**

When cells change their positions relative to each other during tissue and organ development, this change constitutes movement because a displacement in space is involved. Growth itself leads to extension from one part of space to another and can be viewed, therefore, as a form of movement. In effect, living systems are in constant motion in at least some portion of the individual during its entire existence.

This universal property of motion is expressed visibly in various ways, some of which were just mentioned. It has become apparent in recent years, however, that the variety of visible movements are actually based on common structural themes within cells. These concepts of cell movement have developed since the 1950s and 1960s from studies using electron microscopy and biochemical methods that were not available earlier. Movements consistently involve protein fibrous structures classified as **microtubules** and **microfilaments.** In some forms of motion only one kind of fiber may be directly or exclusively involved in producing movements. In other cases both microtubules and microfilaments seem to be needed to achieve displacement in space.

Wherever they occur, microtubules are hollow, unbranched cylinders. They vary in length up to tens of thousands of nanometers, but they generally have a diameter of 18–20 nm. On closer inspection, the cylinder wall can be resolved into 13 adjacent filamentous components, each having a diameter of 4–5 nm (Fig. 7.1). Microtubules are widely distributed cell components, but they are apparently absent from certain cells, such as all prokaryotes, amebae, slime molds, and a few others. Microtubules are most conspicuous in cilia, flagella, the spindle of dividing nuclei, and centrioles, which are usually associated with motility organelles and with the spindle. The tubules are also prominent components of the peripheral cytoplasm in many kinds of cells and within surface specializations of some protozoa (Fig. 7.2).

The second major kind of fibril is the microfilament, which generally occurs in bundles or other groupings rather than singly. Each filament is about 5–6 nm wide, but there are thicker filaments in certain kinds of cells. Microtubules and microfilaments can be distinguished on the basis of width, substructure, and their responses to drugs.

The most complex and highly evolved system of movement is displayed by muscle. At the same time it is the system that has been studied the longest and one which provides models that are useful in explaining and understanding simpler expressions of motility. Because of these considerations, we will discuss muscle first. With this foundation it should be easier to understand how other movements have been interpreted and analyzed.

## MUSCLE FIBERS

Muscle contraction is work that is subsidized by the chemical energy of ATP. In **striated muscle,**

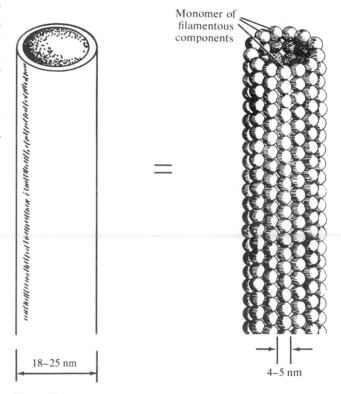

**Figure 7.1**
The microtubule is a hollow unbranched cylindrical structure (left) formed by 13 adjacent rows of filaments which themselves are linear arrays of the globular protein called tubulin (right).

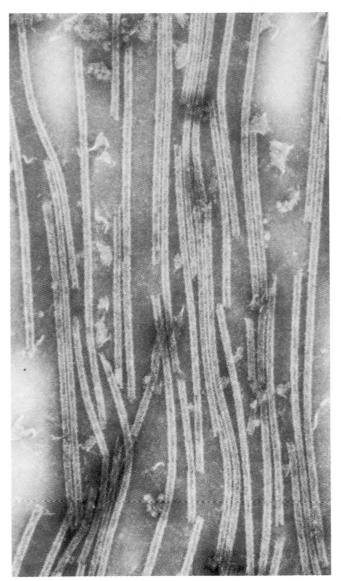

**Figure 7.2**
A group of negatively stained microtubules from the marginal band just beneath the cell surface of newt (*Triturus viridescens*) erythrocytes. The phosphotungstate stain penetrates and accentuates the lumen of the tubules. × 56,000. (Courtesy of J. G. Gall, from Gall, J. G., 1966, *J. Cell Biol.* **31**:639-643, Fig. 1.)

contraction results in voluntary actions by the organism. The structural organization and chemical activities of contraction have been studied for more than 40 years, with increasingly greater understanding during this time. In addition to providing remarkable insight into muscle itself, these studies have allowed us to analyze contractility in nonmuscle systems in more informed ways.

Striated muscle is so named because of its prominent alternating dark and light bands when seen by microscopy. It is also called skeletal muscle because of its close association with parts of the skeleton in vertebrates. The **muscle fiber,** which is a cylindrically-shaped, multinucleate cell, is the whole system (Fig. 7.3). The fiber varies in length but generally has a

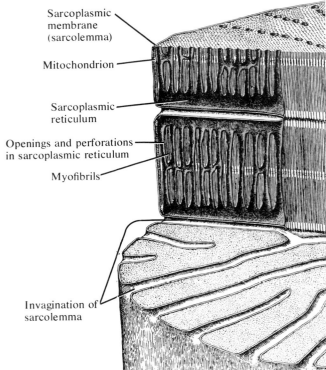

Sarcoplasmic membrane (sarcolemma)

Mitochondrion

Sarcoplasmic reticulum

Openings and perforations in sarcoplasmic reticulum

Myofibrils

Invagination of sarcolemma

**Figure 7.3**
Cutaway view showing some of the major structured components of part of a muscle fiber. The sarcoplasmic reticulum is perforated and the sarcolemma is infolded at various places, forming a system of invaginations that penetrates through the thickness of the cell.

diameter of 50 to 200 $\mu$m. The plasma membrane surrounding the muscle fiber is called the **sarcolemma.** This membrane is bordered by an area of cytoplasm that is called the **sarcoplasm,** which in turn surrounds a bundle of **myofibrils.** Myofibrils are the contractile elements of muscle, running the length of the muscle fiber; each fibril is only about 1–3 $\mu$m in diameter. The sarcoplasm contains organized features similar to any eukaryotic cytoplasm. Nuclei, organelles such as mitochondria, and an endoplasmic reticulum (called **sarcoplasmic reticulum** in this system) are all present.

The striations of muscle myofibrils have been designated by letters of the alphabet in a convention that is readily understood by all investigators (Fig. 7.4). The two major bands are the wider, darker **A-band** next to the narrower, lighter **I-band.** In the center of the A-band there is a lighter region called the **H-zone,** which is bisected by an **M-line.** Each I-band is bisected by a dark, narrow **Z-line.** The region between adjacent Z-lines is called the **sarcomere.** The sarcomere is the contractile unit of the myofibril, and these units are repeated along the length of myofibrils. Each sarcomere occupies about 2.5 $\mu$m of space.

Vertebrate smooth muscle, often called involuntary muscle, has no striations. A single nucleus is found in the center of isolated spindle-shaped cells, and the cells are linked together by connective tissue. Considerable progress is being made in understanding smooth muscle, but it is not nearly as well understood as striated muscle systems.

## Myofibril Ultrastructure and Chemistry

Myofibrils contain two kinds of **myofilaments** in parallel arrays. These types are thick and thin fila-

**Figure 7.4**
The striated muscle system. Striated muscle is made up of cylindrical multinucleated cells called muscle fibers. Each fiber is bounded by a sarcolemma and contains various structured components, including bundles of myofibrils. Each myofibril contains the contractile units, or sarcomeres, of the muscle. Sarcomeres are regions of darker, wider A-bands and narrower, lighter I-bands. Each type of band is bisected by a line. A single sarcomere occupies the area between adjacent Z-lines.

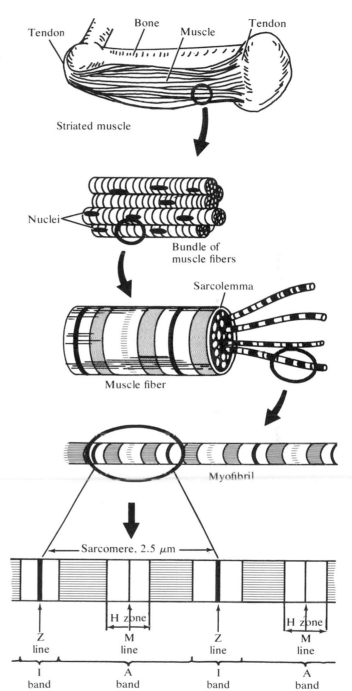

ments arranged in specific spatial patterns. Thick filaments, about 15 nm wide, are arranged hexagonally, with about 45 nm of space between pairs of filaments. Thin filaments are about 6 nm in diameter and are regularly arranged between thick filaments so that each thick filament is encircled by six thin filaments. Thick filaments extend from one end to the other of the dense A-band, while thin filaments extend from their connection to the Z-line through the I-band and into the A-band up to the H-zone boundary (Fig. 7.5). When seen in cross section, the pattern varies according to the particular part of the sarcomere in view. Sections through the A-band will therefore have both thick and thin filaments, I-band sections have only thin filaments, and so forth. High-resolution electron microscopy has further revealed the presence of cross-bridges which are disposed along the thick filaments at regular intervals.

Thick filaments are made of the fibrous protein **myosin.** Each myosin molecule is made up of two identical subunits tightly wound around each other. Myosin molecules aggregate into filaments in a staggered sequence, beginning with the first two molecules lying with their tails side by side and the heads of these asymmetric molecules pointing in opposite directions (Fig. 7.6). This arrangement produces a filament with globular heads projecting from the central core of the aggregate, except for the central region which lacks these heads. This regularity of projecting heads produces the periodic spacing mentioned above, since the cross-bridges are heads of myosin molecules. These cross-bridges connect the thick myosin filaments to the thin filaments, which are composed of the protein **actin.** The head region of a myosin molecule binds specifically to actin. It is also the part of myosin that has ATPase activity and is, therefore, involved in deriving energy from ATP for contraction.

Actin is different from other proteins of myofibrils, since it dissociates into globular monomers (**G-actin**) in the absence of ions, and aggregates into fibrous (**F-actin**) form in the presence of neutral salts. Each G-actin monomer contains one $Mg^{2+}$ ion and one molecule of ATP, both of which are very tightly bound to actin.

Actin and myosin associate and dissociate during contraction; that is, actomyosin is a complex that is not permanent. Sarcomeres are diminished to about 20–50 percent of their resting length during an episode of contraction. But the individual filaments of myosin and actin do *not* contract. Sarcomeres may increase in length if passively stretched, but again, individual filaments are not stretched. These observations were based on exacting measurements of intact muscle sarcomeres. The A-band width remains constant in contracted, stretched, or resting muscle; therefore, the length of its myosin filaments must also be constant. Nor do actin filaments become longer or shorter. This is clear from the fact that the distance between the Z-line and H-zone does not change during contraction. This is the region in which actin filaments occur. The portion of the sarcomere that alters in length is the I-band, which decreases in contraction and increases in stretching.

Taking all these factors into consideration, A. Huxley and H. Huxley first proposed that changes in sarcomere (hence, muscle) length must be due to sliding of thick and thin filaments along each other (Fig. 7.7). According to the **sliding filament model,** muscles contract by a mechanism in which myosin and actin filaments slide past each other, driven by longitudinal forces developed by cyclically operating cross-bridges. The Z-lines are drawn together during contraction because the actin filaments attached to a Z-line are drawn in toward the center of the A-band. This event is repeated in each adjacent sarcomere of a myofibril, since actin filaments have an opposite polarity on either side of a Z-line. Because of this arrangement, actin filaments in each sarcomere can move similarly toward the center of the A-zone.

Myosin molecules can be selectively fragmented into three regional components by proteolytic enzyme activities (Fig. 7.8). The tail end of the molecule (**light meromyosin**) has no ATPase activity and cannot sponsor binding of myosin and actin. The opposite end of the asymmetric molecule (**heavy meromyosin**) binds to actin and displays ATPase activity. If heavy meromyosin is digested further, a globular head portion separates from the remainder. The globular head

(a)

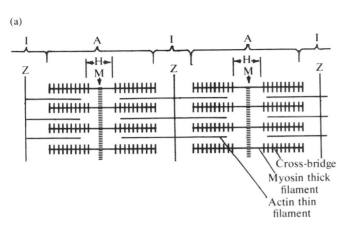

Cross-bridge
Myosin thick filament
Actin thin filament

**Figure 7.5**
Muscle myofilaments: (a) Drawing showing the regular arrangement of myosin thick filaments and actin thin filaments in a sarcomere; (b) electron micrographs of muscle of the fresh water killifish (*Fundulus diaphanus*). (1) longitudal section of a sarcomere, bounded by two Z-lines and including bands and zones lettered according to the convention described in Fig. 7.4 and the text. (2-6) Cross-sectional views through particular parts of the sarcomere: (2) thick and thin filaments in the A-band region of overlap; (3) thick filaments from the H-zone region of the A-band where no overlap occurs; (4) thick filaments from the A-band immediately adjacent to the M-band; (5) thick filaments from the A-band in the region of the M-band, showing each thick filament connected to each of its six neighboring thick filaments by bridges; and (6) thin filaments from the I-band. (Photographs courtesy of F. Pepe, from *Biological Macromolecules Series: Subunits in Biological Systems*. Marcel Dekker, Inc., New York, 1971, pp. 323-353.)

(b)

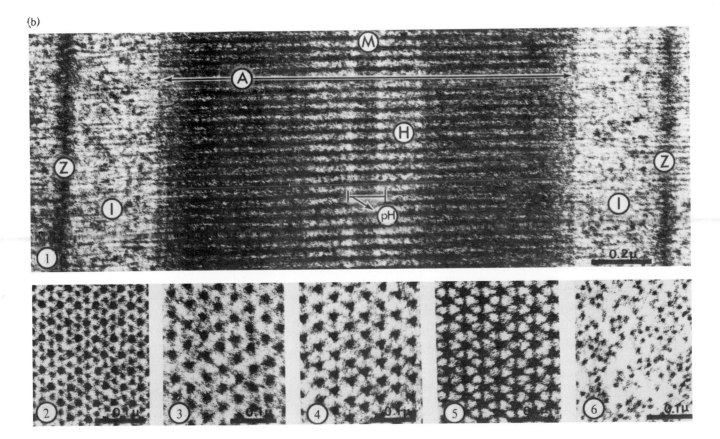

**Figure 7.6**
A thick filament consists of myosin molecules aggregated such that the molecule heads, which act as cross-bridges during contraction, project from the filament core except in the center, where there are only tail portions of myosin molecule aggregates. The region of "tails" is visible as the pseudo H-zone in a sarcomere (see Fig. 7.5b, pH).

is the specific portion of the molecule that binds to actin and has ATPase activity, and it is the active portion of the myosin molecule.

The globular head portion of myosin is believed to be the cross-bridge that occurs in measurable periodicity along the thick filament. These cross-bridges probably break and form continuously during sliding but are never detached from the rest of the myosin

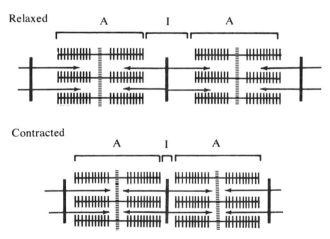

**Figure 7.7**
Diagram illustrating the sliding filament theory of muscle contraction. The thick and thin filaments slide past each other during contraction but do not undergo change in total length within a sarcomere. The sarcomere itself shortens, since adjacent Z-lines are drawn together as actin filaments slide over myosin filaments. Actin filaments on either side of a Z-line have opposite polarity, so that each sarcomere contracts during filament sliding.

molecule. The particular mode of cross-bridging has not yet been resolved, but some models are open to experimental tests.

According to some investigators, the distance involved in an overall cross-bridging movement is 8–12 nm as actin filaments slide over myosin filaments during sarcomere contraction. Because there are two weak regions of myosin molecules which are sites of proteolytic enzyme attack, it is possible that myosin can bend at these two places. If the area between light and heavy meromyosin regions and the area in heavy

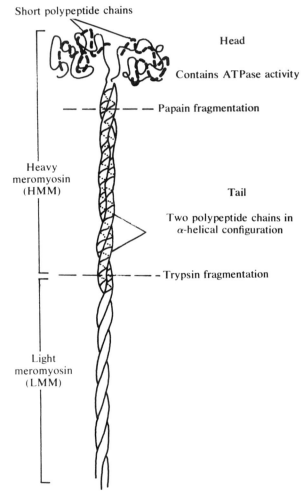

**Figure 7.8**
The myosin molecule. See text for details.

meromyosin between the globular head and the remainder can provide some flexibility to myosin molecules, then bends could occur. These bends could be involved in the interactions between myosin cross-bridges and actin filaments (Fig. 7.9).

About 80 percent of myofibril protein is myosin and actin. The remaining 20 percent consists of four proteins that are all associated with actin filaments: **α-actinin, β-actinin, tropomyosin,** and **troponin.** Tropomyosin accounts for more than 10 percent of myofibrillar protein, and troponin for about 5 percent. Both these proteins have a role in regulating contraction.

Before considering these regulatory proteins we need to have some idea of the stimulus, chemistry, and cellular contributions to episodes of muscle contraction. Afterward these factors can be related to the regulatory processes by which contractions are switched on and off in muscles by control systems.

### Coupled Excitation and Contraction

The concentration of ATP in muscle is relatively high during the resting phase. This observation indicates that ATP itself cannot be the factor that initiates and terminates contraction because it would be present in low amounts in resting muscle if its depletion during contraction was the particular signal to shut off contraction. The stimulus that triggers contraction is produced when an electrical impulse from a nerve is received by the muscle and spreads over the sarcolemma. Since there is a higher positive electrical charge outside than inside the sarcolemma, a potential difference exists across this membrane. As the impulse spreads over the sarcolemma, this transmembrane potential difference suddenly disappears; this phenomenon is called **depolarization.** Depolarization takes place because the transmembrane potential is discharged as the membrane suddenly becomes permeable to cations such as $K^+$, $Na^+$, and $Ca^{2+}$. The flow of these ions through the membrane leads to depolarization. The extremely rapid spread of the electrical impulse is communicated almost instantly to all the myofilaments of a muscle fiber, since they contract simultaneously. The simultaneous contraction occurs even though some myofibrils may be 50

$\mu$m deep within the fiber. Simple diffusion of some chemical from the sarcolemma to myofibrils did not adequately explain this phenomenon because the process of diffusion is too slow. The explanation of rapid and instantaneous communication of the impulse finally was discovered in the 1960s. Using electron microscopy, it was found that the sarcolemma is repeatedly invaginated all around the muscle fiber, in such a way that the tubular infoldings of the membrane make contact with most of the muscle myofilaments (Fig. 7.10). This **T system** of transverse tubular infoldings undergoes depolarization, along with the rest of the sarcolemma, upon excitation by an incoming electrical impulse. The T system explains how an impulse can be communicated almost simultaneously to all the sarcomeres of a muscle fiber.

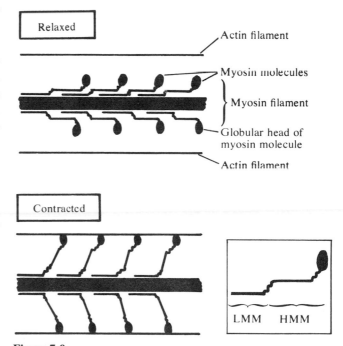

**Figure 7.9**
Postulated mode of cross-bridging during muscle contraction. The "weak" sites (attacked by trypsin and papain) may provide flexibility to the whole myosin molecule. Bending in these two sites may account for alternating attachment and detachment of myosin heads during contraction and relaxation episodes.

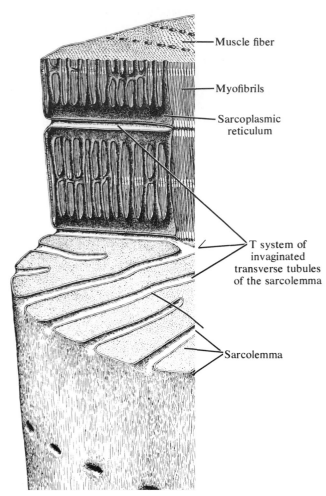

Muscle fiber

Myofibrils

Sarcoplasmic
reticulum

T system of
invaginated
transverse tubules
of the sarcolemma

Sarcolemma

**Figure 7.10**
Three-dimensional reconstruction of part of a muscle fiber
showing T system of transverse tubular invaginations of the
sarcolemma. (From "The Sarcoplasmic Reticulum" by
K. R. Porter and C. Franzini-Armstrong. Copyright © 1965
by Scientific American, Inc. All rights reserved.)

The electrical impulse is translated into chemical or
molecular changes in the myofibrils. This translation is
mediated by the sarcoplasmic reticulum which en-
velops the myofibrils and is closely apposed to them.
Depolarization of the sarcolemma leads to an increase
in permeability of the sarcoplasmic reticulum membrane system, as the change in electrical charge is

transmitted through the sarcolemma T system to the
cisternae and membranes of the reticulum. Calcium
ions, which are stored in the cisternae of the reticulum
in resting muscle, then escape rapidly into the sarcoplasm. Once in the sarcoplasm, $Ca^{2+}$ ions trigger the
interaction of ATP with the actin and myosin fila-
ments. It has been estimated that $10^{-6}$ to $10^{-5}$ M $Ca^{2+}$
is sufficient to initiate a contraction. This concentration of $Ca^{2+}$ is required for activity of myosin ATPase
in the sarcoplasm. In resting muscle only $10^{-7}$ M $Ca^{2+}$
ions are thought to be present in the sarcoplasm,
which is an insufficient concentration for enzyme
activity. $Ca^{2+}$ ions probably are "pumped" back from
the sarcoplasm to the sarcoplasmic reticulum during
muscle relaxation, by an enzymelike molecule within
the reticulum membrane. This "relaxing factor" is
ATP-dependent and can transfer $Ca^{2+}$ ions across the
membrane against a $Ca^{2+}$ concentration gradient, using the free energy generated by ATP hydrolysis. On
receipt of the next electrical impulse, $Ca^{2+}$ once again
is released from its storage vesicles within the
sarcoplasmic reticulum, enters the sarcoplasm and
another cycle of contraction occurs.

Studies using inhibitors of glycolysis and respiration have shown that muscle contraction can be stimulated repeatedly even if metabolic formation of ATP is
prevented. Since there is not enough ATP in muscle to
underwrite contraction and since this amount does not
decrease during contractions, some other high-energy
compounds must contribute to the process. In vertebrate striated muscle, **phosphocreatine** is present in
nearly 5 times the concentration of ATP. Dephosphorylation of phosphocreatine by the enzyme
**creatine phosphokinase** maintains the cellular concentration of ATP at steady levels. Under conditions
known to occur in the sarcoplasm, the reaction leading
to ATP formation is favored. In this way, the high-
energy phosphoryl group is transferred from phosphocreatine to ADP, forming ATP.

$$\text{phosphocreatine} + \text{ADP} \xrightleftharpoons{\text{creatine phosphokinase}} \text{creatine} + \text{ATP}$$

Since the terminal phosphoryl group lost from ATP

during muscle contraction is rapidly replenished at the expense of phosphocreatine, the sarcoplasmic concentration of ATP remains constant. This level of ATP will decrease in poisoned muscles only after the phosphocreatine supply is exhausted and no new ATP is synthesized in glycolysis, respiration, or by creatine phosphokinase action. Various feedback controls regulate ATP formation in metabolic pathways, and these controls influence muscle contraction through the dependency of contraction on ATP.

### Troponin and Tropomyosin: Regulatory Proteins

Cycles of contraction and relaxation in vertebrate muscle are regulated by the intracellular concentration of free $Ca^{2+}$ ions, under sarcoplasmic reticulum control. The *responsiveness* to $Ca^{2+}$ ions, however, involves participation of a **tropomyosin–troponin system** before myosin and actin can interact in response to $Ca^{2+}$ ions introduced from the sarcoplasmic reticulum. Tropomyosin and troponin molecules are precisely arranged along the actin thin filaments (Fig. 7.11).

In effect, troponin is a $Ca^{2+}$-dependent switch for muscle contraction. Troponin detects the signal for muscle contraction because it is able to bind $Ca^{2+}$ ions, which flood into the sarcoplasm after the electrical impulse from the nerve has been received. The troponin switch turns muscle contraction on through a physical interaction between troponin and tropomyosin molecules on the actin filament. In the relaxed state, tropomyosin literally blocks the actin sites required to bind myosin heads. When troponin binds a sufficient amount of $Ca^{2+}$ ions, it actually "pushes" tropomyosin away from myosin and closer inward toward the groove between the two actin chains of an actin filament. Once tropomyosin moves away, actin sites are unblocked, myosin heads make contact with them, ATP can now be hydrolyzed, and muscle contraction occurs. With decreasing $Ca^{2+}$ levels, tropomyosin can move back to the original blocking sites because troponin no longer interferes. Since tropomyosin binds to troponin through one of the troponin subunits, the two regulatory proteins act coordinately in controlling muscle contraction, relative to $Ca^{2+}$ ion concentrations in the sarcoplasm surrounding the myofilaments.

The troponin-tropomyosin control system has been found in every animal species examined so far, except for molluscs (clams, scallops, and others). On the other hand, the system discovered in molluscs seems to be absent in vertebrates, but present in nonvertebrate animals. The mollusc regulatory system has no troponin. In all invertebrates, $Ca^{2+}$ ions trigger muscle contraction by direct interaction with myosin. The $Ca^{2+}$-stimulated myosin then interacts with actin filaments. The actin-linked and myosin-linked muscle contraction control systems have many features in common, such as dependence on appropriate $Ca^{2+}$ ion concentrations. They differ mainly in whether the $Ca^{2+}$ effect (mechanism by which filament interaction is prevented) is regulated at the actin filaments or at the myosin filaments, and in whether troponin is present. It seems, therefore, that all animals have the actin-linked regulatory system, except for molluscs, and all invertebrates have a myosin-linked system that is absent in vertebrate species.

Now that we have examined the bases for muscle contraction we can proceed to look at other and simpler movements. Unlike muscle, in which microfilaments are the underlying components for movement, some other systems are based on microtubular

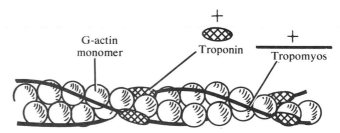

G-actin monomer

Troponin

Tropomyos

**Figure 7.11**
Actin filaments in vertebrate striated muscle are composed of two helically wound rows of G-actin monomers. The regulatory proteins troponin and tropomyosin occur in a regular association along the actin filament.

components. After we have examined some of the microtubular systems of movement we will discuss systems in which both microfilaments and microtubules contribute to cellular motions.

## CENTRIOLES

**Centrioles** appear as small granules when seen by light microscopy, and structural details are not visible at these magnifications. The granules are associated with spindle formation during nuclear division, and they can be recognized by their behavior, location, and small size. When these granules were examined by electron microscopy in the 1950s, it was realized for the first time that centriole structure was essentially identical in various cell types. Centrioles were about 160 to 230 nm in diameter but could vary in length from 160 to 5600 nm. Since the limit of resolution of the light microscope is about 200 nm, it is easy to see why centrioles were not detected in some kinds of cells before electron microscopy was developed. The centrioles would be especially difficult to recognize if they were not in their usual location at the poles of a cell, which is the situation in some cells at some developmental stages.

When seen by electron microscopy in thin sections of cells, there are 9 sets of tubule triplets in the cross-sectional view of a centriole (Fig. 7.12). Each of the 27 (9 × 3) components is a microtubule about 25 nm wide. Delicate strands appear to connect the triplet groups to each other, and other fine fibrils can often be seen radiating from the central hub of the cylinder to the innermost subfiber of each of the 9 sets. This "cartwheel" configuration is not always present, but if it does occur it is usually localized to the denser proximal end of a centriole. In longitudinal section, the cylinder is seen as a heavy-walled structure with a somewhat denser proximal region in its otherwise translucent and amorphous center (Fig. 7.13).

Centrioles have been found in all eukaryotes except for species which never have a flagellated cell type at any time in development. Species lacking centrioles include certain amebae, unicellular red algae, highly evolved gymnosperms such as pines and their relatives, and all flowering plants. This correlation is

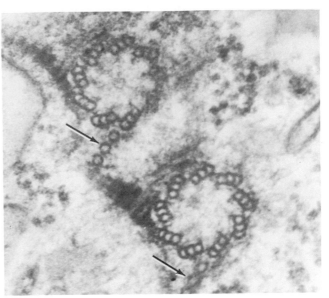

**Figure 7.12**
Cross-sectional view of centrioles with the nine sets of triplet tubules in *Paramecium*, a ciliated protozoan. The "cartwheel" is faintly visible. Microtubules in the nearby cytoplasm (at arrows) are the same in cross-section as the circular microtubules of centrioles. × 105,000. (Courtesy of R. V. Dippell).

interesting in view of the structural identity of the conventional centriole, which acts as a mitotic center, and the **basal body,** from which a cilium or flagellum is produced. All centrioles and basal bodies have the same fundamental 9 subfiber-triplet organization, but the whole organelles differ in their activities and in their locations in the cell. Because they are structurally identical and because a centriole may later become a basal body or a basal body may detach and become a centriole in action and location, the two are considered to be manifestations of the same kind of organelle. A single name identifies the organelles, the preferred term being centriole. We still refer to basal bodies, however, because it is often convenient to use this synonym.

### Centriole Formation

New centrioles usually form in the presence of preexisting centrioles. There are two particular reasons,

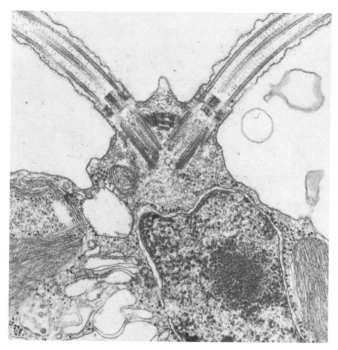

**Figure 7.13**
Longitudinal section of apical portion of *Chlamydomonas reinhardi*. A heavy-walled centriole subtends each of the two flagella that extrude from the cell surface. × 45,000. (Courtesy of U. W. Goodenough and R. L. Weiss)

however, why we do not consider centrioles to be self-replicating structures: (1) the existing and forming centrioles are never attached, but are always separated by a space from the time the new centriole can be detected until it is mature; and (2) centrioles may form in various kinds of cells even when there is no preexisting centriole present. Since centrioles may arise *de novo,* there is no apparent template or parent structure to transfer information and, therefore, no way that we can see for existing centrioles to direct the formation of new centrioles. This is quite different from chromosomes, mitochondria, and chloroplasts, which transmit coded information from preexisting to newly-forming structures, primarily through transfer of replicated DNA during new structure formation. In this connection, we should note that centrioles are not known to have their own DNA.

Whether a centriole arises *de novo* or in the presence of a mature centriole, the sequence of its formation is essentially the same. The first event in centriole formation is the appearance of a circle of 9 singlet fibers that become the inner *A*-subfibers. The *B*-subfibers assemble onto the *A* singlets, and the subfiber triplets are completed when the *C*-subfibers assemble on the *B* set (Fig. 7.14). Once this immature centriole, or **procentriole,** has formed, it grows in length, but usually maintains its original diameter while it matures. According to interpretations of centriole formation from electron microscopy, whether *de novo* or not, it seems that procentrioles develop from various building components by a self-assembly mechanism.

### Centriolar Functions

Centrioles are associated directly or indirectly with two sets of developmental events: (1) formation of cilia and flagella, and (2) formation of a spindle during divisions of the cell nucleus. Microtubules are a principal component in both systems.

FORMATION OF CILIA AND FLAGELLA FROM CENTRIOLES. Eukaryote **cilia** (shorter appendages) and **flagella** (longer appendages) are produced only from centrioles positioned at the cell periphery. These centrioles may have been formed near the cell surface, or they may have migrated there from some place of origin deeper within the cell.

In a very careful study by I. Gibbons and A. Grimstone reported in 1960, the ultrastructural relationship between centriole and flagellum was described as follows. The 9 subfiber triplets in the centriole extend to a transitional region just below the shaft of the flagellum. At some point in this zone near the flagellar shaft, the *C*-subfibers disappear. The remaining *A*- and *B*-subfibers continue as doublets to a place just below the tip of the flagellum (Fig. 7.15). At the same transition region where the *C*-subfibers end, a new pair of microtubules appears in the center of the ring of 9 subfiber doublets in the flagellum. This central pair of microtubules together with the encircling microtubule doublets make up the so-called "9 + 2" organization found in almost all eukaryotic cilia and flagella. The developmental sequence and continuity

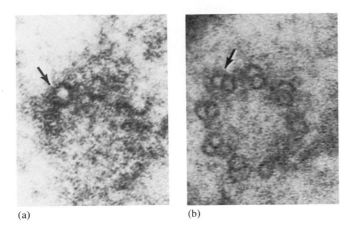

(a)　　　　　　　　(b)

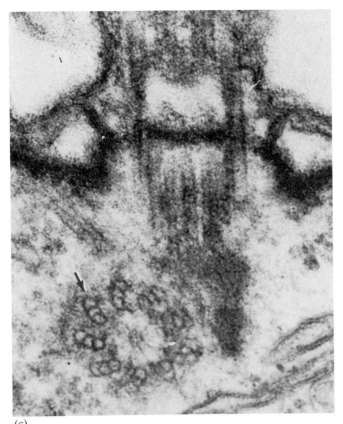

(c)

between centriole and cilium or flagellum has been described in other species and therefore appears to be a common plan of organization.

There is more than enough evidence to state that each cilium and flagellum forms directly from one existing centriole. For example, (1) cilia and flagella never form in the absence of a centriole; (2) if the centriole remains intact, a new cilium will be regenerated when a previous one has been removed; and (3) if the centriole has been destroyed or removed, the flagellum will degenerate and a new flagellum will not form unless a new centriole replaces the one that was removed.

Centrioles display **polarity.** New centrioles form only at the proximal end (farthest from the cell surface) of a mature centriole, while flagellar outgrowth occurs at its distal end (nearest to the cell surface). The organizing events are different in each of these two occurrences. While centriole and flagellum form a continuous structure, mature centriole and procentriole are always separated by a space of about 60–80 nm (Fig. 7.16).

PARTICIPATION OF CENTRIOLES IN NUCLEAR DIVISIONS. Centrioles are associated with microtubular displays called the **aster** and the **spindle** in dividing cells (Fig. 7.17). Centrioles are *not* essential for spindle formation or for nuclear division, according to many lines of evidence and routine observation, including the fact that organisms which never have centrioles are still capable of spindle formation during nuclear divisions. All the flowering plants fall in this category.

In an experimental study using animal cells with

**Figure 7.14**
Stages in development of a new centriole at the proximal end of, and at right angles to, the existing centriole in *Paramecium*. (a) Singlet A-subfibers appear first. × 104,000. (b) B-subfibers assemble onto the initial A-set. × 120,000. (c) Triplet sets are completed when the C-subfibers assemble onto the B-subfibers. × 120,000. (Courtesy of R. V. Dippell)

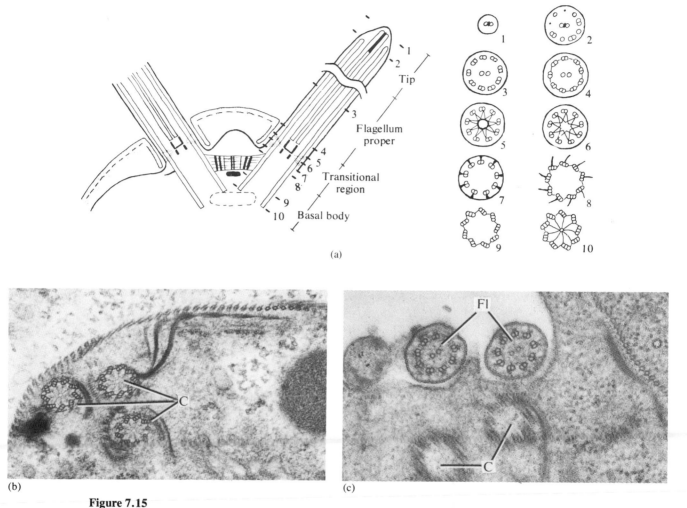

(a)

(b)

(c)

**Figure 7.15**
Centrioles and flagella in unicellular organisms. (a) Schematic drawing of an idealized
longitudinal section through the apex of a *Chlamydomonas* cell showing both centrioles
and flagella. Four regions of the centriole–flagellum system are shown. A series of cross-
sectional views at ten sites along this system shows the subfiber numbers and patterns.
(Reproduced with permission from Ringo, D. L., 1967, *J. Cell Biol.* **33**:543, Fig. 30.) (b)
Electron micrograph of three centrioles in the protozoan *Hypotrichomonas acosta*,
showing a distinct "cartwheel" interior in each. × 49,400. (c) Cross-sectional view of the 9
+ 2 microtubule architecture of flagella of the protozoan *Pentatrichomonas*. × 56,000.
(Both photographs courtesy of C. F. T. Mattern and B. M. Honigberg)

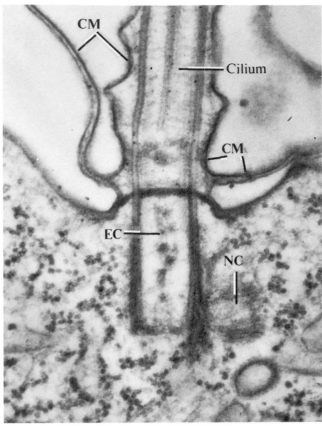

**Figure 7.16**
Centriole formation in *Paramecium*. The new centriole (NC) forms at right angles to the existing centriole (EC), which is attached to the cilium. A distance of about 70 nm separates the new and existing centrioles at all times. The cell membrane (CM) is continuous around the cell periphery and each cilium of the cell. × 78,000. (Courtesy of R. V. Dippell)

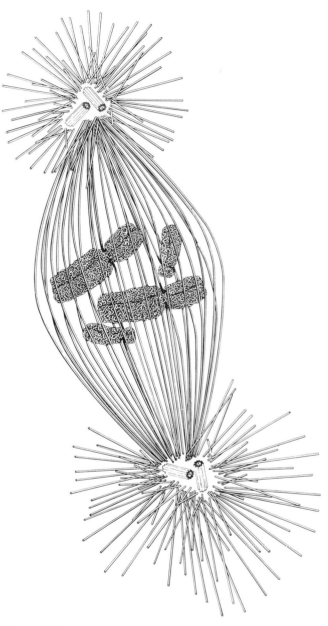

**Figure 7.17**
An interpretation of mitotic apparatus ultrastructure. A pair of centrioles occupies a clear zone surrounded by short microtubules of the aster at each pole of the spindle. Some microtubular spindle fibers extend from pole to pole, while others extend from one pole to their site of attachment on the chromosome. (Reproduced with permission from DuPraw, E. J., *Cell and Molecule Biology*, 1968, Academic Press. Copyright © 1968.)

typical centrioles, asters, and spindle, the cells were manipulated so that an aster developed only at one pole of the cell and not at the other. Spindle formation continued normally even though one pole lacked an aster and also lacked centrioles. At the end of the division, one daughter cell had centrioles and aster while the other daughter cell lacked these components. In the next division cycle, the cell lacking centrioles and an aster was still able to produce an essentially complete spindle.

Many lower organisms have centrioles that never

participate in nuclear divisions, but act only in ciliogenesis (formation of cilium or flagellum). In other species, however, centrioles may first act as a mitotic center and later migrate to the cell periphery where they produce a motility organelle. These and other observations lead to the conclusion that centrioles are essential for ciliogenesis but not for spindle formation, even in cells where centrioles regularly participate in nuclear divisions. It is premature to speculate at present on the relationship between spindle microtubules and centrioles.

## CILIA AND FLAGELLA

Cilia and flagella are specialized surface structures whose movements propel a cell through a liquid medium. Some ciliated cells are not motile, and in these cells, cilia sweep materials past the cell surface. **Flagella** are long whiplike appendages measuring over 150 $\mu$m in some cases, while **cilia** average 5–10 $\mu$m in length (Fig. 7.18). Cilia occur in relatively large numbers per cell. Flagella are usually less numerous, often only 1 or 2 per cell; but there may be thousands in certain protozoa and some other cell types. As we

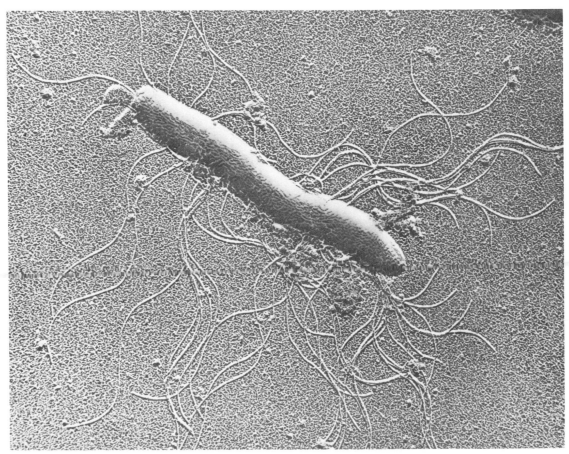

**Figure 7.18**
Electron micrograph of *Bacillus*, a flagellated bacterium. Numerous flagella are evident.
(Reproduced with permission from Brown, W. V., and Bertke, E. M., *Textbook of Cytology*, 2nd ed., St. Louis, 1974, The C. V. Mosby Co. Courtesy Dr. J. Swafford, Laboratory of Electron Microscopy, Arizona State University, Tempe, Arizona.)

mentioned earlier, cilia and flagella have the same ultrastructure and are both formed from centrioles.

The movement patterns of cilia and flagella are generally distinctive. Cilia beat coordinately in groups or rows of individual hairs, in a rowing-like motion. The beat is transmitted progressively along a row of cilia, not synchronously within a group. Ciliary beating is very similar in cells as different as protozoa and vertebrate epithelial tissue. The major differences are in the form and magnitude of the stroke in the beating pattern. Flagella undulate in a complex pattern apparently based on a progressive, sharp transition from straight to curved areas along the length of the structure. Sperm tails are the most frequent material examined in studies of flagella.

### Structure and Chemistry

In addition to electron microscopic studies of ciliary structure, biochemical analysis can be made using cilia and flagella detached from cells by mechanical agitation. Such isolated motility organelles can be used in motion studies, in addition to examining whole moving cells.

The membrane covering can be removed by detergents, and the remaining fibrous and matrix materials of the **axial shaft** of cilia and flagella can be collected in separate fractions by centrifugation. It is entirely possible to isolate purified preparations of subfiber doublets, singlet microtubules from the center of the axial shaft, and other proteins (Fig. 7.19). After chemical analysis, each component can be related to ciliary movements.

Three particular protein components of cilia have so far been identified and named. The major protein of the doublet and singlet microtubules is called **tubulin.** Tubulin resembles muscle actin in amino acid composition, but differs in several important features: (1) tubulin is not responsive to $Ca^{2+}$ ions, while actin is; (2) tubulin binds GTP (guanosine triphosphate), while actin binds ATP; and (3) the two proteins are immunologically unrelated as shown by the failure of each protein to cross-react with the other's antiserum. Tubulin is the microtubule monomer protein.

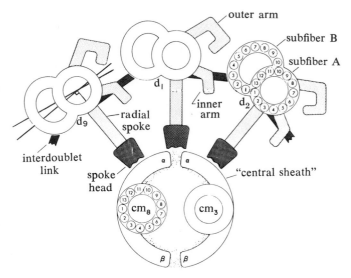

**Figure 7.19**
Diagram of a cross section through the 9 + 2 axoneme or axial shaft of a cilium or flagellum as viewed from the base toward the tip of the structure. Only three of the nine doublets are shown here, or one-third of the circular display of a complete cross-sectional view of an axoneme. See text for details. (Reproduced with permission from Warner, F. D., and P. Satir, 1974, *J. Cell Biol.* **63**:35-63, Fig. 3.)

A second kind of protein, called **dynein,** has been extracted from purified subfiber doublets. This protein is a $Mg^{2+}$-activated ATPase, and it is an important molecule, since it sponsors hydrolysis of ATP and therefore makes energy available for ciliary movement. (Muscle ATPase is $Ca^{2+}$-dependent.) Dynein has been localized to the arms of subfiber-*A*. When dynein is extracted from purified doublets, the arms disappear; when dynein is added back, they reappear.

A third protein, **nexin,** is located in the links between subfiber doublets. There are other proteins that have been identified in general by their sensitivity to digestion by trypsin and similar enzymes, but few other details are available. One other feature that is important to verify is the tentative localization of ATPase activity in a second place within the axial shaft; specifically, in association with the thickened knobs at the ends of the radial spokes that connect subfiber doublets to the central sheath surrounding the two singlet microtubules. This location is pertinent to the postulated mechanism of ciliary movement.

### Ciliary Movement: Sliding Microtubule Mechanism

Cilia can move even after they are detached from cells; the cells themselves stop moving. Amputated cilia continue to swim until their ATP supply is exhausted. These and previous observations allow us to draw at least three conclusions:

1. Cells are moved by their cilia (or flagella), which are the primary organs of motility.
2. ATP is the energy source and its hydrolysis by ATPase makes free energy available for ciliary movement.
3. Ciliary fibers probably serve as the contractile machinery for locomotion.

We may also infer that the source of ATP is within the cell rather than in the cilium itself, otherwise ATP would be made in isolated cilia and they would move for long periods of time. The ATP supply must therefore be cut off when cilia are detached from cells.

When isolated cilia are suspended in glycerol in the cold for a long time (a method called glycerination), ATP and other soluble components are removed, but the structural integrity of the cilia is maintained. If ATP is supplied to such glycerinated cilia, they are reactivated. The frequency of ciliary beating is directly proportional to the amount of ATP provided, showing that energy for movement is obtained by ATP hydrolysis. The muscle contraction system works in a similar way.

There is no apparent effect on movement of glycerinated cilia if $Ca^{2+}$ ion levels are severely reduced, showing that the ATPase is insensitive to this ion. Since myofibril contraction is inhibited when $Ca^{2+}$ ions are removed, there must be somewhat different contractile mechanisms underlying muscle contraction and ciliary movement.

Recent studies by Peter Satir and others have led to the proposal that ciliary movement is based on a sliding microtubule mechanism, which resembles sliding filament models for muscle contraction in certain ways (Fig. 7.20). From exacting measurements of electron micrographs showing cuts through different regions of bent and relaxed cilia, it has been proposed that ciliary subfiber doublets must slide past one another without contracting. As the microtubule doublets slide past one another, their movement causes a localized bend of the cilium. Doublet sliding results from cycles of attachment and detachment of the ATPase-containing dynein arms to the *B*-subfiber of the adjacent outer tubule doublet. This mechanism resembles sliding filaments in muscle where actin slides over myosin filaments by cyclic attaching and detaching of myosin heads to actin filaments. In cilia, interdoublet *sliding* is apparently converted into ciliary *bending* by cyclic interactions between the radial spokes and the central sheath. Individual tubules do not change in length or width, which shows that they themselves do not contract. Similar observations showed that muscle myofilaments remained constant in length and width, but sarcomeres contracted as sliding took place.

In addition to ATPase activity of the dynein arms of *A*-subfibers, it is important to know whether there actually is some ATPase activity as has been reported to occur at the swollen ends of the radial spokes where they make contact with the central sheath. It seems reasonable to expect ATPase activity there, since energy must be needed for intermittent interactions with the central sheath.

The sliding microtubule model satisfies conditions for movement in conventional 9 + 2 cilia and flagella and could provide a simple explanation for nonmotility of modified cilia with a 9 + 0 pattern. Without the central pair of microtubules and the surrounding central sheath in 9 + 0 cilia, there is no place for radial spokes to undergo attachment-detachment cycles. There are other known modified fiber patterns, however, and it remains to be seen whether the sliding microtubule mechanism can explain ciliary movement in various types of modified cilia and flagella as neatly as it explains the 9 + 2 structural pattern.

The most likely source of ATP for ciliary movement are mitochondria, which are usually located very close to ciliary basal bodies (centrioles). ATP could easily be transferred from neighboring mitochondria to the motility organelles. Sperm tails generally have a modified mitochondrion wrapped around the region between the head and tail of the sperm cell (Fig. 7.21). Sperm tails stop moving when their ATP is exhausted

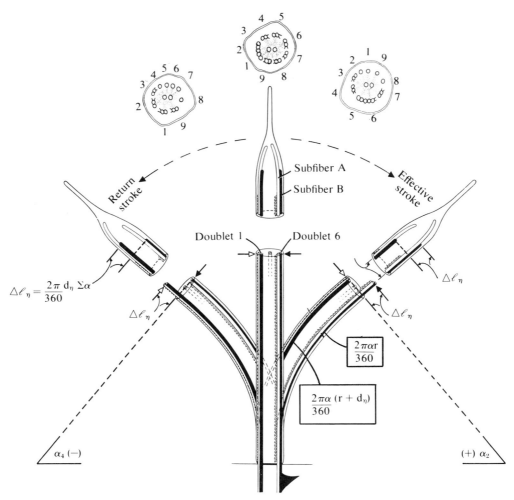

**Figure 7.20**
Diagrammatic illustration of the sliding microtubule hypothesis of ciliary motion. The behavior of subfiber doublets numbers 1 and 6 provides information relating microtubule sliding to cilium bending. Subfiber doublets are shown in cross section near the tip of the cilium. The cilium is in a straight neutral position, bent in the direction of the effective stroke and bent in the return-stroke direction. In the neutral position subfibers-B of all 9 doublets are visible, indicating these terminate at the same level across the tip of cilium. In the bent ciliary positions, subfiber-B of doublet 1 or of doublet 6 (as well as others) is missing from the outer side of the bend. Since the microtubules are flexible, these cross-sectional views indicate that the microtubule doublet on the inner (concave) side of a bend must slide tipward. When microtubules slide past one another, shear resistance in the cilium changes sliding to bending. Displacement at the tip ($\Delta\ell\eta$) can be measured using geometrical formulae based on characteristics of the arc produced by bending of the cilium. (Reproduced with permission from Satir, P., 1968, *J. Cell Biol.* **39**:77.)

or if ATP synthesis is inhibited when glycolysis and respiration are poisoned.

### Modified Cilia and Flagella

It was mentioned earlier that sperm tail ultrastructure varied from the conventional eukaryotic 9 + 2 pattern in some species. Many animal sperm have a 9 + 2 pattern of doublets and singlets but also contain another 9 components that are called **accessory fibers** (Fig. 7.22). These longitudinal components lie outside the ring of doublets and may be dense, massive structures. They are particularly prominent in mammalian sperm, but also occur in some other vertebrates and invertebrate groups. Little is known of the origin or function of accessory fibers, but they are usually found in species characterized by internal fertilization. Because of this correlation, studies were conducted to see if accessory fibers increased sperm motility in the viscous environment of the female reproductive tract. Evidence that accessory fibers contain an ATPase and a protein similar to one known in striated muscle has provided support for the suggestion that these structures make an essential contribution to sperm motility.

Specialized cells that respond to environmental stimuli such as light, sound, and odor are called **sensory receptors.** A modified cilium, subtended by a centriole, is found in many but not all of these cell types (Fig. 7.23). The central pair of singlet microtubules are usually missing or are extremely short in the ciliary shaft. Visual receptor cells of both rod and cone types have a ciliary structure that connects the outer and inner segments of the cell. A ring of 9 doublets encircles a hollow center where the singlet tubule pair would otherwise be located. In vertebrate visual

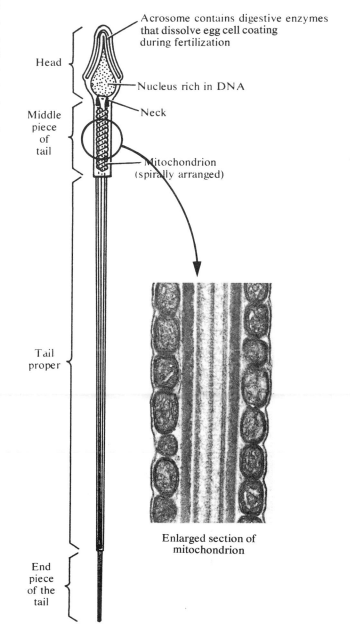

Acrosome contains digestive enzymes that dissolve egg cell coating during fertilization

Head

Nucleus rich in DNA

Neck

Middle piece of tail

Mitochondrion (spirally arranged)

Tail proper

End piece of the tail

Enlarged section of mitochondrion

**Figure 7.21**
Electron micrograph of longitudinal section through the middle piece of the tail of a mammalian spermatozoan showing cut views of the spirally shaped mitochondrion (note cristae). The drawing of a spermatozoan provides reference for the parts of the cell and the location of the mitochondrion. (Photograph reproduced with permission from Bloom, W., and Fawcett, D. W., *A Textbook of Histology*, 10th ed., 1975, W. B. Saunders Co., Philadelphia.)

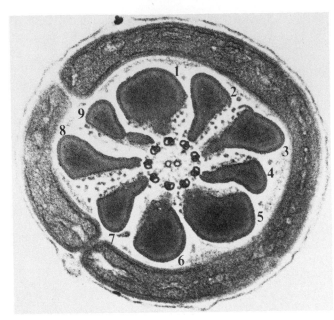

**Figure 7.22**
Electron micrograph of guinea pig sperm in cross section. The nine dense accessory fibers surround the typical 9 + 2 tubules of the sperm tail. This section was taken from the middle piece of the spermatozoan. × 66,400. (Courtesy of D. W. Fawcett)

receptors there may be up to 1000 membrane-limited discs arranged transversely in the outer segment. The discs are apparently formed from extensively folded ciliary membrane, which is derived from the cell plasma membrane just as in any cilium. The folded membrane system in visual receptors provides increased surface area for the conversion of light into electrical impulses. This represents an evolutionary advance in vertebrate animal eye organization.

### Bacterial Flagella

The bacterial version of a flagellum is very different from the 9 + 2 eukaryotic organelle. In bacteria each flagellum is made up of 2 to 5 individual filaments which are only 4–5 nm wide but up to several $\mu$m long. The width of the flagellum (about 20 nm) depends entirely on the bundle of fibrils. Each filament in a flagellum is made up of a single kind of protein called **flagellin.** Globular monomers of flagellin aggregate end

to end to form a filament, like beads on a string. No membrane encloses the flagellum, except in spirochetes, which have an unusual sheathing as well as very different internal components.

Bacterial flagella grow at the tip. The bacterial appendage is attached to a granule embedded within the protoplast, but it bears no resemblance to a centriole or other known structure (Fig. 7.24). Flagellar insertion can be readily seen in electron micrographs; flagella remain attached to a protoplast after the bacterial cell wall has been removed. A bacterium is pro-

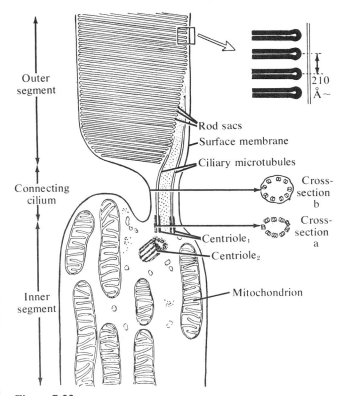

**Figure 7.23**
Diagram of a mammalian retinal rod cell showing the centriole pair associated with the modified cilium of the outer segment, which contains rod sacs that are transversely arranged membranous discs formed from the folded ciliary membrane. The cross-sectional views through the cilium show (a) the usual subfiber triplets of a centriole and (b) nine subfiber doublets of a ciliary axoneme, but lacking the central pair of microtubules.

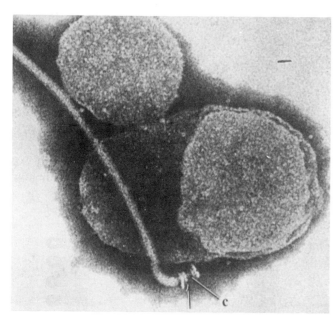

**Figure 7.24**
Negatively-stained preparation of lysed cells of the
bacterium *Rhodospirillum molishianum*. The basal portion
of the flagellum consists of two pairs of discs connected by a
narrow collar (c) to each other. Globular subunits make up
discs, as shown in one of the inner discs (arrow). × 160,800.
(Reproduced with permission from Cohen-Bazire, G., and J.
London, 1967, *J. Bacteriol.* **94**:458-465, Fig. 3.)

pelled through liquid medium by rotation of its
flagellum around its base, unlike the undulating motion
of eukaryotic flagella.

Special staining methods are required to see 20 nm-
wide flagella in the light microscope where the limit of
resolution is 200 nm. A cementing substance is added
first to increase the flagellum width, and stain is then
applied. The organelles require no special treatment to
be visible by electron microscopy where 1 nm resolution is easily achieved.

Purified flagellin can be obtained from flagella,
which are easily broken off when cells are agitated. As
the pH of the medium is varied, globular monomers of
flagellin readily aggregate and disaggregate in what is
obviously a self-assembly process. Fibrils form spontaneously under proper conditions, but there must be
an organizing site in the structural components in

order to produce long polymers. Monomers of flagellin
have a molecular weight of 20,000 to 40,000 daltons,
which is lower than the weight of eukaryotic tubulin
(50,000 to 60,000 daltons).

Flagellin and muscle actin are chemically similar in
many ways, especially in having a high proportion of
nonpolar amino acids such as aspartic and glutamic
acids. No ATPase activity is associated with bacterial
flagellar protein, however, while eukaryotic myosin
and dynein provide ATPase for motility directly in the
structural system which moves.

## THE MITOTIC APPARATUS

Chromosome movement during nuclear divisions
will not take place unless there is an organized system
of microtubules to form a **spindle.** In addition to the
prominent **spindle fibers,** the mitotic apparatus includes other components in animals and many other
groups of organisms. An **aster** is found at each pole.
The aster surrounds a pair of **centrioles,** but is
separated from them by a clear zone called the
**centrosome.** The spindle and astral fibers are similar in
construction, and neither set of microtubules actually
touches the centrioles. Species lacking asters have a
spindle figure that is different in shape from astral
types. Where there are no asters, the spindle fibers are
arranged in a cylindrical pattern with a bulging
midline, rather than the more spindle- or lens-shaped
system which is tapered at the poles (Fig. 7.25) Although it is called a "mitotic" apparatus, similar
fibrous configurations are present in meiotic division
stages.

### Structure

The detailed structural organization and composition of the aster was not described until it was studied
by electron microscopy. Many components are below
the limit of visibility of the light microscope, but the
centrioles can be seen as a barely visible pair of
granules separated by a clear zone from radiating
astral fibers. In electron micrographs it is possible to
see the precise tubular construction of the centrioles,
the microtubules that form the astral rays, and various

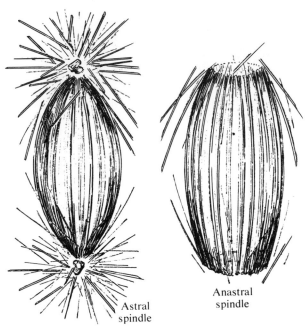

**Figure 7.25**
Comparative aspects of astral and anastral spindles.

structural components within the "clear" centrosome region. Ribosomes and small vesicles are dispersed among the astral microtubules and within the centrosome.

The spindle fibers are conventional microtubules about 18–22 nm wide and many micrometers in length. There may be as many as 3000 microtubules in a spindle figure. The wall of the cylindrical hollow tubules is made from globular **tubulin** monomers arranged into 13 parallel filaments that run the entire length of the microtubule (see Fig. 7.1). According to chemical studies, each tubulin monomer has two binding sites normally occupied by guanosine triphosphate (GTP). In the presence of the drug colchicine, GTP is displaced and its binding site is taken by the drug. These tubules resemble the central pair of singlets in cilia in their construction and their sensitivity to colchicine.

Microtubule monomer units assemble spontaneously *in vitro* to produce recognizable structures essentially identical to spindle fibers that are isolated from various kinds of cells. Groups of spindle fibers have been assembled *in vitro* and these have been shown to be functional, since chromosomes moved toward the poles of these fiber assemblies. These observations indicate that formation of functional microtubules in organized patterns takes place by a self-assembly mechanism. Additional evidence also points to a self-assembly process; for example, microtubules will form even when the system is poisoned, showing that new protein synthesis is not required.

Polymerization of tubulin monomers is inhibited by the drug colchicine because it binds to monomer units and blocks the binding sites needed to attach to existing tubulin polymers. The polymers, themselves; are insensitive to colchicine. This particular observation has led to suggestions for a mechanism of chromosome movement during anaphase of nuclear divisions.

### Spindle Function

During metaphase of nuclear division, replicated chromosomes congregate at the midpoint of the spindle. The two parts of each replicated chromosome then separate, and each moves to an opposite pole at anaphase of mitosis. The situation is similar in meiosis, at least for chromosome movement to the poles. The separation of sister chromosomes or half-chromosomes is independent of their subsequent anaphase movement. If the spindle is disrupted, chromosomes will separate but will not move to the poles. The result is formation of a **polyploid** nucleus, having twice the number of chromosomes as the premitotic nucleus.

Before the 1950s there were serious doubts that a spindle figure actually existed. It was considered to be an artifact of cell preparation because well-prepared cells often seemed to have no spindle in dividing tissues. Electron microscopy put these doubts to rest, but evidence in support of the existence of a spindle had also come from studies which used a light microscope equipped with polarizing optics. In this optical system, parallel, ordered arrays of fibrous molecules have a brightness due to their property of **birefringence.** Disordered arrays of molecules lack this quality. In the early 1950s, Shinya Inoué showed that birefringence in the spindle region disappeared if colchicine was added to the cell suspension. The birefringence reappeared, however, when colchicine was washed out. Since colchicine had long been known to inhibit anaphase movement of chromosomes and cause the formation of polyploid nuclei, the polarizing microscope studies clearly related chromosome movement to spindle presence and function. The nature of the spindle fibers and their associations with the chromosomes, however, were not clarified until the mid-1960s. At this time important improvements in methods for preparing materials for electron microscopy were developed. As the quality of the images improved, fragile components such as microtubules could be consistently seen to be integral parts of the cell and not artifacts.

Many studies have shown that there are at least two kinds of spindle microtubules (see Fig. 7.17):

1. Those extending from pole to pole.
2. Those attached to a chromosome and terminating near one pole of the spindle.

As early as the 1940s, Hans Ris had produced information from careful measurements which showed that spindle fiber lengths changed during anaphase. The chromosome-connected fibers shortened, at the same time that the length of the entire spindle increased. Any model of anaphase movement of chromosomes must account for these simultaneous but opposite changes in the two kinds of spindle fibers. Electron microscopy has verified the occurrence of these changes in length of individual spindle fibers.

Energy for chromosome movement is apparently derived by ATPase-catalyzed hydrolysis of ATP. A dyneinlike catalyst has been found in preparations of isolated spindle fibers. Using a specific method which reveals the inorganic phosphate released during ATP conversion to ADP and phosphate, the highest levels of enzyme activity (greatest deposition of phosphate) were found near the spindle poles.

### Anaphase Movement of Chromosomes

Early ideas about chromosome movement centered around physical features in the cell, such as electromagnetic field or sol-gel changes. Current notions center on the connections between chromosomes and spindle microtubules and changes these might undergo during anaphase. Three main hypotheses have been proposed to explain the mechanism of anaphase movement: **assembly-disassembly, contraction,** and **sliding microtubules.**

The **contraction mechanism** has the least support. According to this proposal, spindle microtubules would shorten and fold or contract. Instead, the evidence shows that spindle fibers undergo no changes in diameter or in the thickness of the microtubule wall. Another argument against this hypothesis is that contraction fails to explain why pole-to-pole fibers lengthen at the same time that chromosome-to-pole fibers shorten.

The **tubule assembly-disassembly mechanism** was first proposed by Inoué in 1967. It is based on the existence of a pool of tubule monomers and polymerized microtubules of the spindle in an equilibrium mixture. Inoué suggests that chromosomes move toward the poles because spindle fibers disassemble at their poleward end. The disassembly produces shortened fibers and the chromosomes are pulled to the poles. The monomers released from the polar ends of chromosome-connected fibers either return to the cellular pool or add on at the ends of the pole-to-pole fibers, which increases their length. The poles would be pushed farther apart by the lengthening spindle figure made of continuous pole-to-pole microtubules. The "push-pull" system accounts, therefore, for both increasing spindle length and decreasing distance between chromosomes and their respective poles.

A major problem is that the mechanism explains changes in lengths of fibers, but it does not explain how chromosomes move when their fibers are disassembled at the poleward end. It does not necessarily follow that the attached chromosome would be pulled closer to the pole because the fiber shortens at one end. On the other hand, evidence in support of this hypothesis comes from the presence of an active ATPase at the poles, which could liberate the energy to generate the required pulling force.

The **sliding microtubule mechanism** for chromosome movement is derived from analogies with muscle contraction, as is the mechanism for ciliary movement discussed earlier (see Fig. 7.20). According to this hypothesis, chromosomes would move as chromosome-to-pole and pole-to-pole fibers slide past each other. The force required for movement would be generated by alternative breaking and reforming of bridges or chemical bonds during fiber sliding. The spindle fiber system is different in at least one significant way, however, when compared with either the cilium or the muscle systems. No precise spatial relationship between the two kinds of fibers is found in the spindle, while the architecture of the fiber displays in muscle and cilia lend themselves to specific filamentous associations and disassociations.

In some organisms, the chromosome fibers are arranged around the periphery of a "hollow" spindle, while the continuous fibers occur as a separate bundle in the axis of the spindle. In certain protozoa the intact nucleus encircles a bundle of pole-to-pole fibers, while the other spindle fibers extend from each pole to the chromosomes within the membrane-bounded nucleus. Anaphase movement of chromosomes still takes place in these systems even when the two kinds of fibers are in physically separated regions.

While the sliding microtubule mechanism proposed in 1969 by J. McIntosh is an attractive idea, its major advantages are that the mechanism fits the general pattern for ciliary and muscle movements. There is no particular reason to expect all motility systems to have a sliding fiber mechanism. At the present time, therefore, it is difficult to choose between the assembly-disassembly and sliding microtubule hypotheses for anaphase movement of chromosomes.

## MICROFILAMENTS AND NONMUSCLE MOVEMENTS

So far we have been discussing systems that are either entirely or predominantly microtubular *or* microfilamentous in underlying structure. Ciliary and chromosomal movements rely primarily on a microtubular apparatus, while muscle contraction seems to work with a system made of microfilaments. Many other forms of cellular motion have been studied, and in most cases these seem to depend on contributions made by both kinds of fibrillar components. In other cases, the movements take place in the total absence of microtubules. This great variety of structural systems has not yet been placed into a single comprehensive framework. We will consider examples of cellular movements that are under study and then look for the common features, if any, that link together seemingly different expressions of motility.

Actin or actinlike filaments have been found in almost all eukaryotic cells. These fibrils are usually 5–7 nm in diameter, they occur in almost all regions that are capable of movement, and they can be recognized by their specific interaction with heavy meromyosin from muscle. Actin filaments can form polarized complexes with heavy meromyosin to produce a "decorated" or "arrowhead" longitudinal display (Fig. 7.26). The more general term microfilament has been used because the earlier identifications were based on measured dimensions. Chemical and functional criteria have been used more recently to identify microfilaments of different kinds.

In addition to conventional microtubules and microfilaments, other classes of filaments have been described in certain kinds of cells. Myosin thick filaments are one obvious example of a special kind of filament that differs from the conventional actinlike microfilament in all or most cell types. Many kinds of animal cells contain filaments about 8 to 12 nm in diameter. These filaments can be recognized and distinguished by other criteria in addition to width. They show different responses to physical and chemical agents intended to disrupt microfilaments, and their distribution in the cell is distinctive. In some cases a specific name has been applied to these filament types.

All these filamentous components have been related to some movement phenomenon involved in cell locomotion, transport and translocation of organelles and particles within the cell or cell extension, or developmental modifications of cell and organ shape. These relationships have been studied by electron microscopy and other optical methods, effects induced by changes in temperature and pressure, and by responses to **colchicine** and **cytochalasin B** (Fig. 7.27).

Colchicine binds to tubulin monomers and disrupts

(a)

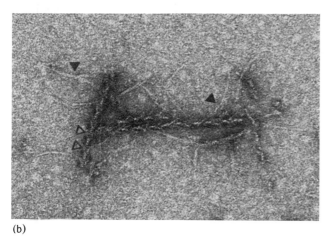

(b)

**Figure 7.26**
Negatively stained actin and actomyosin filaments from the slime mold *Physarum polycephalum*. (a) Purified actin filaments show an "arrowhead" display after reaction with heavy meromyosin from rabbit muscle. × 218,000. (From Nachmias, V. T. *et al.*, 1970, *J. Mol. Biol.* **50**:83-90, Plate Vic) (b) Purified *Physarum* actomyosin filaments are indistinguishable from those obtained from muscle preparations, once "enriched" for myosin by removal of part of the actin in the preparation. Closed arrows point to long myosin "tails" that project off the "arrowheads"; open arrows point to the visible repeat periods, or cross-overs, on the actin filament. × 92,600. (Photographs courtesy of V. T. Nachmias)

microtubules under certain conditions. High concentrations of colchicine do not disrupt doublet microtubules in cilia, but are effective against the singlet tubules and the spindle fibers in dividing nuclei. The different responses may be related to the mode of formation of the several kinds of microtubules and the conditions for their maintenance. If continual

Colchicine
($C_{22}H_{25}O_6N$)

Cytochalasin B
($C_{29}H_{37}O_5N$)

**Figure 7.27**
Structural formulae for the alkaloid drugs colchicine and cytochalasin B.

assembly and disassembly is required to maintain microtubules, colchicine would deplete the pool of available monomers and eventually lead to microtubule disruption. If microtubules reach some stable end point and do not require continued replacement of tubulin monomers, colchicine would have no disruptive effect. At the moment these are only suggestions to explain the differential colchicine effect.

While colchicine has been used for many years to produce polyploid plants with larger flowers or fruits, cytochalasin B only came into widespread use in 1969. Thomas Schroeder reported at that time that cytochalasin B treatment led to disruption of microfilaments in dividing marine eggs, but left the spindle microtubules unaffected. The observations spurred a number of independent studies in various laboratories on the contribution of microfilaments to cell movements, using cytochalasin B as a selective disruptor of these fibrils. Microtubules are insensitive to the drug, but there is some controversy about whether the effects of cytochalasin B are direct or indirect in producing alterations in microfilaments.

### Movements of Protoplasm

All particles exhibit continuous motion in water because they are randomly bombarded by water molecules. The irregular displacements that result from bombardments depend on the relative amounts of kinetic energy of the water molecules which hit particles in suspension. Such random movement is called **Brownian motion,** in honor of Robert Brown who described the phenomenon in 1820. In contrast with undirected Brownian motion, particles, organelles, and the whole cytoplasmic content of cells can move in directed pathways that are far from random. These directed movements are produced by activities of microfilamentous assemblies, or by a combination of microtubules and microfilaments.

SALTATORY MOVEMENTS. Particles suspended in the cytoplasm often undergo directed migrations which result in a displacement of up to 30 $\mu$m in a few seconds. The rapid, directed displacement of particles over long distances is called **saltatory movement.** This form of movement has been studied in several species, with different results and explanations. In the giant fresh-water alga *Nitella,* no microtubules are found in the moving regions of the cytoplasm, yet these are the regions in which jumplike saltatory movements have been observed. Since microfilaments are present, these fibrils have been implicated in saltations. Particles adjacent to each other do not necessarily move together. In some episodes some particles may move at a constant rate, while their immediate neighbors are totally unaffected. Various explanations for the way in which microfilaments produce particle saltatory motions have been offered. The major difficulty with all these hypotheses lies in explaining the means by which the motive force is generated. If microfilaments are attached at one end to some cellular structure or region, they may "whip" around and cause particles to be rapidly displaced in some particular direction. An energy supply, which can be transduced to provide the force needed to activate microfilament movement, must be present, however. Little is known about these matters at present.

PROTOPLASMIC STREAMING. In plants, which have rigid cell walls, the flow of cytoplasm in circular pathways is commonly observed. The large cylindrical cells of the freshwater algae *Nitella* and *Chara* serve as excellent materials for various studies (Fig. 7.28). The outer, gel-like layer of **ectoplasm** encloses the fluid, sol-like **endoplasm.** In these algae, the chloroplasts are embedded in the relatively stationary ectoplasm, while numerous granules flow with the endoplasmic stream and serve as indicators of **cyclosis.**

Several lines of investigation have shown that the motive force which produces streaming is located at the interface between the stationary ectoplasm and the moving endoplasm. L. Rebhun and others have found twisted bundles of 50–100 microfilaments at this interface parallel to the direction of cyclosis. Microtubules were located in the stationary ectoplasm immediately beneath the plasma membrane. Microtubules seem, therefore, to be unrelated to streaming events.

Thick myosin filaments have been observed and a myosinlike protein must be present since actomyosin complexes can be isolated from these algal cells. Furthermore, using glycerinated materials it was shown that ATPase activity could be demonstrated if $Ca^{2+}$ ions and ATP were provided.

Using cytochalasin B, N. Wessells and colleagues were able to stop cytoplasmic streaming in *Nitella* within 1 hour of treatment. These cells resumed vigorous streaming at original rates once the drug was washed out of the system. Cycloheximide had no effect at any stage in treatment or recovery, indicating that new proteins are not needed for these events. Microtubules also appeared to be unrelated to streaming, since colchicine treatment did not stop streaming. Similar results were obtained using oat cells. These studies further emphasize the participation of microfilaments in plant cell cyclosis.

### Displacement of Cells

In addition to using ciliary motility, which is based on a microtubular system, single cells can move from one place to another by means of gliding, ameboid motion, "ruffled membrane" movement, and other ways.

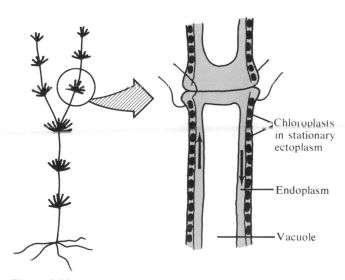

**Figure 7.28**
Drawing of a *Nitella* plant and a detail of the system showing the distribution of protoplasm and vacuole in the large cylindrical cells of the "stem" region of this alga.

Two of these alternatives have been especially well studied. Many amebae, macrophages, and leucocytes exhibit **ameboid motion,** while **"ruffled membrane" movement** characterizes locomotion of vertebrate cells in culture. Each of these systems has provided insights into the participation by microfilaments and microtubules in movement.

AMEBOID MOTION. Amebae move in a particular direction by forming one or more pseudopods and retracting their posterior region from an attached surface. A continuous forward flow of endoplasm takes place within an enclosure of relatively stationary ectoplasm. Two major hypotheses have been proposed and subjected to more comprehensive tests than have other possible ideas. Each of the two major hypotheses postulates a connection between the forward surge of endoplasm and ultimate cell displacement. The hypotheses differ in the site suggested as providing the motive force for the forward flow of endoplasm.

As endoplasm moves forward, it changes from a gel to a sol state, and when endoplasm is converted back to ectoplasm at the end of the surge, sol-to-gel conversion takes place. Either site could theoretically provide the motive force for endoplasmic flow (Fig. 7.29). According to Robert Allen, force is generated at the frontal zone where sol-to-gel change occurs. R. Goldacre, however, has provided evidence showing that the force is generated in the rear of the cell in conjunction with gel-to-sol change in protoplasm consistency. Each model requires contractile fibril assemblies that are coupled to ATP hydrolysis for protoplasmic flow.

Amebae contain numerous actinlike filaments that complex with heavy meromyosin to produce "arrowhead" displays (see Fig. 7.26). These cells also have occasional thick myosinlike filaments within the cell periphery. Despite the increasing amount of information, the system is extremely difficult to analyze and interpret with confidence. Little evidence which would relate the component steps of the contractile phase with the resulting flow phase of movement has been found. Assembly and disassembly of filaments

Chloroplasts in stationary ectoplasm

Endoplasm

Vacuole

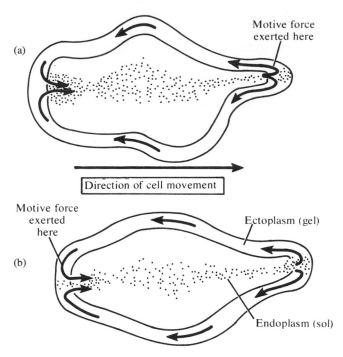

**Figure 7.29**
Ameboid movement. The forward surge of endoplasm leads to cell displacement, which may be propelled by a motive force exerted in (a) the frontal zone or (b) the rear of the cell, according to different theories.

occurs with astonishing speed, and these materials are not easily prepared for the necessary studies using electron microscopy. Cytochalasin does interrupt the forward flow of endoplasm and pseudopod extension, indicating the participation of microfilaments in these aspects of movement.

RUFFLED MEMBRANE MOVEMENT. The term "ruffled membrane" has been used to describe the appearance of moving mammalian cells when grown in tissue culture in single layers. By time-lapse photography, electron microscopy, and other methods, it has been observed that transient waves of undulating movement by the membrane develop at the leading end of the motile cell. The ruffled membrane is actually a surface protrusion that makes intermittent contact with the glass surface of the tissue culture flask. At these times

of contact, the cell moves forward by advancing its unattached membrane regions (Fig. 7.30). A forward flow of cytoplasm is not required for ruffled membrane movement, as it is in amebae, and protoplasmic streaming seems to occur in the peripheral regions of these cells rather than in a central channel.

These membranes are also important in intercellular interactions, and in control over cell movements by contact with other cells. The term **contact inhibition** was coined in 1954 to describe the inhibition of ruffled membrane movement in fibroblast cells grown in glass (or plastic) culture vessels. Once the borders of individual cells make physical contact, there is a cessation of movement so that cells do not pile up on one another. The cells are then distributed as a confluent layer, one cell thick. Contact inhibition of locomotion is often confused with the phenomenon of inhibition of cell multiplication, which occurs in many but not all such cell monolayers, but the two inhibition phenomena are separable and may even be based on different mechanisms. Inhibition of cell growth and division in a layer of confluent cells has been called **density-dependent** or **post-confluence inhibition.**

It has been postulated that the cell surface membrane is somehow least stable along the leading

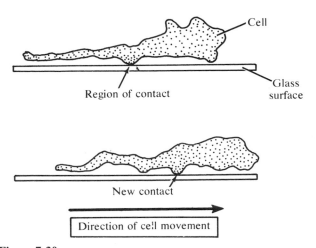

**Figure 7.30**
Forward movement of cells in culture has been postulated to occur by alternating attachment and detachment of the leading margin ("ruffled membrane") of the cell surface.

cell margin, which in some way sponsors ruffling, endocytosis, and other cell surface activities. Perhaps the membrane becomes more stabilized upon contact with another cell surface, causing locomotor inhibition. The experimental evidence provides support for the idea that the locomotor machinery is "turned off" by contact so that traction is no longer exerted at the site of cell contact and the cell stops moving. Since little is known about the machinery of locomotion or the propulsive force for cell movement, the phenomenon of contact inhibition has been difficult to analyze experimentally.

Circulating blood cells do not display contact inhibition *in vivo* or *in vitro,* for reasons that are not known. Nor is it known why some malignant tumor cells lose the capacity for contact inhibition which they possessed before their transformation from normal to neoplastic growth. The basic questions can be answered when more is learned about the properties of the plasma membrane and about the locomotor machinery of the animal cell.

Cultured mammalian cells of all types contain microfilaments and microtubules, and some also have an intermediate class of thicker filaments (8–12 nm diameter). Despite observed variation, the one consistent generalization is that microtubules and microfilaments contribute to movement in different degrees in different cells and situations. Microtubules provide a framework or cytoskeletal component while microfilaments are more directly associated with the contractile machinery itself. Two examples illustrating these points will be discussed briefly.

Norman Wessells has reported one set of experiments in which embryonic nerve cells were exposed to colchicine and cytochalasin (Fig. 7.31). Cytochalasin stopped cell movement and the leading edge of the undulating membrane was retracted, but the axonal portion of the cell did not change in length. Microfilament assemblies were disrupted, but microtubules and 10-nm filaments were not affected. Cells recovered after the drug was washed out of the medium. In the presence of colchicine, axon growth continued for about 30 minutes. Afterward, the axon gradually shortened until it collapsed into the body of the nerve

cell. Neither system was affected by cycloheximide, which indicated that protein synthesis was not required in the recovery phase. New fibrils self-assembled from existing monomer pools. These results may be interpreted as showing that locomotion by the ruffled membrane mode of movement at the growing tip was due to microfilament integrity, whereas

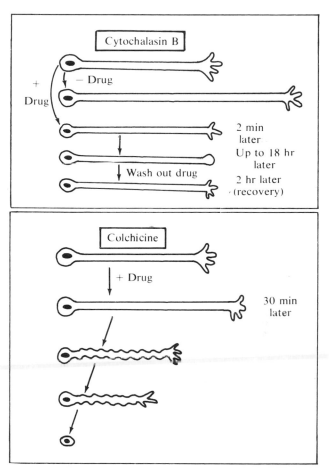

**Figure 7.31**
Different effects are observed in embryonic nerve cells exposed to cytochalasin B and colchicine. Movement stops in each case, but changes in cell morphology vary according to the drug used. See text for details. (Adapted with permission from Wessells, N. K. *et al., Science* **171**:135-143, 15 January 1971, Fig. 5. Copyright 1971 by the American Association for the Advancement of Science.)

maintenance of cell shape during growth depended on intact microtubules. The role of the 10-nm neurofilaments was not resolved.

In a similar study, using fibroblasts in cell culture, Robert Goldman found that colchicine treatment prevented formation of the long cellular processes of typical fibroblasts (Fig. 7.32), but had no effect on cell

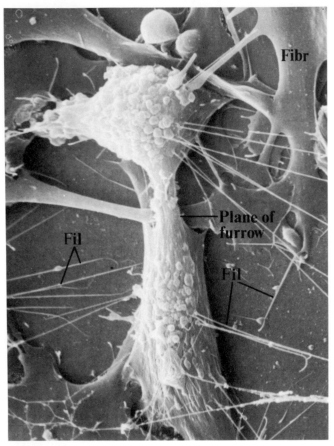

**Figure 7.32**
Photograph of normal rat fibroblasts growing in explant culture, taken with the scanning electron microscope. Cell in center is just completing cytokinesis. Blebbing of surface is characteristic of cells entering G₁ of the cell cycle. Other cells in the neighborhood are in a later stage of interphase and have relatively smooth surfaces. Slender filopodia connect the dividing cell to the substrate below. × 2,700. (Courtesy of K. R. Porter)

attachment or movement. Microfilaments were present in the usual places near the plasma membrane, but microtubules had been disrupted and were not visible. The thicker filaments remained near the nucleus instead of spreading to other parts of the cell. He concluded from these observations that microtubules provided a cytoskeletal framework essential to maintain normal cell shape and to aid in dispersion of the 10-nm filaments within the cell. Since microfilaments were not affected by colchicine and cell movement continued to occur, these fibrils were assumed to be responsible for locomotion. When fibroblasts were treated with low concentrations of cytochalasin B, no effect on microfilaments was observed, but ruffled membrane movement and cell division were inhibited. If higher levels of the drug were provided, microfilaments disappeared. Microtubule and microfilament formation continued to occur during recovery even in the presence of cycloheximide, again indicating that reassembly did not require protein synthesis.

The two sets of experiments produced essentially similar results. But, the differences point out one problem in current studies. There has been little standardization of experimental systems and conditions. One set of results may differ from another because the drug concentration was not the same or the duration of treatment varied. Different types of cells may respond in different ways to identical treatments. Until these technical problems are overcome, one set of experiments may not be directly comparable to another set.

MORPHOGENETIC MOVEMENTS. During development the shape of an organ changes, as it proceeds from its embryonic stage to a differentiated form. Changes during morphogenesis usually involve displacement of cells from one relative position in the three-dimensional organ to a different position within that space. Movement of cells which are parts of a tissue or organ, rather than occurring singly, also comes under the heading of cell displacement or motion. The same fibrillar motility components are found at work in these expressions of cell movement, as in the others we have been discussing.

An early clue to the role of microfilaments in

changes in organ or cell shape was contained in a study reported by T. Schroeder in 1969. He found that cytochalasin caused cleaving eggs to "round up" as the cleavage furrow disappeared. At the same time, the "contractile ring" of microfilaments in the peripheral cytoplasm just under the furrow also disappeared. When cytochalasin was removed from the medium, the furrow and underlying microfilaments reappeared during the recovery phase. The formation of a furrow has been viewed as an effect induced by contraction of microfilaments, analogous to the drawing of a pursestring (Fig. 7.33).

These observations have formed the basis for the study of cytochalasin effects on embryonic organs grown in culture. Organs which develop lobes have been shown to contain bundles of microfilaments in the areas of depression between lobes, usually lying just under the plasma membrane. The same kind of "pursestring" effect seems to be responsible for cell displacement in the developing organ, as it was in cleaving eggs, but changes in organ shape are a consequence of contractility in this case.

Similar invaginations and evaginations have been reported to occur in several kinds of embryonic organs with microfilaments appropriately positioned just under a kink in an organ. These studies have been interpreted as showing that microfilaments serve as a contractile apparatus in morphogenesis. Further support for contractility has come from studies using calcium ions. When $Ca^{2+}$ ions are injected into young toad embryos, contraction occurs very rapidly. When these embryos were examined by electron microscopy, it was found that a dense region of microfila-

ments was present in the place where calcium had entered. If cytochalasin was applied before injection of calcium, contraction did not occur and the dense region of filaments was absent. This response is similar to the response to calcium seen in muscle contraction. On the basis of results in this and other systems, it is thought that the actinlike filaments in the embryo and actin filaments in muscle have similar functions.

### SUMMARY

1. Microfilaments and microtubules are fibrous protein structures found in eukaryotes. Singly or in combination they contribute to all directed motion of living cells and their parts. The best studied system is striated muscle in which the work of contraction and relaxation is subsidized by ATP, which provides free energy for sliding of thin actin filaments past thick myosin filaments. Microtubules are not present in the multinucleated muscle fiber. The sliding filament mechanism of striated muscle contraction is regulated by troponin and tropomyosin, which are proteins bound to actin filaments. Troponin acts as a $Ca^{2+}$-dependent switch, since troponin binds $Ca^{2+}$ ions that flood into the cytoplasm (sarcoplasm) after an electrical impulse from a nerve has been received by the muscle fiber. After binding $Ca^{2+}$ ions, troponin "pushes" tropomyosin molecules away from the active sites on the actin filaments which are blocked in relaxed muscle. When the actin sites become exposed they make contact with the myosin heads of myosin filaments, myosin head ATPase hydrolyzes ATP, and muscle contraction occurs as filaments slide over one another. Striated muscle is recognized by its alternating wider, darker A-bands and its narrower, lighter I-bands. The sarcomere, which extends for about 2.5 $\mu$m from one Z-line to another Z-line in an adjacent I-band, is the unit of contraction.

2. Cylindrically-shaped centrioles contain 9 sets of microtubule triplets, and are found in many kinds of cells at the base of cilia or flagella and at the poles of dividing cell nuclei. Centrioles are essential for formation of cilia and flagella, but they are not essential for formation of the mitotic spindle during nuclear division. Centrioles appear to assemble spontaneously from molecular precursors, and may form in the presence or absence of preexisting centrioles.

3. Cilia and flagella are specialized surface structures whose movements propel a cell through a liquid medium or sweep molecules past the surface of nonmotile cells lined with these organelles. Cilia and flagella have the same ul-

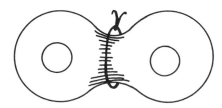

**Figure 7.33**
Schematic illustration of cell furrow formation as analogous to the tightening of a pursestring around the cell.

trastructure, in which 9 sets of microtubule doublets encircle a central sheath within which there are two singlet microtubules. This is the common "9 + 2" cross-sectional display. Ciliary movement appears to be based on a sliding microtubule mechanism in which the dynein arms of subfiber-A make and break contact with the adjacent doublet's subfiber-B. In addition to ATPase activity of the dynein arms of subfiber-A there may be ATPase activity associated with the swollen knob at the end of each radial spoke that extends from the microtubule doublets to the central sheath in the center of the motility organelle. Hydrolysis of ATP provides the free energy for the work of movement in this system, as it does in muscle.

4. Modified cilia and flagella may contain additional components, as in mammalian sperm which have 9 accessory bodies surrounding the microtubule doublets, or they may lack components, usually the two singlet microtubules in the center of the cylinder. Bacterial flagella are quite unlike eukaryotic flagella in their structural organization, mode of insertion in the cell, and mechanism for propulsion; and in having filaments composed of the protein monomer flagellin rather than the tubulin found in eukaryotic cilia and flagellar microtubules.

5. The mitotic apparatus consists of a microtubular spindle and, in many but not all species and cell types, a pair of centrioles within an aster formed from short microtubules. Spindle microtubules assemble and disassemble spontaneously during nuclear divisions, and are sensitive to colchicine. Chromosome movement to the poles at anaphase of nuclear division will not take place in the absence of a spindle. The mechanism for anaphase movement of chromosomes has been proposed to be by assembly and disassembly of pole-to-pole and chromosome-to-pole spindle fibers, respectively, or by some mechanism of sliding microtubules.

6. Various examples of nonmuscle movements have been described as having a system of microfilaments acting as a contractile mechanism, or a coordinated system of microfilaments acting as contractile machinery which in turn acts in conjunction with a microtubular structural framework that provides rigidity and shape to a system. The drug colchicine leads to microtubule disassembly while the drug cytochalasin B disrupts microfilaments. Using either or both of these drugs has provided one approach to analyzing the particular contributions made by each fibrous component to cell movements.

7. Protoplasmic movements include sudden displacements, or saltatory motion, and streaming. Both movements appear to depend on microfilament systems and not on microtubules, according to colchicine and cytochalasin B experiments. In contrast, displacement or movement of whole cells may depend either on microfilaments alone (as in ameboid motion) or on a coordinated microfilament–microtubule system (as in "ruffled membrane" movement of mammalian cells in culture). Cell displacements during organ or tissue growth also appear to rely on the presence and activities of microfilament and microtubule components, but specific details are not yet available.

### STUDY QUESTIONS

1. What kinds of movements characterize (a) whole-cell locomotion, (b) displacements of parts of a cell, and (c) cell growth? What are the underlying structural components in all cellular movement systems? How do we recognize and distinguish microfilaments and microtubules?

2. How is striated muscle organized? What is a muscle fiber, and how is it constructed? What are myofibrils? myofilaments? What is the chemical and physical difference between thick and thin myofilaments? How do thick and thin filaments interact during muscle contraction and relaxation? What is a major line of evidence in support of the sliding filament theory of muscle contraction? What is the contractile unit of striated muscle? What is found in the A-band? the I-band? the H-zone? the Z-line?

3. Why do we say that ATP is not the factor that initiates and terminates striated muscle contraction? What happens when an electrical impulse from a nerve is received by the muscle? What is depolarization and what are its consequences? How does the T system of the sarcolemma help us to understand almost simultaneous contraction of all the sarcomeres of a muscle fiber? What is the location and function of $Ca^{2+}$ ions in contracted and in relaxed muscle? Why does ATP continue to be produced in muscle even after respiration has been poisoned? How long will ATP continue to be produced in poisoned muscle?

4. How do troponin and tropomyosin interact to regulate muscle contraction? What is the function of troponin in muscle contraction? How do $Ca^{2+}$ ions fit into this picture of regulation of muscle contraction?

5. What are centrioles? What is their ultrastructure? How do centrioles act in ciliogenesis? in mitotic cells? Are centrioles essential for both phenomena? What is the evidence in support of your statement for centriole requirement? What cell types or organisms never have centrioles?

sometimes have centrioles? always have centrioles? How do centrioles form? What is a procentriole?

6. What is the ultrastructure of a cilium? a flagellum? How can we analyze the components of cilia and flagella to isolate and identify specific proteins? What and where are tubulin, dynein, and nexin in cilia and flagella? How do we know that dynein has ATPase activity? What is the significance of this observation?

7. Why do we think that cells are propelled by their cilia? What is the energy source for ciliary movement? What is the proposed mechanism for ciliary movement? What is the difference between cilia of sensory receptor cells and other cell types? In what ways are bacterial and eukaryotic flagella different? What similarities are there, if any, between these two kinds of locomotor organelles?

8. What is the function of spindle fibers in anaphase movement of chromosomes during nuclear division? How do we know that spindle fibers are microtubules? What mechanisms have been proposed to explain shortening of chromosome-to pole fibers at the same time that pole-to-pole fibers lengthen in later stages of nuclear division? What is the supporting evidence and what is the conflicting evidence for these proposed mechanisms?

9. How can we identify actin filaments chemically? by electron microscopy? using drugs with selective activity? How do microfilaments contribute to protoplasmic movements, such as saltation and streaming? What evidence supports these notions? What is the difference between movement of a cell by ameboid motion and by ruffled membrane displacement phenomena? What kinds of experimental evidence support the idea that ruffled membrane movement depends on interacting microfilament and microtubule systems? Why are there conflicting results using similar experimental procedures?

10. How do cell displacements during growth and development in multicellular systems take place? What is the effect of cytochalasin treatment on change in shape or position of developing cell systems?

## SUGGESTED READINGS

Berg, H. C. How bacteria swim. *Scientific American* **233:**36 (Aug. 1975).

Bryan, J. Microtubules. *BioScience* **24:**437 (1974).

Cohen, C. The protein switch of muscle contraction. *Scientific American* **233:**36 (Nov. 1975).

Dippell, R. V. The development of basal bodies in *Paramecium. Proceedings of the National Academy of Sciences* **61:**461 (1968).

Gibbons, I. R., and Grimstone, A. V. On flagellar structure in certain flagellates. *Journal of Biophysical and Biochemical Cytology* **7:**697 (1960). (Journal is now entitled *Journal of Cell Biology.*)

Harris, A. Contact inhibition of cell locomotion. In *Cell Communication,* R. P. Cox, ed., pp. 147–185. New York: John Wiley (1974).

Huxley, H. E. The mechanism of muscle contraction. *Science* **164:**1356 (1969).

Lester, II. A. The response to acetylcholine. *Scientific American* **236:**106 (Feb. 1977).

Murray, J. M., and Weber, A. The cooperative action of muscle proteins. *Scientific American* **230:**58 (Feb. 1974).

Porter, K. R., and Franzini-Armstrong, C. The sarcoplasmic reticulum. *Scientific American* **212:**72 (Mar. 1965).

Satir, P. How cilia move. *Scientific American* **231:**44 (Oct. 1974).

Sleigh, M. A. (ed.) *Cilia and Flagella.* New York: Academic Press (1974).

Wessells, N. K. How living cells change shape. *Scientific American* **225:**76 (Oct. 1971).

# Chapter 8

# Cellular Genetics: DNA, RNA, Protein

In this and the next two chapters we will examine basic questions about cell continuity from one generation to another, and about how the genetic machinery sponsors and regulates the expression of cell structure and function. In this chapter we want to know: (1) What is a gene? (2) How is genetic information stored in cells? (3) How is this stored information expressed?, and (4) What regulates gene expression? Afterward we can look at the cellular organization of its genetic machinery and at gene expression at the level of chromosome structure (Chapter 9). In Chapter 10 we will explore the ways in which genetic information is distributed during cell reproduction, and the significance of the two major modes of nuclear division, mitosis and meiosis.

## DNA: THE GENETIC MATERIAL

Long before we learned that genes were made of DNA there was a great deal of evidence about the gene and its manner of inheritance. Mendel's studies of 1866 went unappreciated in his time, but we can only marvel at his conceptual achievement in relating visible characteristics of garden peas to invisible and hypothetical factors that were responsible for height, seed color, seed shape, and four other traits. Many did marvel when Mendel's work was ''rediscovered'' in 1900, and an incredible amount of information was collected in the 50 years or so before we entered the era of molecular biology of the gene.

Mendel interpreted his breeding experiments as showing that unit factors (now called **genes**) were transmitted, unchanged, from one generation to another. These unit factors *segregated and reassorted* so that new combinations of characteristics appeared in progenies of the parental strains that were crossed. Furthermore, unit factors *existed in alternative forms* (now called **alleles**) that behaved as *dominants* or as *recessives*. When tall plants were crossed with dwarf plants, all their progeny (the $F_1$ generation) were tall. But the capacity for dwarf had not disappeared because the next generation, or $F_2$ progeny, produced by interbreeding $F_1$ plants, contained both tall and dwarf individuals. One of Mendel's great perceptions was to apply simple statistical methods of analysis. Not satisfied to simply note that both parental types were recovered in the $F_2$ generation, he counted the different types in the progeny. He found that three-fourths were tall and one-fourth were dwarf. He obtained this same **3:1 ratio** in each of the other six breeding experiments involving six particular traits. Since these were consistent results he reasoned there must be some common explanation for all these inheritance patterns, regardless of the particular trait being studied.

The explanation that best fit the results was that each unit factor segregates and reassorts independently, producing all possible combinations in strictly chance proportions. It is the same as tossing two coins and finding two heads 25 percent of the time, two tails 25 percent of the time, and one head and one tail the remaining 50 percent of the time. Each coin acts independently of the other and has a 50–50 chance of landing as heads or as tails. When both coins are tossed, we get two heads in 25 percent of the attempts because each coin has a 50 percent chance and for both coins to land as heads the chances were 50% × 50%, or 0.5 × 0.5 = 0.25, or 25%. In Mendel's breeding experiments, the 3:1 ratio was derived on the basis of the percentages of each allele (50% of each type) and the chance of each combination occurring (Fig. 8.1). Because of dominance, the ratio of observed types is 3:1. If we imagine that one or more ''heads'' in

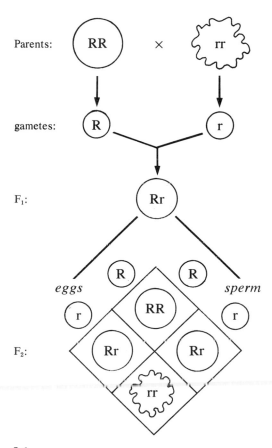

**Figure 8.1**
A typical Mendelian inheritance pattern for two alleles of one gene, showing dominance and recessiveness. The homozygous round-seeded (*RR*) and wrinkled-seeded (*rr*) parents produce heterozygous (*Rr*) $F_1$ progeny, all having round seeds. When $F_1$ plants interbreed to produce the $F_2$ generation, the progeny occur in a ratio of 3 round : 1 wrinkled. This ratio occurs when each parent produces 50% *R* and 50% *r* gametes, and gamete fusions take place at random, leading to all possible combinations in particular proportions. Since *R* is dominant, *RR* and *Rr* plants all have round seeds.

the coin tossing analogy leads to the expression of "heads" as the dominant form, then we would find a 3:1 ratio for "heads" and "tails." The chance for two independent events occurring simultaneously is the product of their separate probabilities, such as $0.5 \times 0.5 = 0.25$. In these ways, Mendel proposed that the basis for inheritance was *particulate*, that is, a unit factor. Today we know that the gene is a unit of information, made of DNA.

Two particularly important experimental studies showed that DNA was the genetic material, one reported in 1944 and the other in 1952. The 1944 study by O. Avery, M. McCarty, and M. McCleod used a pneumonia-causing bacterium and one of its mutant forms that did not cause disease (an avirulent strain). Extracted and purified DNA from the virulent strain was added to avirulent cells, and these avirulent cells were **transformed** to virulent ones (Fig. 8.2). The newly-transformed cells maintained their virulent characteristic and transmitted it to all their des-

cendants. Clearly they had acquired a new genetic characteristic along with the added DNA, and the simplest conclusion was that they had acquired new genes, which must therefore be made of DNA. But there was a lack of enthusiasm for this view in 1944, and since there was very little possible follow-through by other investigators the results were not accorded their proper significance.

In 1952 A. Hershey and M. Chase provided striking information in support of DNA as the genetic material and against the prevailing notion that genes were proteins. They used newly-available isotopes to label the experimental viruses with $^{32}$P for their DNA and $^{35}$S for proteins. Since these bacterial viruses are made of approximately equal amounts of DNA and protein, with no other constituents, they were well suited to the experimental design. The prediction was that the genetic material would go *into* the bacterial host cells (where the viruses multiply), whereas nongenetic material would remain outside for the most part. When

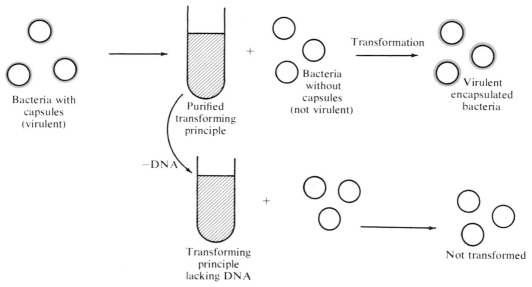

**Figure 8.2**
Diagram illustrating the results of a bacterial transformation experiment. DNA in the extract of transforming principle from donor bacteria is the agent responsible for transformation of an avirulent receptor strain to a virulent strain like the donor cells. No capsule surrounds avirulent cells in these pneumococcus bacteria.

$^{32}$P-DNA and $^{35}$S-protein-labeled viruses were added to their bacterial hosts, $^{32}$P-DNA was found inside and most of the $^{35}$S-protein remained outside (Fig. 8.3). These and other results from a series of experiments led to the conclusion that DNA was the genetic material.

Shortly afterward, James Watson and Francis Crick proposed their double-helix model of the DNA molecule and pointed out how well this molecule suited the requirements for genes. There was potential for mutation, precisely accurate replication, and the stability expected for molecules with genetic functions. As soon as the model was reported in 1953, the scientific community initiated hundreds of different experiments and molecular genetics became the focus for future gene studies.

### The Double Helix

Each of the two long polynucleotide chains in the duplex, helical DNA molecule is linear. The double helix is a remarkably stable conformation because of strong internal bonds: hydrogen bonds between each pair of nitrogenous bases along the entire length of the duplex, and hydrophobic interactions between stacked bases in each of the chains (Fig. 8.4). The polynucleotide sugar-phosphate backbone has considerable stiffness, and the two chains are **antiparallel** (pointed in opposite directions). The adenine–thymine and guanine–cytosine base pairs are exactly the same diameter and shape, so *any* sequence of base-pairs may occur in a duplex molecule. This permits enormously high numbers of different possible sequences, and easily accounts for genetic variety.

**Figure 8.3**
Diagrammatic illustration of the main features in the experiments of Hershey and Chase. T2 phages carrying radioactively-labeled protein or DNA were allowed to infect *E. coli*. After infection was initiated, the viruses were mechanically separated from their infected host cells. The whole mixture was then centrifuged to separate and recover the viruses and *E. coli*. Only labeled viral DNA was found in the infected host cells, and only labeled viral DNA was later recovered in progeny viruses formed in the isolated bacterial cells allowed to complete the infection cycle.

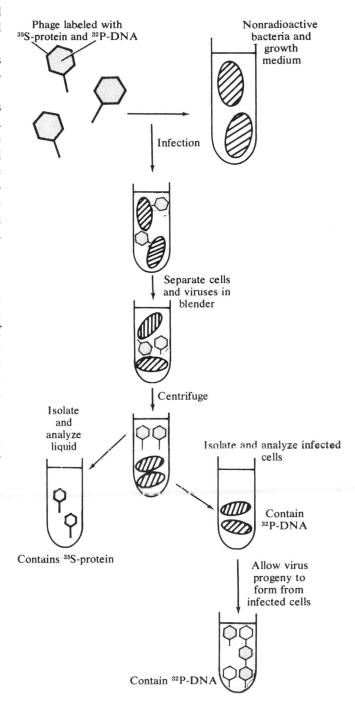

Phage labeled with $^{35}$S-protein and $^{32}$P-DNA

Nonradioactive bacteria and growth medium

Infection

Separate cells and viruses in blender

Centrifuge

Isolate and analyze liquid

Isolate and analyze infected cells

Contains $^{35}$S-protein

Contain $^{32}$P-DNA

Allow virus progeny to form from infected cells

Contain $^{32}$P-DNA

Because base-pairing is specific and essentially invariant, each chain of the duplex is **complementary** to its partner. This feature, along with the others noted above, provides the framework for precise replication of DNA (Fig. 8.5). It also ensures that any alteration due to mutational change will also be replicated precisely, since complementary base-pairing accounts for the fidelity of replication. The consistency of inheritance, the fact that parents resemble their progenies and these in turn produce other progenies like themselves, can be explained by accurate replication of genes (DNA) in every generation.

### Semiconservative Replication

One compelling feature of the Watson-Crick model for DNA was that it provided a basis for a model of

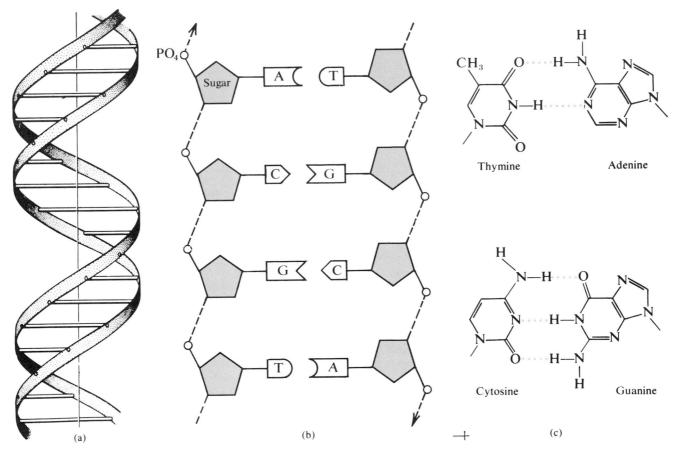

**Figure 8.4**
The molecular structure of DNA: (a) The intertwining strands of the double helix have "steplike" connections at regular intervals along the length of the molecule; (b) complementary pairs of nitrogen-containing bases, each base attached to the pentose sugar of the twisting sugar—phosphate "backbone," keep the two strands of the double helix at a constant distance (diameter) for the entire length of the molecule; and (c) the pairing is specific between each kind of purine and pyrimidine, but there is no restriction on the vertical sequence of base-pairs that can be incorporated into a DNA molecule. Two regions of H-bonding hold thymine and adenine together, while three H-bonded regions occur in a cytosine-guanine pair.

replication guided by an inherent property of the molecule itself, namely, complementary base pairing. Watson and Crick suggested that the two strands of the duplex could each serve as a template for synthesis of a new complementary partner, thus producing two identical molecules from the original duplex. Since only one strand of the old duplex is conserved in each new duplex, this mode of replication is **semiconservative;** the two other possible modes are **conservative** (old duplex intact and entirely new duplex formed) and **dispersive** (pieces of old and new duplex mixed in the next generation of molecules).

In 1958, Matthew Meselson and Franklin Stahl provided the most convincing evidence in support of the semiconservative mode, simultaneously making the other two possibilities remote or void. Using the method of equilibrium density gradient centrifugation to separate DNA in cesium chloride (CsCl) gradient (Fig. 8.6), Meselson and Stahl designed their experiments to obtain data which would clearly point out the correct mode and exclude the other two. These studies are examples of elegant experimental design, which we are able to achieve only too rarely.

The *E. coli* cultures had been previously grown in nutrient media containing the heavy isotope $^{15}$N, so their DNA was fully labeled at time zero of the experiment. These $^{15}$N-labeled cells were transferred afterward to media containing the ordinary $^{14}$N isotope. In the first generation after doubling of cells and their DNA, the DNA was extracted and centrifuged to identify the kinds of isotope-marked molecules that had been synthesized.

There were three different predictions, each unique to one of the three possible modes of DNA replication (Fig. 8.7). If replication was semiconservative, all first-generation DNA molecules should be half-heavy, that is, $^{15}$N-$^{14}$N. If replication was conservative, half the DNA would be heavy ($^{15}$N-$^{15}$N) and half the molecules would be unlabeled, or light ($^{14}$N-$^{14}$N). Dispersive replication would lead to DNA with varying amounts of $^{15}$N and $^{14}$N in the molecules, depending on how many of the original $^{15}$N pieces were incorporated along with newly-synthesized $^{14}$N pieces in the same structures.

Sedimentation of extracted and purified DNA by high-speed centrifugation in CsCl density gradients was carried out until equilibrium, when DNA had settled in specific regions of the gradient that corresponded to the *buoyant density of DNA in CsCl*. The

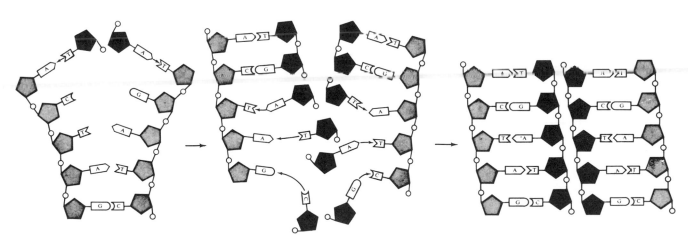

**Figure 8.5**
The duplex DNA molecule replicates to produce a new complementary strand from each of the original parental strands. At the end of the process, two duplex molecules, identical to each other and to the original parental molecule, have been formed. Faithful replication ensures genetic continuity from generation to generation.

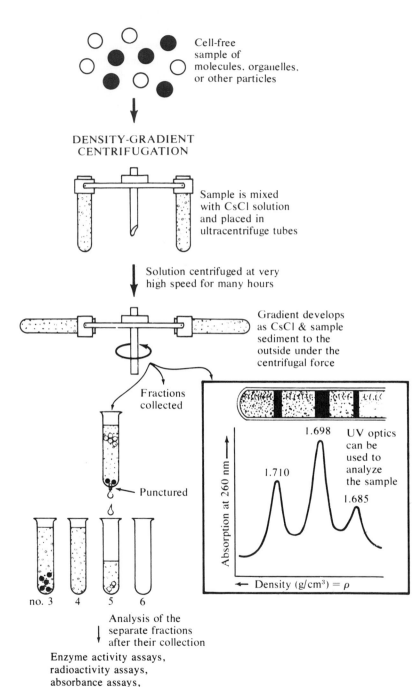

Cell-free sample of molecules, organelles, or other particles

DENSITY-GRADIENT CENTRIFUGATION

Sample is mixed with CsCl solution and placed in ultracentrifuge tubes

Solution centrifuged at very high speed for many hours

Gradient develops as CsCl & sample sediment to the outside under the centrifugal force

Fractions collected

Punctured

no. 3  4  5  6

Analysis of the separate fractions after their collection

Enzyme activity assays, radioactivity assays, absorbance assays, and others

1.698  UV optics can be used to analyze the sample

1.710

1.685

Absorption at 260 nm

← Density (g/cm³) = ρ

**Figure 8.6**
Diagram illustrating the protocol for equilibrium density gradient centrifugation. After materials have been centrifuged to their equilibrium positions in the gradient, fractions may be collected dropwise from a hole punctured in the bottom of the tube (or may be withdrawn in regular amounts beginning at the top of the tube). Fractions may also be photographed using ultraviolet optics. Densitometer tracings may then be made to show the positions and amounts of DNA or RNA in the gradient (inset diagram). The buoyant density of the solute ($\rho$) is expressed in g/cm³, and reflects the density of the solute relative to the density of CsCl or other gradient material.

$^{15}$N-$^{15}$N, $^{15}$N-$^{14}$N, and $^{14}$N-$^{14}$N duplexes each settled in a distinct and nonoverlapping region of the gradient. The contents of the centrifuge tubes were photographed and interpreted (Fig. 8.8). Since all DNA molecules of generation-1 were half-heavy, $^{15}$N-$^{14}$N, the observations supported the semiconservative mode.

These observations and interpretations were confirmed by examining DNA extracted at various times during growth. The distribution of labeled DNA always coincided with the predictions based on semiconservative replication. For example, second generation DNA was 50 percent half-heavy and 50 percent unlabeled, as expected only by semiconservative replication (see Fig. 8.7).

These results do not tell us *how* DNA synthesis occurs, but they do tell us that each parental strand acts as a template for a new, complementary partner strand. The details of DNA replication and the

particular enzymes and other proteins known to be involved in synthesis are shown and described in Figure 8.9. We will have occasion to refer to some of these details in Chapter 10.

## INFORMATION STORAGE AND FLOW

DNA has three main functions: **replication, storage of genetic information** that specifies the characteristics of cells and organisms, and **information transfer** to molecules that can bring the genetic blueprints to sites of protein synthesis. In addition, DNA has the capacity to undergo **mutation,** which leads to new genetic information that forms the basis for evolutionary change. Information is stored in coded form, transferred to complementary messenger RNA molecules, which then direct synthesis of encoded proteins at the ribosomes in the cytoplasm surround-

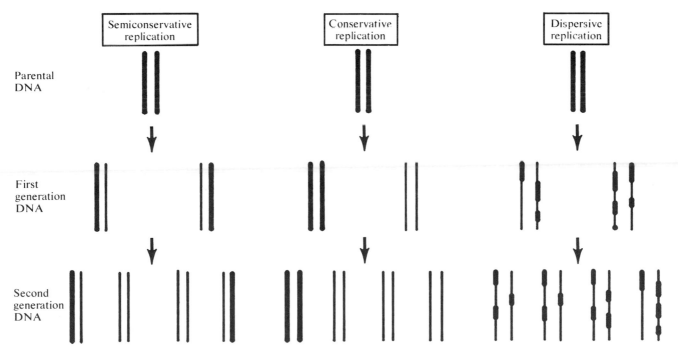

**Figure 8.7**
The distribution of first and second generation progeny molecules follows different predictions according to the three modes of DNA replication set out as hypotheses before 1958.

ing DNA areas in the cell. We will look at these events in the remainder of this chapter.

### The Genetic Code

On the basis of the Watson-Crick molecular model for DNA, it was realized that there was a great variety of possible nucleotide combinations and, therefore, of genes, as Watson and Crick had suggested in their 1953 publications. The only way such a store of information could be miniaturized in DNA molecules was by some form of coding. The presumed coding unit, or **codon,** was a triplet of nucleotides that specified a particular amino acid in proteins. Since there are 20 naturally occurring amino acids included in the genetic code, at least 3 nucleotides per codon are needed. Taking 4 kinds of nucleotides in all possible combinations of three, or $4^3$, 64 unique codons can be assembled. If there were only 2 nucleotides per codon then only $4^2$, or 16, unique combinations would be possible and this was not enough to code for 20 amino acids.

The first codon to be identified was UUU, which specified the amino acid phenylalanine. The genetic code is conventionally described according to the messenger RNA transcripts of DNA nucleotides, because RNA is easier to work with than DNA in experiments and because it is messenger RNA that is decoded directly during protein synthesis. While all information ultimately is referred back to DNA, the usual discussions concern RNA transcripts and protein translations (Fig. 8.10).

In 1961, Marshall Nirenberg and Henry Matthaei showed that an artificial messenger RNA made up only of uracils bound to the sugar-phosphate backbone of the polynucleotide chain, or poly(U), could direct synthesis of a polypeptide consisting entirely of phenylalanine residues. The *in vitro* test system

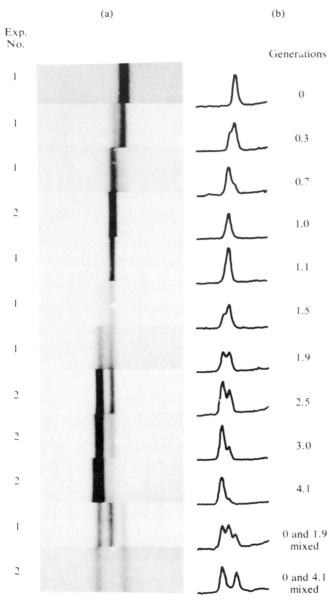

(a)    (b)

Exp.
No.

Generations

| | |
|---|---|
| 1 | 0 |
| 1 | 0.3 |
| 1 | 0.7 |
| 2 | 1.0 |
| 1 | 1.1 |
| 1 | 1.5 |
| 1 | 1.9 |
| 2 | 2.5 |
| 2 | 3.0 |
| 2 | 4.1 |
| 1 | 0 and 1.9 mixed |
| 2 | 0 and 4.1 mixed |

**Figure 8.8**
Some of the results of the Meselson and Stahl 1958 experiment showing (a) ultraviolet absorption photographs of DNA bands resulting from density gradient centrifugation of bacterial preparations sampled at different times after development in $^{14}$N-labeled medium of $^{15}$N-labeled cells, and (b) densitometer tracings of the DNA bands shown in the adjacent photographs. The bottom photograph and tracing, which show fully unlabeled and labeled DNA from a mixed preparation of known content, serve as a reference for DNA density. (From Meselson, M., and F. W. Stahl, 1958, *Proc. Natl. Acad. Sci.* **44**:675.)

contained a mixture of 18 kinds of amino acids (the other two are relatively uncommon) and all the other necessary ingredients for protein synthesis in a test tube. Using 18 different tubes, each with a different *one* of the 18 amino acids radioactively labeled and the other 17 unlabeled, it only required looking for the one test tube out of 18 in which a *radioactively-labeled polypeptide* had been produced from poly(U) instructions. On collecting the labeled polypeptide and hydrolyzing it to its constituent amino acids, only phenylalanines were present. As the codon for phenylalanine was UUU, similar studies soon showed that proline was coded by CCC (using poly(C) in a comparable experiment), and lysine by AAA.

Various coding studies were reported between 1961 and 1964. In 1964, Nirenberg used a new test system he had devised and showed that selected triplets of messenger RNA specified certain amino acids. Many codons were assigned according to these tests. By 1967 the last of the 64 codons had been specified, an astonishing accomplishment for such a brief period of time. The codon dictionary consisted of 61 codons that were specific for the 20 amino acids, and 3 punctuation-mark codons that were "stop" signs for polypeptide synthesis. In addition to finding **punctuations** as a feature of the genetic code, there were two other characteristics that became obvious: (1) the code was **degenerate,** that is, more than one codon usually specified the same amino acid; and (2) the code was **universal,** since the same codewords spell out the same amino acids in all life forms tested to date. Also, the code is **consistent,** since each codon spells out only one amino acid, although from 1 to 6 different codons may code for the same amino acid (Fig. 8.11).

### Co-linearity of Gene and Protein

The code is read in a continuous reading frame beginning at one end of the message and continuing to the other end where a punctuation codon signals

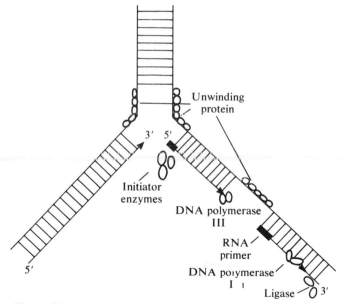

**Figure 8.9**
Summary diagram of the requirements and events relative to replication of duplex DNA molecules. The initiating pieces of RNA primer are the shaded areas. (From *DNA Synthesis* by Arthur Kornberg. W. H. Freeman and Company. Copyright © 1974.)

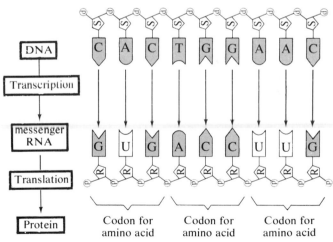

**Figure 8.10**
The messenger RNA copy of genetic information contained in DNA is translated into a co-linear protein. Each 3-letter codeword in the nucleic acid is translated into one of the 20 kinds of amino acids in proteins. The precise kinds and sequence of amino acids in the protein are specified by the particular codons in a particular sequence in the genetic material.

Second nucleotide

| First Nucleotide | A or U | G or C | T or A | C or G | Third Nucleotide |
|---|---|---|---|---|---|
| **A or U** | AAA *UUU*<br>AAG *UUC* Phenylalanine<br>AAT *UUA*<br>AAC *UUG* Leucine | AGA *UCU*<br>AGG *UCC*<br>AGT *UCA* Serine<br>AGC *UCG* | ATA *UAU*<br>ATG *UAC* Tyrosine<br>ATT *UAA*<br>ATC *UAG* "Stop" | ACA *UGU*<br>ACG *UGC* Cysteine<br>ACT *UGA* "Stop"<br>ACC *UGG* Tryptophan | A or U<br>G or C<br>T or A<br>C or G |
| **G or C** | GAA *CUU*<br>GAG *CUC*<br>GAT *CUA* Leucine<br>GAC *CUG* | GGA *CCU*<br>GGG *CCC*<br>GGT *CCA* Proline<br>GGC *CCG* | GTA *CAU*<br>GTG *CAC* Histidine<br>GTT *CAA*<br>GTC *CAG* Glutamine | GCA *CGU*<br>GCG *CGC*<br>GCT *CGA* Arginine<br>GCC *CGG* | A or U<br>G or C<br>T or A<br>C or G |
| **T or A** | TAA *AUU*<br>TAG *AUC* Isoleucine<br>TAT *AUA*<br>TAC *AUG* Methionine | TGA *ACU*<br>TGG *ACC*<br>TGT *ACA* Threonine<br>TGC *ACG* | TTA *AAU*<br>TTG *AAC* Asparagine<br>TTT *AAA*<br>TTC *AAG* Lysine | TCA *AGU*<br>TCG *AGC* Serine<br>TCT *AGA*<br>TCC *AGG* Arginine | A or U<br>G or C<br>T or A<br>C or G |
| **C or G** | CAA *GUU*<br>CAG *GUC*<br>CAT *GUA* Valine<br>CAC *GUG* | CGA *GCU*<br>CGG *GCC*<br>CGT *GCA* Alanine<br>CGC *GCG* | CTA *GAU*<br>CTG *GAC* Aspartic acid<br>CTT *GAA*<br>CTC *GAG* Glutamic acid | CCA *GGU*<br>CCG *GGC*<br>CCT *GGA* Glycine<br>CCC *GGG* | A or U<br>G or C<br>T or A<br>C or G |

**Figure 8.11**
The genetic code. The DNA codons are shown in roman and the complementary RNA codons are in italics. The 20 amino acids are specified by 61 of the 64 triplet codons. The remaining 3 codons are "punctuation marks" that signal the end of a genetic message in a sequence of nucleotides.

"stop". From studies reported by Charles Yanofsky, Sidney Brenner, Francis Crick, and others, it is clear that the polypeptide is a **co-linear translation** of the messenger RNA copy of informational DNA. Yanofsky's work with *E. coli* showed that a specific codon change in a mutant DNA led to a specific amino acid change in the translated polypeptide, *at precisely corresponding sites* on the two linear molecules.

In a different set of studies Brenner, Crick, and others showed that the head protein of phage T4 was co-linear with the phage gene for this protein. A series of mutants was obtained, each of which had a terminator (punctuating) codon somewhere within the DNA sequence rather than the normal amino acid-specifying codon. During protein synthesis, polypeptide chains of varying lengths were made in the different mutants. The length of the polypeptide corresponded to the known position of the mutated codon in the gene (Fig. 8.12).

The underlying basis for gene expression can be summarized as follows. Information coded in DNA is **transcribed** to a messenger RNA copy. The messenger binds to ribosomes, where **translation** of the genetic message into polypeptide is accomplished. The basic feature common to all these activities, and to DNA

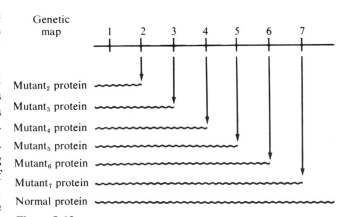

**Figure 8.12**
Demonstration of co-linearity of gene and protein in T4 phage. The known position of an inserted termination codon in a mutant DNA sequence corresponded to the observed site of premature termination in the protein chain translated from the altered DNAs.

replication as well, is a recognition system based on complementary base pairing. The correct DNA strands are replicated, the correct messenger-RNA copied instructions are made, the correct amino acids are incorporated into polypeptides by a recognition system involving transfer RNA molecules (to be discussed shortly), all based on complementarity and the restriction of pairing between guanine and cytosine, and adenine with thymine (in DNA) or uracil (in RNA) (Fig. 8.13).

## REGULATION OF GENE EXPRESSION

We have already discussed certain aspects of this topic in Chapter 3, where regulation of enzyme synthesis was described. Some background information will be repeated here, but the emphasis will be on gene regulation of gene expression. By **gene expression** we refer to translation of the coded information during polypeptide synthesis.

As our information broadened during the 1940s and 1950s when biochemistry became an essential approach to genetics, the concept of genes and gene products moved from *one gene–one enzyme* to *one gene–one protein* and finally to *one gene–one polypeptide*. In every case it was still a system in which some protein or polypeptide chain of a protein, usually but not always an enzyme, was the direct expression of gene action. In 1961 Sidney Brenner presented evidence for the existence of messenger RNA, introducing the concept of an intermediary between DNA and polypeptide. This made sense since DNA remains in its place in the cell whereas proteins are synthesized in the cytoplasm, in regions which might be far removed from the DNA.

Also in 1961, a profoundly different and more sophisticated concept of gene action in the expression of coded information was proposed as the **operon hypothesis,** by François Jacob and Jacques Monod. They presented convincing evidence that some genes regulated the transcription of other genes in coordinated gene groups, or **operons.** We no longer think in terms of one gene–one polypeptide because we know it may take several different kinds of genes to act

before a polypeptide-specifying gene is actually transcribed into messenger RNA prior to translation.

### The Operon Concept

Briefly, since these points were discussed at greater length in Chapter 3, inducible enzymes and repressible enzymes *fluctuate in amount* according to the presence or absence of the substrate they act on or the end product of the reaction they catalyze. **Inducible enzymes** occur in trace amounts most of the time, but quickly increase in amount when their substrate is provided. **Repressible enzymes** are present and active only when there is a low concentration of the end product of the catalyzed reaction.

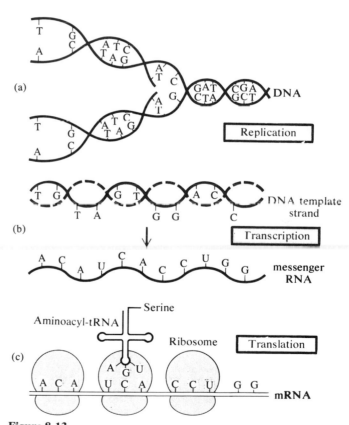

**Figure 8.13**
Complementarity is the underlying theme for recognition between nucleotides during (a) replication, (b) transcription, and (c) translation of genetic information.

The particular system studied by Jacob and Monod was the inducible enzyme β-galactosidase, which catalyzes breakdown of lactose and other β-galactosides to simpler sugars, in *E. coli*. The *lac* operon, or gene system, includes three genes coded for proteins needed to metabolize lactose in *E. coli*. These genes are situated next to one another, and each gene is responsible for a different protein (Fig. 8.14). Since gene *y* and gene *a* protein products are not as easily studied, and any mutations in the *z* gene lead to modified lactose metabolism anyway, studies were restricted to the *z* gene and to other genes that regulated the *z y a* cluster coordinately. In fact, because the *z y a* cluster was affected coordinately by various metabolites (all three protein products increasing or decreasing in unison), it was concluded that some common control system governed coordinated synthesis of the three proteins.

Based on these and other considerations, Jacob and Monod postulated the existence of two classes of genes: structural and regulatory. **Structural genes** coded for the proteins of lactose utilization, and **regulator genes** were responsible for turning the structural genes "on" and "off" so that proteins were made or were not made in the cells. Mutations in any one of the *z y a* cluster affected only the one protein specified by that mutant gene, but mutations in regulator genes affected all three inducible proteins equally and coordinately.

Putting together a large variety of experimental results, Jacob and Monod proposed that the product of the regulator gene was a **repressor** protein, which interacted with the **inducer** metabolite (lactose) and

regulated enzyme synthesis at the level of transcription (Fig. 8.15). When the inducer was absent, the repressor blocked transcription of the structural genes. Without messenger RNA, proteins would not be synthesized. When lactose was present, on the other hand, the inducer combined with the repressor and prevented the repressor from blocking transcription. Transcription was turned on when the inducer was present, and turned off when the inducer was absent.

In the case of repressible enzymes, the same operon concept could be applied to explain fluctuating amounts of these catalysts. The enzyme is synthesized when there is little of the reaction end product in the free metabolite pool of the cell, but enzyme synthesis is repressed when this end product accumulates. In this system, the repressor protein of the regulator

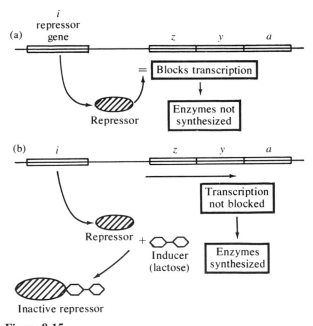

**Figure 8.15**
Diagram illustrating the interaction between the repressor product of the regulator gene *i* and transcription of the structural genes that specify the three inducible enzymes of the *E. coli lac* operon in the (a) absence and (b) presence of the inducer metabolite lactose.

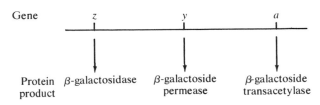

**Figure 8.14**
Genetic map showing the genes that specify enzymes in the *lac* system of *E. coli*.

gene cannot block transcription all by itself. When repressor and metabolite (end product) combine, the complex is able to block transcription. The end product, or metabolite, in these cases is usually called a **corepressor,** since it aids in repressing transcription (Fig. 8.16). Most repressible enzymes function in biosynthetic pathways, and most inducible enzymes function in degradation reactions.

### Operators and Promoters

How does the repressor protein block transcription? From their genetic studies, Jacob and Monod proposed the existence of an **operator,** just to the left of the *z y a* gene cluster. The operator component of the operon signals "on" and "off" for transcription. When repressor binds to the operator it blocks the signal and transcription is turned off. When the repressor is removed, the operator is free to signal and transcription proceeds. Direct confirmation of these predictions was obtained in 1966 in different studies reported by Walter Gilbert and Mark Ptashne. They isolated specific repressors, showed that they were

proteins, and that they bind with DNA as well as with their specific inducer metabolites. Specifically, they showed that repressors bind to the operator site of their operon, but become detached from the DNA when added inducer combines with repressor protein. A change in shape of the repressor protein probably occurs when it binds with inducer, and in its altered conformation the repressor binding with DNA is weakened enough to be detached from the operator DNA site.

Another **recognition site** in the operon, in addition to the operator, was discovered in the late 1960s. This site is called the **promoter,** and it is also situated next to the operator but on the opposite side from the structural genes. The promoter is the DNA site recognized by the enzyme that catalyzes messenger RNA synthesis during transcription, an **RNA polymerase.** Since the operator sits between the promoter and the structural genes, the RNA polymerase can only move toward the structural genes if the operator site is not blocked by a bound repressor (Fig. 8.17). In both inducible and repressible enzyme syn-

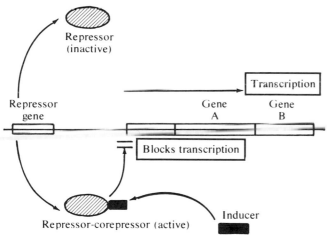

**Figure 8.16**
Diagram illustrating the negative control of repressible enzyme synthesis in prokaryotes. Transcription of mRNA is blocked when corepressor binds with repressor at the operator site of the operon. Transcription proceeds when the corepressor metabolite is absent since repressor alone cannot bind to the operator in such a system.

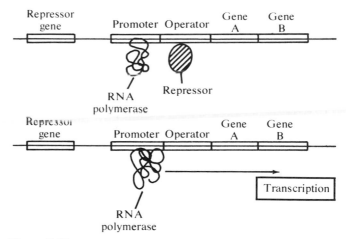

**Figure 8.17**
Relationship of the promoter site to the other components of the operon. RNA polymerase binds to the promoter DNA. The polymerase can then direct transcription of messenger RNA from the structural gene sequences. Transcription occurs only if its progress is not blocked by repressor bound at the adjacent operator site.

thesis systems, the same components regulate gene expression by controlling messenger RNA transcription. In the repressible system, of course, it is a repressor-corepressor complex that binds to the operator, but the outcome and actions are all parts of a common system of **negative control** of transcription. The repressor has to be detached before transcription can take place. There are other transcriptional control systems which are positive, that is, which require something to bind to DNA and thereby stimulate transcription. In such systems positive control works together with negative control, both regulating transcription of messenger RNA from structural gene regions of the operon. With positive and negative controls working together there is a finer regulation of gene expression.

It should be clear from this discussion that there are functionally distinct kinds of DNA sequences, or genes. Regulator genes apparently transcribe all the time and repressor protein products are always present. The target of the repressor proteins is the operator site of DNA in the operon. Both the promoter and operator are recognition sites that are not transcribed. It is the structural genes that correspond to the classical genes of the one gene–one polypeptide concept, now outmoded. The protein products of structural gene expression are primarily enzymes, which in turn control synthesis and degradation reactions that comprise cell metabolism and are responsible for the diverse characteristics of the organism. Some structural gene products are not catalysts; for example, the polypeptide chains of hemoglobin are gene products. In the overall view, however, all the protein products of gene action, whether catalysts or not, direct the outcome of differentiation and development of each individual organism.

The operon concept has been successfully applied to bacterial systems, but is not quite this simple in eukaryotes. There are regulatory and structural genes in eukaryotes, but there is a superimposed complexity due to chromosomal organization and compartmentation in eukaryotic cells. Some of these complicated control systems have been analyzed and have been interpreted in terms of models, but we still are uncertain about many basic features of complex regulation systems in eukaryotic gene expression during differentiation and development. After all, there is a great deal more to cellular differentiation in multicellular organisms than just turning operators on and off. We will mention some of these special qualities of eukaryotic gene action in the next chapter.

## RIBOSOMES AND PROTEIN SYNTHESIS

During all of our discussions of enzymes and genes and regulation there has been the implication that all cellular molecules, *except for DNA*, are constantly made and degraded during the life of a cell. In other words, there is continual **turnover** of molecules and structures. Since cellular activities are specific, replenished molecules must be like the old molecules they replace to explain continuing and recognizable characteristics of each cell type. A very important component of molecular continuity despite molecular turnover is the precision with which new proteins are made, particularly enzymes, during cellular existence. Amino acids are not put together in some helter-skelter manner; they are made according to exact genetic instructions. What regulates and ensures such precision of protein construction? It is one thing to have a lot of bricks and a blueprint, and quite another thing to put these bricks together so that they conform to the blueprint. Our next topic concerns the nature of the cellular machinery that virtually guarantees the proper construction of thousands of different proteins specified by coded DNA. The events of protein synthesis absolutely require the participation of ribosomes.

### Ribosomes

Every living cell contains **ribosomes** during all or part of its existence. Ribosomes are too small to be seen with the light microscope, but they are preserved and visible in materials suitably prepared for electron microscopy. Ribosomes provide the surface on which translation takes place, that is, where messenger RNA transcripts are decoded and proteins are synthesized according to the coded information. They also *actively* *coordinate* and *catalytically assist* in the processes of polypeptide synthesis. In fact, ribosomes are the activity center for many interacting processes and molecules required for synthesis of virtually every protein in the cell.

Prokaryotic cells contain abundant ribosomes distributed throughout the cytoplasm. In eukaryotic cells, however, ribosomes may occur free in the cytoplasm or they may be bound to various cellular membranes. One of the striking ribosomal displays in eukaryotes is the system called the **rough endoplasmic reticulum** (Fig. 8.18). Ribosomes are bound to the outer surface of the membrane sheets of the endoplasmic reticulum. When ribosomes are absent from these membranes, the system is called **smooth endoplasmic reticulum.** The endoplasmic reticulum was discussed in Chapter 6.

Ribosomes are also present on other membrane surfaces, and in mitochondria and chloroplasts. Protein synthesis takes place on all these ribosomes in all their locations in the cell. Only the acellular viruses lack ribosomes. Viruses therefore cannot synthesize their own structural and catalytic proteins. Instead they subvert the ribosomal machinery of the host cell so that host ribosomes translate virus genetic information into virus-specific proteins.

Ribosomes are made up of two different subunits which act in concert during protein synthesis. Neither subunit alone can guide synthesis, but each performs a set of functions that is complementary to the other's. The entire functional unit is the **ribosome monomer,** which can be described and identified according to its physical and chemical characteristics. The two kinds of subunit of each monomer are also distinguishable in these ways.

PHYSICAL CHARACTERIZATION OF RIBOSOMES. The commonest way to identify ribosome monomers and subunits is by their properties of sedimentation during high-speed centrifugation. Under standard conditions of centrifugal speeds in centrifuge rotors, monomers and subunits settle at equilibrium (the point at which no further change takes place) in particular regions of the graduated solution in which the particles have been suspended. The solution of sucrose or some other appropriate solute is graduated across a range of concentrations, or densities of the solute. Particles and molecules settle in different parts of such **density gradients** according to size, shape, and molecular weight. The particles or molecules are described by their **S value,** which is their sedimentation coefficient expressed in **Svedberg units,** as determined under standard conditions in previous analytical centrifugation tests. Ribosomes from the colon bacillus *Escherichia*

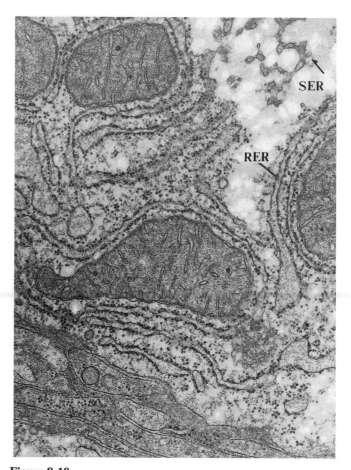

**Figure 8.18**
Thin section of rat liver cells showing rough (RER) and smooth (SER) endoplasmic reticulum. × 25,000. (Courtesy of M. Federman)

*coli* are 70S, and these particles are frequently used as a reference standard for other monomers and subunits from various sources and species. The subunits of *E. coli* ribosomes are 50S and 30S particles. Since sedimentation behavior of particles and molecules is influenced by shape as well as size and molecular weight, the individual subunit values are higher than one might expect. The combined values for the two subunits are always higher than the S value of the monomer particle.

Ribosomes of prokaryotes other than *E. coli* are also 70S. On the other hand, ribosomes from cytoplasm of eukaryotic cells are 80S. The subunits of the eukaryote cytoplasmic ribosomes are 60S and 40S. Common reference standards for eukaryotic 80S monomers and their subunits are particles from yeast and rat liver cells. It is very useful to include an established standard reference particle in ribosome studies, since S values may vary because of slight changes in centrifugation conditions.

Mitochondrial and chloroplast ribosomes also consist of two unequal-sized subunits. Mitochondrial ribosomes range from 55S in multicellular animals to about 80S in some protozoa and fungi. Chloroplast ribosomes appear to be 70S in all green cells that have been studied so far. We will focus the discussion in this chapter on cytoplasmic ribosomes.

Although ribosome dimensions may vary according to the method of preparation used for electron microscopy of these particles, isolated ribosomes from prokaryotic cells measure about 17 nm wide by about 20 nm long. Eukaryote cytoplasmic ribosomes are larger, measuring about 25 nm wide and between 20 and 27 nm long in materials prepared by different methods. Using centrifugation, electron microscopy, and other methods, a consistently larger eukaryotic than prokaryotic ribosome has been reported, regardless of the differences in the absolute measurements obtained.

CHEMICAL CHARACTERIZATION OF RIBOSOMES. Ribosomes are made up of some 50 to 80 different proteins and 3 to 4 kinds of RNA molecules. Each subunit has a unique and different set of these macromolecular constituents. Prokaryotic ribosomal proteins are far better characterized than eukaryotic ones, including the absolute number of proteins per subunit. The ribosome is not as simple as its relatively small size might indicate, since 50 or more different kinds of proteins are assembled into each monomer, along with at least 3 RNA molecules.

In contrast with its protein constituents, **ribosomal RNA (rRNA)** is relatively well characterized in prokaryotes and eukaryotes. The small subunit contains one RNA molecule, but at least two different RNA molecules occur in the large subunit of every ribosome known. Like monomers and subunits, rRNA molecules are identified and discussed according to their S values. Each type of rRNA molecule is a single polynucleotide chain, but some regions fold back and become effectively double-stranded because of local hydrogen bonding between complementary bases (adenine and uracil, guanine and cytosine). Some molecular features of ribosomes are given in Table 8.1.

Little is known at present about the function of ribosomal RNAs. We know quite a bit, however, about their formation and their incorporation into subunits, along with ribosomal proteins. Ribosomal proteins assemble, in rather sequential fashion, on a basic framework of ribosomal RNA. The process is one of **self-assembly,** which implies that the geometry and interactions between constituent molecules is sufficient information for each component to be installed in its correct place in the whole particulate structure. Subunit assembly from individual protein and RNA molecules has been achieved in the laboratory. This achievement has opened the way to further explorations of the functions of each macromolecular constituent, and to more detailed information about the capacities of each of the two kinds of subunit.

### Polysomes

During the 1950s and 1960s it was generally believed that proteins were synthesized on single, free ribosomes. This notion was shown to be incorrect by elegant studies reported in 1963 by Alexander Rich and his colleagues. Their investigations showed that proteins were made at groups of ribosomes, called

**TABLE 8.1   Some characteristics of cytoplasmic ribosomes**

| Source | Intact Ribosome | Ribosome Subunits | rRNA in Subunit | No. Proteins in Subunit |
|---|---|---|---|---|
| Prokaryotes | 70S | 30S | 16S | 21 |
|  |  | 50S | 23S, 5S | 32–34 |
| Eukaryotes | 80S | 40S |  | ~30 |
|  |  | 60S |  | ~50 |
| Animals |  | 40S | 18S |  |
|  |  | 60S | 28S, 5S, 5.8S |  |
| Plants |  | 40S | 18S |  |
|  |  | 60S | 25–26S, 5S, 5.8S |  |
| Fungi |  | 40S | 18S |  |
|  |  | 60S | 25–26S, 5S, 5.8S |  |
| Protozoa (and some other protists) |  | 40S | 18S |  |
|  |  | 60S | 25–26S, 5S, 5.8S |  |

polyribosomes, or **polysomes,** rather than on free ribosome monomers. The polysome is a complex that includes a messenger RNA strand which holds together a variable number of ribosomes. The number of ribosomes in a polysome is proportional to the length of the messenger RNA transcript coded for a polypeptide.

Rich studied rabbit reticulocytes because almost the only protein being made there is hemoglobin, and it simplified the problem of looking for the particular ribosomes engaged in hemoglobin chain synthesis. Almost all ribosomes were making hemoglobin chains. In addition to convincing biochemical studies which showed that new polypeptides were present in 5-ribosome aggregates and not in single ribosome populations isolated from these cells, electron microscopy provided strong supporting evidence as well.

The particular polypeptide chains making up hemoglobin have about 145 amino acids in a polymer molecule. A messenger RNA transcript of a gene coding for this size polypeptide should be about 145 nm long, since each amino acid is coded by a triplet of nucleotides measuring 1 nm in length. At 1 nm per codon and 145 codons for the 145 amino acids, there should be $1 \times 145 =$ about 145 nm for the messenger (I have rounded off some numbers, so they are slightly dif-

ferent from those reported). When polysomes were examined by electron microscopy, the predominant aggregate contained 5 ribosomes and measured about 145 nm from end to end. A thin thread connected the 5 ribosomes in a polysome and seemed to be the messenger RNA, since it could be specifically digested by the enzyme ribonuclease (Fig. 8.19).

In eukaryotic systems, ribosome subunits and messenger RNA are made in the nucleus and transported from the nucleus to the cytoplasm, where polypeptide synthesis then takes place. They arrive in the cytoplasm independently of each other and become associated only at the time translation is begun. Rich was able to study polysomes in rabbit reticulocyte cytoplasm, even though the nucleus has been lost in mature cells. In fact, this simplified the study because there was no need to treat the cells specially to remove nuclei and then get at the cytoplasm for biochemical analysis. Bacterial systems are somewhat different, since transcription and translation take place concurrently and in association with the cellular DNA. Ribosomes become attached to messenger RNA while the messenger is still being transcribed from the DNA molecule. Translation along the ribosome groups begins as soon as the first ribosomes are attached to the growing messenger molecule "peeling" off the

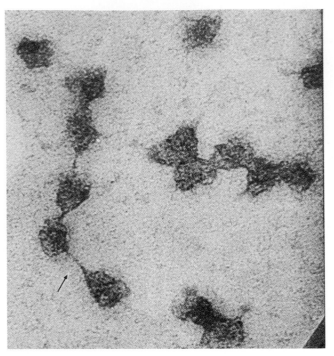

**Figure 8.19**
Polysomes from rabbit reticulocytes. Two pentamers are shown, each of the five ribosomes in the aggregates held together by a messenger RNA strand presumably coding for a hemoglobin polypeptide. × 382,000. (Reproduced with permission from Slayter, H. S. *et al.,* 1963, *J. Mol. Biol.* 7:652-657, **Plate III**a.)

DNA. In 1969, Oscar Miller produced striking electron micrographs showing polysomes in *E. coli* (Fig. 8.20). Here too, the size of the polysome varies in proportion to the length of the coded message and the polypeptide which is translated from it.

### Amino Acids and Transfer RNA

If amino acids are to be linked together to form a particular polypeptide, there must be an **energy source** and a **recognition mechanism.** Energy is needed to sponsor the "uphill" reaction of amino acid addition as the polypeptide grows in length. If the correct amino acids are to be incorporated during synthesis, there must be some way in which the properly coded units recognize the messenger RNA codons during

translation. It is also obvious that messenger RNA (mRNA) must be in an accessible conformation, and that ribosome surfaces also must be available for interactions to take place.

Energy for polypeptide chain growth is made available by amino acid interaction with ATP. When an amino acid is activated (raised to a higher energy level) after reacting with ATP, it has undergone one necessary step in preparation for protein synthesis. The second preparatory step involves attachment of the activated amino acid to its **transfer RNA (tRNA)** carrier. Both activation of the amino acid and its binding to tRNA are catalyzed by the same enzyme, called an **aminoacyl-tRNA synthetase.** There is a different and specific synthetase for each of the 20 amino acids represented in the genetic code.

The synthetase drives the first of the two reactions by coupling it to the hydrolysis of ATP to AMP (adenosine monophosphate). A high-energy intermediate is produced, when the amino acid's carboxyl group binds with the phosphate of AMP, which is called an aminoacyl–adenylate complex. This complex remains bound to the enzyme until a second reaction attaches the aminoacyl to the terminal nucleotide of a tRNA. The enzyme basically interacts with a particular amino acid and tRNA, and couples them to produce the functional **aminoacyl-tRNA** complex which participates in polypeptide synthesis. These reactions can be stated as

$$aa + ATP + enzyme$$
$$\downarrow \qquad\qquad [8.1]$$
$$aa\text{-}AMP\text{-}enzyme + pyrophosphate\ (PP)$$

followed by:

$$aa\text{-}AMP\text{-}enzyme + tRNA$$
$$\downarrow \qquad\qquad [8.2]$$
$$aa\text{-}tRNA + AMP + enzyme$$

or, summarized as:

$$aa + tRNA \xrightarrow[ATP]{enzyme} aa\text{-}tRNA \qquad [8.3]$$

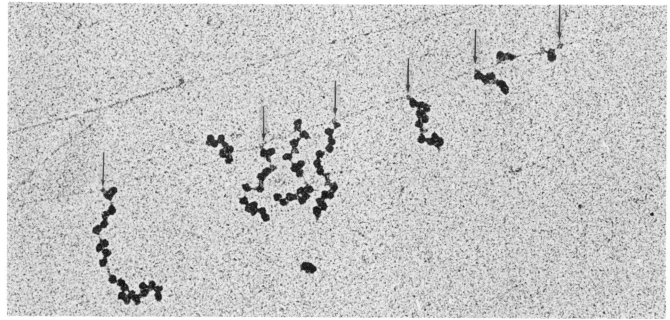

**Figure 8.20**
Electron micrograph of transcription-translation complexes isolated from *Escherichia coli*.
The polysomes assemble while messenger RNA is being transcribed from active DNA
template. Molecules of RNA polymerase (arrow), which catalyze synthesis of messenger
RNA, are present on the DNA strand next to each mRNA being transcribed.
× 115,300. (Courtesy of O. L. Miller, Jr., from Miller, O. L., Jr. *et al*., 1970, *Science*
**169**:392-395, Fig. 3.)

How do the aminoacyl-tRNA units recognize the correct mRNA codon during translation? Is it the amino acid or its tRNA that recognizes the mRNA codon? A classic experiment of the early 1960s provided the answer to this question. It is the tRNA, not its amino acid, which recognizes and specifically interacts with mRNA codons. In the experiment, the amino acid cysteine was linked to its proper tRNA and the complex was then exposed to a nickel catalyst. The catalyst removed the sulfur atom from cysteine, converting it to the amino acid alanine. Nothing else was altered by the treatment, so the catalyst-treated complex was an alanyl-tRNA$^{cys}$, that is, an altogether functional complex in which the usual tRNA for cysteine now carried the amino acid alanine. In a test-tube system that allowed the investigators to see whether the altered complex was inserted into a growing polypeptide as alanine or as cysteine, it was dis-

covered that the alanine was incorporated as if it were cysteine. These results clearly showed that the tRNA and not its amino acid was responsible for recognizing mRNA codons (Fig. 8.21). It was therefore established that an amino acid is carried passively to the mRNA codon, which is recognized by the region of a tRNA molecule called the **anticodon.** Recognition is achieved through pairing between complementary bases in mRNA codon and tRNA anticodon nucleotide triplets. For example, the anticodon UUG would only bind with the mRNA codon AAC with which it is exactly complementary. Each of the 61 codons is unique, and only the correct and unique anticodon will bind regularly and specifically through hydrogen bonds between all three complementary base pairs. **Complementary base pairing** underlies all specific interactions between nucleotides or nucleic acids (see Fig. 8.13).

The tRNAs are relatively small, single-stranded

molecules containing 75–85 nucleotides. The first tRNA to be completely analyzed and sequenced was an alanine-carrying tRNA from yeast; the primary structure was reported in 1965 by Robert Holley. From studies of primary structure in this and many other tRNAs now known, models of secondary and tertiary structure have been developed (Fig. 8.22). It is from such models that we hope to understand how tRNA interacts with enzymes and ribosomes during translation. If all that was needed from tRNA was a nucleotide to bind an amino acid and three more for an anticodon, why are tRNAs 75–85 nucleotides long? We hope to have some answers very soon.

### Polypeptide Chain Elongation

Synthesis of a polypeptide can be divided into three phases: **initiation, elongation,** and **termination** of the chain. We will begin with elongation since it will be

useful to establish the basic principles and processes of the repeated additions as a chain gets longer. Afterward we can look at the special conditions required to initiate and terminate chain growth.

It was in 1961 that experimental evidence was first reported showing clearly that amino acids are added *sequentially* from the start to the finish of a polypeptide chain. Before 1961 it was considered possible that amino acids could be inserted randomly, but that each amino acid somehow correctly entered its place in a polymer undergoing synthesis.

We have already mentioned some of the requirements for polypeptide synthesis: ribosomes, amino acids, tRNAs, aminoacyl-tRNA synthetases, and messenger RNA. In addition, various protein factors, several inorganic ions ($Mg^{2+}$, $K^+$, and $NH_4^+$), and guanosine triphosphate (GTP) must be available.

Various lines of evidence show that ribosome

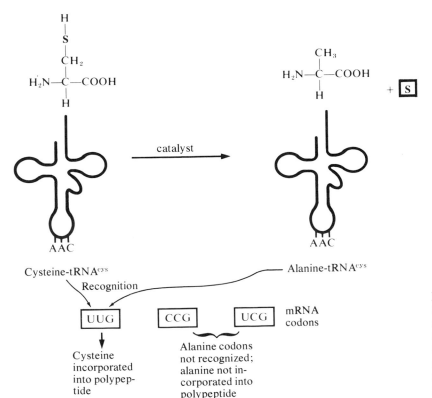

**Figure 8.21**
Diagrammatic illustration of the experiment showing that tRNA$^{cys}$ recognized the mRNA codon for cysteine even though it was carrying alanine instead of its usual amino acid. Alanine-carrying tRNA$^{cys}$ installed the alanine it was carrying into polypeptide sites meant for a cysteine residue.

subunits have specific activity sites, which together allow overall ribosome participation in protein synthesis. From such lines of experimental evidence we believe that specific sites on the two subunits interact to (1) accept incoming aminoacyl-tRNA, (2) add it on to the growing peptidyl chain of amino acids, and (3)

prepare the ribosome for the next incoming aminoacyl-tRNA in the sequence until synthesis is completed.

In the current view, only two sites on a ribosome are thought to accommodate tRNAs at any one time and, therefore, only two molecules of tRNA can be attached simultaneously to an active ribosome. The **A**

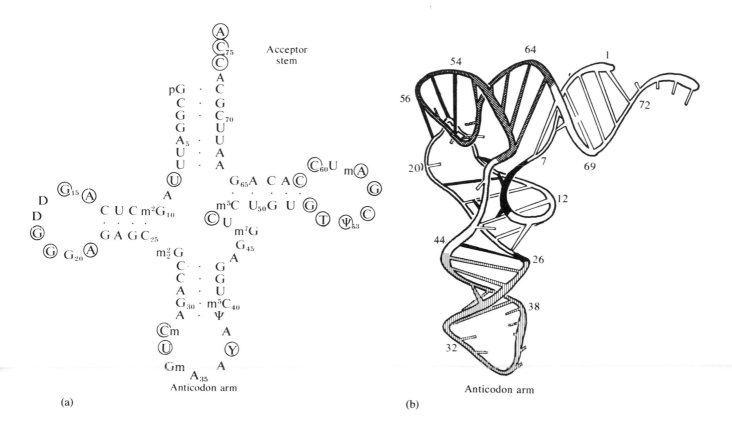

(a)
(b)

**Figure 8.22**
Transfer RNA molecular structure.
(a) Generalized model of secondary structure in tRNA, showing bonding between some of the bases along the molecule, as well as the acceptor stem and anticodon loop. The amino acid binds to the acceptor stem. The site that recognizes the appropriate mRNA codon during translation is located in the anticodon arm of the molecule. (Reproduced with permission from Kim, S. H. *et al., Science* **185**:435-440, Fig. 1, 2 August 1974. Copyright © 1974 by the American Association for the Advancement of Science.)

(b) Model of the tertiary structure of yeast phenylalanine tRNA. The numbers refer to nucleotides in their primary structure locations in the sequenced tRNA molecule. (Reproduced with permission from Kim, S. H. *et al., Science* **185**:435-440, Fig. 3, 2 August 1974. Copyright © 1974 by the American Association for the Advancement of Science.)

site is the place of entry, where incoming aminoacyl-tRNA is accepted at the ribosome. The **D site** (also known as the P site) is the place of release for free tRNA which is discharged from the ribosome after depositing its peptidyl chain onto the incoming aminoacyl-tRNA at the A site. The tRNA is released after a sequence of actions in which the peptidyl chain has been lengthened by one amino acid unit (Fig. 8.23). The same sequence of actions takes place for each amino acid unit added during polypeptide chain elongation, regardless of the particular amino acid to be added. Because the processes are the same for any polypeptide to be synthesized, variety of proteins is achieved on the sole basis of amino acid combinations, all of which are synthesized by a single system of components.

According to the model just presented, when a ribosome which is engaged in protein elongation carries the peptidyl-tRNA at its D site, the ribosome can accept an incoming aminoacyl-tRNA at its A site at the same time. The particular aminoacyl-tRNA accepted depends on the exposed codon of the mRNA. Once in place, peptide bond formation occurs to link the incoming amino acid to the existing peptidyl chain. Bond formation occurs by transfer of the peptidyl chain from its tRNA at the D site to the aminoacyl-tRNA just brought to the A site. The reaction is catalyzed by the enzyme **peptidyl transferase.** Once the peptide bond has been formed, the lengthened peptidyl-tRNA at the A site is translocated back to the D site. Translocation requires **GTP** and a **translocase** protein called the **G factor.** Before the peptidyl-tRNA can return to the D site, however, the now uncharged tRNA must be expelled from the place where it has remained after unloading its peptidyl chain. Once all this is accomplished, the ribosome is restored to its previous state of readiness to accept an aminoacyl-tRNA at the A site. But the polypeptide is now one amino acid unit longer and is temporarily attached to a different tRNA than it was at the beginning of the episode.

The ribosome moves to the next triplet codon after every cycle of peptide bond formation, along the mRNA strand attached to the 30S (smaller) subunit. If GTP is not hydrolyzed, further synthesis stops be-cause the A site remains blocked until peptidyl-tRNA is translocated back to the D site. There is no place of entry for another aminoacyl-tRNA until the A site is opened again. G factor (translocase) must also be available if translocation is to occur and the A site be made accessible to incoming aminoacyl-tRNA.

In addition to GTP and G factor, one or more protein **elongation factors** (EF) must be present if polypeptide chain elongation is to proceed normally (Fig. 8.24). Elongation factors combine with GTP, and the EF-GTP complex combines with incoming aminoacyl-tRNA. In the form of EF-GTP-aminoacyl-tRNA, the incoming unit is brought to the A site of the ribosome where it can then participate in polypeptide chain elongation.

The sequence of events in polypeptide chain elongation proceeds, therefore, as a result of close interaction among the various molecular and structural constituents described. Once it is complexed with EF-GTP, an aminoacyl-tRNA will bind *only* to the A site of a ribosome that has its D site filled with a peptidyl-tRNA. Once bound to the ribosome, GTP of this complex is hydrolyzed so that EF-GDP is released and the GTPase (guanosine triphosphatase, the enzyme that catalyzes the hydrolysis of GTP) site on the ribosome becomes open to binding of G factor and a free second molecule of GTP. This second GTP is hydrolyzed during translocation of the peptidyl-tRNA from the A site to the D site. At the same time, GDP and G factor are released and the A site is opened again to entry of the next aminoacyl-tRNA. From these considerations, it is believed that the ribosome must have at least four active sites or centers:

1. The **D site** situated largely on the 30S subunit.
2. The **A site** located largely on the 50S subunit.
3. A **peptidyl transferase center** found exclusively on the 50S subunit.
4. A **GTPase site** situated on the 50S subunit.

All these closely coordinated features of chain elongation provide greater insurance that few mistakes will take place during polypeptide synthesis. Enzymes and other proteins would be less effective or nonfunctional if they were put together inaccurately, and cell sur-

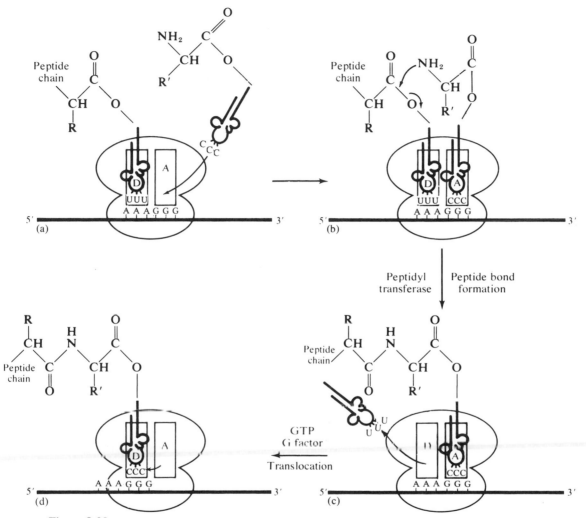

**Figure 8.23**
The sequence of events that takes place at the ribosome during peptide chain elongation:
(a) Incoming aminoacyl-tRNA enters the A site on the ribosome, recognizing the mRNA
codon *GGG*. Peptidyl-tRNA carrying the nascent polypeptide chain occupies the D site. (b)
Peptide bond synthesis takes place by transfer of the polypeptide chain attached to the
tRNA at the D site to the aminoacyl-tRNA at the A site. (c) Formation of the peptide bond
is completed, catalyzed by peptidyl transferase. The A site is now occupied by a peptidyl-
tRNA, while the freed tRNA at the D site is released to recycle. (d) The peptidyl-tRNA is
translocated to the D site, a process that requires GTP and G factor. The A site is again free
to accept the next amino-acyl-tRNA, the ribosome is one codon farther along the message
(moving toward the 3′ end of mRNA), and the polypeptide chain is one amino acid longer.

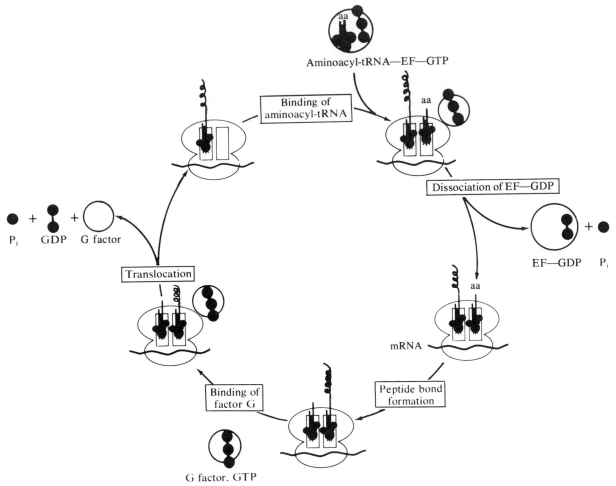

**Figure 8.24**
Summary diagram showing the elongation factors and other components involved in lengthening the peptide chain by one amino acid residue during an elongation episode.

vival would be uncertain. Since even one amino acid substitution is known to lead to protein malfunction in some cases, the cell has a very narrow limit of tolerance for defective proteins. The closely-knit system we have described is fine-tuned in such a way as to make very few mistakes. Since the same basic set of operations and components occurs in all cellular life, it must have been developed to its successful levels early during evolution and been maintained with minor modifications perhaps for several billion years.

**Polypeptide Chain Initiation**

Until 1963 there was a general belief that any amino acid could begin a new polypeptide. J. Waller noted, however, that methionine was the amino-terminal amino acid in 45 percent of *E. coli* proteins studied, and only a few other amino acids occurred in this first position in other *E. coli* proteins. This observation indicated that there might be some special conditions for initiation of protein synthesis which differed from those required to add on amino acids during elongation of a polypeptide. When **N-formylmethionyl-tRNA (fmet-tRNA)** was discovered in extracts from *E. coli* cells, studies were begun to determine its role in polypeptide chain initiation. This compound cannot participate directly in chain elongation because its amino group is blocked, and it therefore cannot undergo peptide bond formation at this part of the molecule. *N*-formylmethionine can act as the first amino acid, however, since its carboxyl group is free for peptide bond formation (Fig. 8.25). This modified amino acid can easily be converted to conventional methionine, the amino acid in the first position in almost half the proteins of *E. coli*. A deformylase can remove the formyl group and leave methionine, and a protein-digesting enzyme known as an aminopeptidase can cleave methionine itself from a chain and leave some other amino acid in the amino-terminal position

of a finished polypeptide. The **initiator mRNA codon** is AUG, which specifies methionine.

There is a slight difference between the tRNA that carries *N*-formylmethionine and the tRNA that carries methionine, although both kinds of tRNA have the same anticodon, UAC. Because of this difference and other reasons, fmet-tRNA interacts only with initiator AUG, and met-tRNA interacts with any other AUG codon in a mRNA molecule. Interestingly, fmet-tRNA is not found in eukaryote cytoplasm although it is typical in prokaryotes. In eukaryotes the initiator codon is also AUG, but it is filled by met-tRNA with the same special tRNA that occurs in prokaryotes. Because the initiating met-tRNA is special, eukaryotic polypeptides are initiated in the same way as in prokaryotes. The main difference is that the initiating aminoacyl-tRNA in eukaryotes is not formylated (and

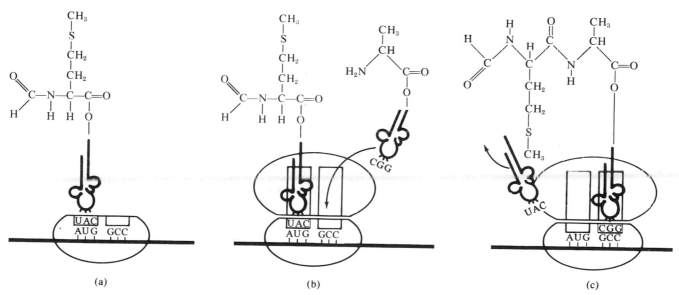

(a)   (b)   (c)

**Figure 8.25**
Polypeptide chain initiation at the ribosome: (a) fmet-tRNA binds at the D site of the 30S ribosome subunit, the anticodon UAC recognizing the AUG initiator codon in mRNA. (b) The aminoacyl-tRNA carrying alanine is entering the A site where the tRNA anticodon CGG recognizes the complementary mRNA codon GCC. (c) Peptide bond synthesis occurs producing a dipeptidyl-tRNA at the A site. Elongation continues according to the steps shown in Fig. 8.23.

there is no formylase in eukaryote cytoplasm). The meaning of these differences is not yet certain.

Discrimination between initiating and internal AUG codons is twofold:

1. Ribosomes must bind to messenger RNA at or near the initiator AUG codon and not to AUG located elsewhere in the message.

2. The initiating fmet-tRNA or met-tRNA must respond to a ribosome engaged in initiation, whereas other met-tRNAs must respond to ribosomes engaged in elongation of a polypeptide chain.

It is the smaller ribosome subunit that acts discriminately, not the larger one. Initiation is mediated by the small subunit of the ribosome once it has accepted and bound a messenger RNA. At this stage the large subunit is not involved and is not attached to the small subunit. Initiation begins when fmet-tRNA or its equivalent in eukaryotes is accepted by the 30S–mRNA complex at the AUG codon which begins the message. Only this aminoacyl-tRNA is accepted in the presence of **initiation factors** associated with the 30S subunit. Afterward, the 50S subunit adds on, the ribosome monomer is formed, and chain elongation can take place. In elongation, the ribosome will not accept initiating fmet-tRNA or met-tRNA, but will accept any noninitiating met-tRNA at AUG codons, in the presence of elongation factors. Protein factors, therefore, are part of the discrimination system for AUG codons.

In test-tube studies of polypeptide synthesis it has been shown that initiation depends only on the small subunit and can occur even if the system lacks the larger ribosome subunits. Elongation, on the other hand, takes place only if complete ribosome monomers are present. These observations make it very likely that both subunits contribute to elongation events.

The **initiation complex** consists of a 30S subunit, messenger RNA, and initiator aminoacyl-tRNA. Its formation requires GTP and at least three kinds of protein initiation factors, only some of whose functions are known. When initiation is complete, the 50S subunit binds to the 30S—mRNA—fmet-tRNA initiation complex (Fig. 8.26). Although we have made constant reference to 30S and 50S subunits, and to 70S ribosome monomers, this was merely for convenience. In every case we could as easily have referred to 40S and 60S subunits, and to 80S ribosome monomers, which occur in eukaryotes. The events are essentially the same for prokaryotes and eukaryotes, except for a few differences as described above.

It should be obvious from this discussion that the pool of inactive ribosomal particles in the cytoplasm consists of subunits rather than monomers. Subunits come together as monomers *after* initiation of polypeptide synthesis, and subunits dissociate after the polypeptide is completed. Under normal conditions of active protein synthesis in cells, whole ribosomes are found mainly as units in polysome groups. These ribosomes are actively engaged in polypeptide chain elongation. Once released from the polysome after the message has been translated, ribosome subunits separate from each other. They come together again in another round of protein synthesis, but continue to recycle in successive syntheses (Fig. 8.27).

**Polypeptide Chain Termination**

Before the finished polypeptide can be released from the ribosome, the link must be broken between the last amino acid of the chain and the tRNA to which it is attached. Separation does not take place spontaneously. Nor does it take place efficiently even when **terminator codons** UAA, UAG, or UGA are present in the messenger. Proper release of tRNA and the completed polypeptide requires the terminator codon signal, and an active termination process aided by specific protein **termination factors** that act in response to such a signal in the mRNA sequence.

The most likely situation is that the termination factor enters the ribosome at the A site, otherwise used by an incoming aminoacyl-tRNA. The exact mechanism of the release action at the ribosome is not known, however.

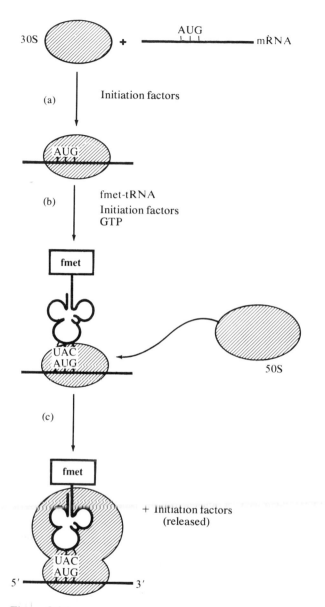

## Overall View

In the overall view of gene action leading to polypeptide synthesis, there is a requirement for DNA, at least three kinds of RNA (mRNA, rRNA, tRNA), regulatory proteins, and enzymes (Fig. 8.28). There must be enzymes for DNA replication so that instructions are available in every cell generation, enzymes for synthesis of different kinds of RNA, enzymes for catalyzing synthesis of each of the fifty or more ribosomal proteins, twenty or more aminoacyl-tRNA synthetases, and enzymes associated with ribosomal coordination of polypeptide synthesis. The incredibly well-coordinated systems work efficiently and consistently every minute of a cell's existence, and in virtually every one of the trillions of cells that make up a human being and all other organisms regardless of their cell numbers and size. The similarity or identity of the system in all cellular life is clear evidence of the common descent of life during billions of years of evolution.

### SUMMARY

1. Genes are units of inheritance made of DNA. Inheritance patterns are based on segregation and reassortment of alleles of a gene, one allele often behaving as a

**Figure 8.26**
Formation of the initiation complex in protein synthesis at the ribosome: (a) a 30S subunit binds to mRNA, stimulated by initiation factors; (b) the binding of fmet-tRNA to the 30S-mRNA complex is promoted by GTP and initiation factors; and (c) the 50S subunit binds to the 30S-mRNA-fmet-tRNA initiation complex, and all the initiation factors are released from the 70S ribosome.

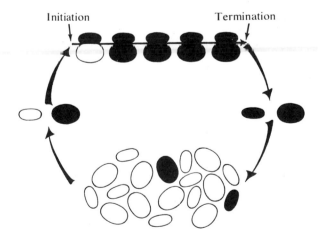

**Figure 8.27**
The subunit cycle in prokaryotic and eukaryotic cells.

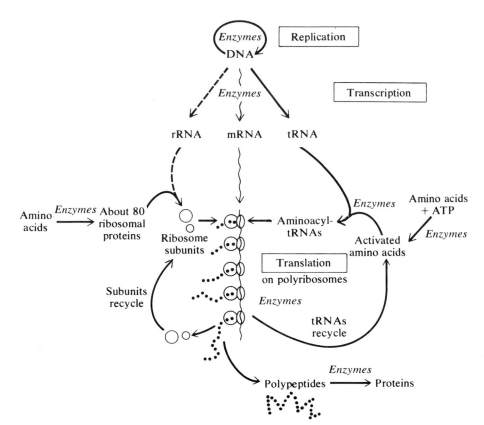

**Figure 8.28**
Summary scheme showing information flow from coded DNA to the final protein products of gene action.

dominant and the alternative allele behaving as a recessive in relation to one another in the development of the organism.

2. The DNA molecule is a double helix composed of two long, linear, polynucleotide chains held together by hydrogen bonding between the complementary base pairs adenine and thymine or guanine and cytosine. Particular features of DNA molecules underlie gene properties of mutation, precise and faithful semiconservative replication, storage of genetic information in coded form, and transfer of information via complementary messenger RNA molecules to the sites of protein synthesis at ribosomes. The specificity of complementary base pairing is responsible for recognition between DNA and DNA, RNA and RNA, or DNA and RNA in the replication, transcription, and translation of genetic information.

3. Genetic information is encoded in DNA. The genetic code consists of 64 unique triplet codons which specify 20 different amino acids in proteins made at the ribosomes during translation of the messenger RNA copy of coded DNA.

The genetic code is universal (the same in all organisms) and consistent (each codon specifies only one amino acid), but it is also degenerate (more than one codon may specify the same amino acid). Three of the 64 codons act as termination signals which "stop" translation when the end of the genetic message has been reached. The sequence of codons in DNA (or its messenger RNA complementary copy) is co-linear with the sequence of amino acids in the polypeptide or protein translation product of gene action. Any alteration in a site in the gene leads to an amino acid alteration at the corresponding site along the length of the polypeptide chain.

4. Gene expression or gene action, that is, transcription of coded information during RNA synthesis, is regulated by certain genes of coordinated gene groups called operons. Structural genes of the operon are the genes of classical genetics, which code for enzyme and other functional proteins. Transcription of structural genes is turned on and off by repressor proteins which are products of regulator gene action. In addition to regulator and structural genes, there

are two other DNA regions in an operon: the operator and the promoter sites. The operator specifically binds repressor protein, while the promoter binds the RNA polymerase which catalyzes transcription of DNA into complementary messenger RNA. When repressor is bound to the operator, transcription is turned off. When repressor is removed from the operator, transcription is turned on since the polymerase can then move past the unblocked operator site and continue on to the structural genes where RNA synthesis then takes place. Inducible enzyme synthesis is turned on when inducer metabolite is present, since inducer and repressor bind together, leading to release of repressor from the operator DNA. Repressible enzyme synthesis is turned on when the corepressor metabolite is absent, since repressor can only bind to operator DNA if it is in combination with its corepressor. In either case, transcription is under negative control since it is turned on only when repressor is removed from its binding site at the operator region of the operon. The operon concept of gene regulation and action has been studied most intensively in bacterial systems, and appears to be considerably more complex in eukaryotic cells.

5. Accurate construction of new proteins to replace those lost in cell turnover is based on the accuracy of translation at the ribosomes, which actively coordinate and catalytically assist in the processes of polypeptide chain synthesis.

6. Ribosomes are made up of two unequal-sized subunits, each having 20 or more different proteins and one or more molecules of ribosomal RNA (rRNA) in its construction. Ribosome monomers, ribosome subunits, and ribosomal RNA molecules are usually described according to their sedimentation constant or S value. Prokaryote cytoplasmic ribosomes are 70S, made up of one 50S and one 30S subunit; eukaryotic cytoplasmic ribosomes are 80S, made up of a 60S and a 40S subunit. The larger subunit of a ribosome monomer has a 5S rRNA and a 23S (prokaryote) or 28S (animal) rRNA, while there is a 16S (in prokaryotes) or a 18S (in eukaryotes) rRNA molecule in the smaller subunit.

7. Polypeptide synthesis takes place at groups of ribosomes, called polysomes, and not on individual monomer ribosomes. The cluster of ribosomes in a polysome is held together by a messenger RNA molecule to which the ribosomes bind by their smaller subunit. Synthesis of a polypeptide from free amino acids requires a source of energy and a mechanism by which the correct amino acid is inserted in its proper place and joined to the growing polypeptide chain during synthesis. Free amino acids are activated (raised to a higher energy level) and attached to a specific transfer RNA (tRNA) molecule through the catalytic activity of aminoacyl-tRNA synthetases. The aminoacyl-tRNA product of these catalyzed reactions interacts with messenger RNA by recognition between the complementary messenger RNA codon and transfer RNA anticodon. Base pairing between codon and anticodon ensures accurate positioning and sequence of the amino acids specified by the genetic code in the messenger RNA.

8. Initiation of polypeptide synthesis requires interaction between messenger RNA, small subunits of ribosomes, and either fmet-tRNA or met-tRNA as the initiating component that binds to the AUG initiator codon of the messenger, all in the presence of initiating protein factors. Elongation of the polypeptide chain then takes place, with each amino acid being added on in proper sequence as the chain grows by one amino acid unit in each successive and repeated set of events at the ribosome surface. An aminoacyl-tRNA enters at the A site of the ribosome, the peptidyl chain is linked to the newly-arrived aminoacyl-tRNA, and this complex is then translocated to the D site of the ribosome surface, which is empty because the previous tRNA is discharged after giving up its peptidyl chain, and the A site is now ready for the next aminoacyl-tRNA as specified in the messenger coded sequence. Termination is signaled by one of the three "stop" codons, and the new polypeptide is released.

## STUDY QUESTIONS

1. What were the principles of inheritance discovered and described by Mendel in 1866? Why does a 3:1 ratio in an $F_2$ progeny indicate that the two alleles of a gene segregate and reassort at random? What would the ratio be in the $F_2$ if the two alleles of a gene were co-dominant, that is, if neither one were dominant over the other's expression.

2. What were the experimental contributions made by Avery, McCarty, and McCleod in their transformation studies and by Hershey and Chase in their bacteriophage studies to the proposition that DNA was the genetic material? How does the molecular construction of the DNA double helix provide a basis for explaining gene properties of diversity and faithful replication in every generation? What was the convincing evidence presented in 1958 by Meselson and Stahl showing that DNA replicated semiconservatively rather than by conservative or dispersive processes? If we start with fully unlabeled $^{14}N$-$^{14}N$ duplex DNA at time-zero, what kinds of DNA duplexes would we predict to occur in first generation and second generation cells grown in $^{15}N$-

containing media according to conservative replication? semiconservative replication?

3. What are three main functions of DNA? What is the genetic code and how was it deciphered? What are "punctuation" codons and how do they work? What do we mean by stating that the genetic code is degenerate? universal? consistent? What was some of the convincing experimental evidence showing that the informational DNA molecule was co-linear with its protein translation product? How does complementary base-pairing underwrite faithful replication of DNA and recognition between nucleic acid regions or whole molecules?

4. What's wrong with the concept of one gene–one polypeptide? What is an operon? How is it organized and what are the functions of its constituent sequences of DNA? How is β-galactosidase synthesis regulated in *E. coli*? What happens in systems of repressible enzyme synthesis when corepressor is present? is absent? What is the function of repressor proteins? How do they differ in inducible and repressible enzyme synthesis systems? Are promoters and operators correctly called genes? In what ways do these differ from regulator and structural genes, and how are they similar? What is transcription? What is translation?

5. What is a ribosome? How is it put together and from what kinds of molecules? What is the difference between cytoplasmic ribosomes in prokaryotes and eukaryotes? Where do ribosomes occur in prokaryotic and eukaryotic cells? What do we mean by a 70S ribosome monomer? What is self-assembly? What is the evidence showing that polypeptides are made at polysomes rather than at ribosome monomers in the cell? What holds the units of a polysome together in the aggregate? How do we know this is the case? Why do polysomes occur in different aggregate sizes rather than always as 5-ribosome groups?

6. What is the energy source and what is the recognition mechanism required to link together amino acids during translation into a specific polypeptide chain? How does an aminoacyl-tRNA form from free amino acids and free tRNA molecules? What is the basis for specificity of aminoacyl-tRNA molecules—in their formation, and their action?

7. What are the steps in polypeptide elongation that lead to growth of the molecule? What regulates the specificity of adding amino acids of specific kinds in a specific sequence during polypeptide chain elongation? What would happen if the peptidyl-tRNA did not move to the D site after reactions were completed at the A site? What are the roles of elongation factors, GTP, and G factor in polypeptide chain elongation? Why are whole ribosome monomers essential for polypeptide elongation, but only small subunits of ribosomes needed to initiate polypeptide chain synthesis?

8. How does the initiation complex form and what are its component parts? Are smaller subunits needed for initiation? are larger subunits needed? How do we know? What happens at termination of polypeptide chain synthesis? How do all three kinds of RNA participate in polypeptide synthesis at the ribosome?

## SUGGESTED READINGS

Cohen, S. N. The manipulation of genes. *Scientific American* **233**:24 (July 1975).

Dickson, R. C., Abelson, J., Barnes, W. M., and Reznikoff, W. S. Genetic regulation: The *lac* control region. *Science* **187**:25 (1975).

Ingram, V. M. How do genes act? *Scientific American* **198**:68 (Jan. 1958).

Jacob, F., and Monod, J. Genetic regulatory mechanisms in the synthesis of proteins. *Journal of Molecular Biology* **3**:318 (1961).

Kornberg, A. *DNA Synthesis*. San Francisco: W. H. Freeman (1974).

Kriegstein, H. J., and Hogness, D. S. Mechanism of DNA replication in *Drosophila* chromosomes: Structure of replication forks and evidence for bidirectionality. *Proceedings of the National Academy of Sciences* **71**:135 (1974).

Lane, C. Rabbit hemoglobin from frog eggs. *Scientific American* **235**:60 (Aug. 1976).

Lewin, B. *Gene Expression,* vols. 1 and 2. New York: John Wiley (1974).

Maniatis, T., and Ptashne, M. A DNA operator-repressor system. *Scientific American* **234**:64 (Jan. 1976).

Meselson, M., and Stahl, F. W. The replication of DNA in *E. coli. Proceedings of the National Academy of Sciences* **44**:671 (1958).

Nirenberg, M. W. The genetic code. II. *Scientific American* **208**:80 (Mar. 1963).

Nomura, M. Assembly of bacterial ribosomes. *Science* **179**:864 (1973).

Ptashne, M., and Gilbert, W. Genetic repressors. *Scientific American* **222**:36 (June 1970).

Raacke, I. D. *Molecular Biology of DNA and RNA. An Analysis of Research Papers*. St. Louis: Mosby (1971).

Rich, A. Polyribosomes. *Scientific American* **209**:44 (Dec. 1963).

Sanger, F., *et al.* Nucleotide sequence of bacteriophage ϕX174 DNA. *Nature* **265:**687 (1977).

Stent, G. S. Prematurity and uniqueness in scientific discovery. *Scientific American* **227:**84 (Dec. 1972).

Swanson, R. F., and Dawid, I. B. The mitochondrial ribosome of *Xenopus laevis. Proceedings of the National Academy of Sciences* **66:**117 (1970).

Watson, J. D. *Molecular Biology of the Gene,* 3rd ed. Reading, Mass.: W. A. Benjamin (1976).

Watson, J. D., and Crick, F. H. C. Molecular structure of nucleic acids: A structure for deoxyribonucleic acid. *Nature* **171:**737 (1953).

Yanofsky, C. Gene structure and protein structure. *Scientific American* **216:**80 (May 1967).

# Chapter
# 9

# Cell Control Center: The Nucleus

The nucleus is the control center of the cell. DNA containing instructions for the expression of cellular developmental potential is organized into chromosomes. Chromosomes act in regulating gene action, distributing genes during nuclear division, and as the bodies which house the genes. All RNA is transcribed at DNA sites within the nucleus and must be channelled to different parts of the cell for their different activities. Ribosome subunits are made in the nucleolus, a nuclear structure essential to eukaryotic life. The membranes that envelop the nucleus participate in the dynamic exchanges of particles and

molecules that allow coordination in the entire **nucleocytoplasmic system.** We will see how each of the major, structured nuclear features contributes to cellular activities.

## NUCLEAR ORGANIZATION

The membrane-bounded nucleus is a trademark of eukaryotic cells. Except for the nuclear envelope, however, there are no permanent membranes in the nucleus. There may be one nucleus per cell (uninucleate), two nuclei (binucleate), or more than two (multinucleate). Most cells in higher organisms are characteristically uninucleate. In certain mature cells the nucleus is no longer present. For example, mammalian red blood cells (erythrocytes) typically lose their nuclei at maturity and survive for only a few months afterward. Certain food-conducting phloem cells in flowering plants continue to function for years after their nuclei degenerate. Usually, however, cell death follows very soon after nuclear degeneration.

The three structured components of a nucleus are the **nuclear envelope, chromosomes,** and one or more **nucleoli,** which are bathed in **nucleoplasm,** an amorphous suspension of assorted particles and molecules of a primarily protein nature (Fig. 9.1). Of the major types of biological macromolecules in the nucleus, DNA is confined to chromosomes, RNA occurs mainly in the nucleolus, lipids (mostly phospholipids) are parts of nuclear envelope construction, and proteins are everywhere within the nucleus. Little carbohydrate has been found in the nucleus.

### The Nuclear Envelope

The **nuclear envelope** is a two-membrane system that separates the nuclear interior from surrounding cytoplasm. The cytoplasm-facing surface of the outer membrane is lined with ribosomes, which are not present anywhere on the inner membrane. The inner membrane, however, is closely associated with condensed chromosomal material (Fig. 9.2). The two membranes of the nuclear envelope are separated by an intermembrane space, or lumen. Within the nuclear envelope there are openings, or "pores," which have a highly ordered border structure. These pores are passageways between the nucleus and cytoplasm.

Continuities can often be seen between the nuclear envelope and the rough endoplasmic reticulum in the cytoplasm (Fig. 9.3). Because of this, as well as similarities in structure, chemical composition, ribosomes attached to the membrane, and staining patterns, the nuclear envelope can be viewed as part of the continuous endomembrane system in cells. The nuclear membranes are most like ER membranes in other ways as well. Both kinds of membranes give rise to vesicles and act as transport channels that guide secre-

tory proteins to the Golgi apparatus for processing and packaging. In some organisms with minimal ER, such as many fungi, the nuclear envelope assumes a number of the functions normally associated with ER systems in cells that are more abundantly endowed.

### Nuclear Pore Complexes

The pore opening is formed by a local fusion between the outer and inner membranes of the nuclear envelope. The perimeter of the opening is bordered by a ringlike structure called the **annulus,** which is found on both the outer surface of the outer membrane and the inner surface of the inner membrane. The annulus has an 8-fold symmetry because of 8 symmetrically distributed particles. The opening and the two annuli comprise the **nuclear pore complex.** A surface view of the nucleus reveals many nuclear pores, with a relatively uniform diameter of 70–75 nm, all essentially circular in outline (Fig. 9.4).

The pore opening is usually filled with a "plug" of material, which is distinct from the fibrils that can also be seen within the pore complex. The fibrils are ribonucleoprotein, but their precise affiliations with other nuclear components are uncertain. Various models of nuclear pore complex organization have been proposed (Fig. 9.5).

The pore opening itself is about 15 nm wide, which would permit some particulate materials to pass through directly. These openings may be especially significant in movement of ribosomal subunits from the nucleus to the cytoplasm, where ribosomes engage in protein synthesis. Strands of messenger RNA must also pass from the nucleus, where transcription takes place, to the cytoplasm where proteins are made according to the copied instructions. Most of the molecular exchanges between nucleus and cytoplasm, however, probably involve the nuclear membranes directly. The flow of materials within the nucleocytoplasmic system undoubtedly requires the entire spectrum of diffusion, passive and active transport, and endo- and exocytosis involving nuclear membranes; and transport of fibrous and particulate materials through the nuclear pore complexes in these membranes.

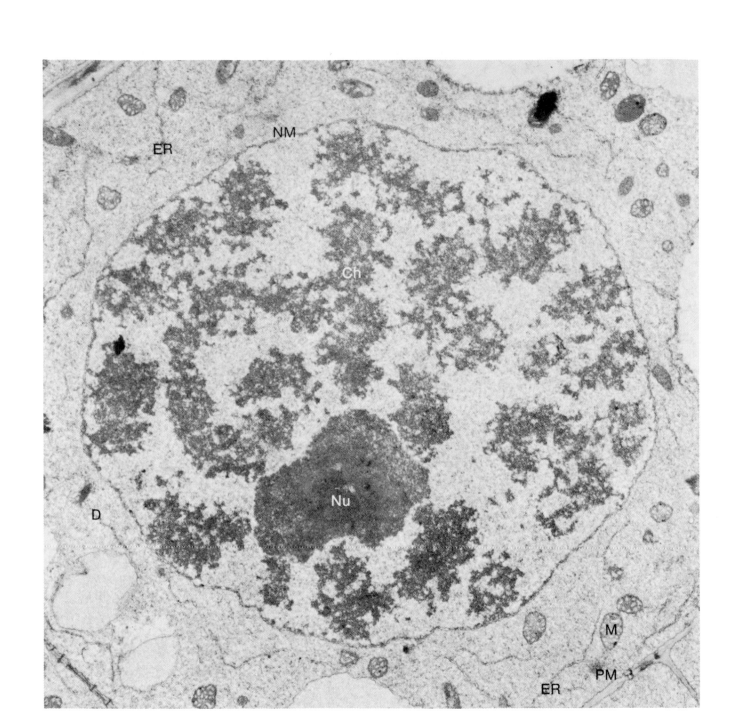

Interestingly, the nuclear pore complexes retain their shape, numbers, and distribution even after the phospholipid membranes of the nucleus have been dissolved away by detergents. The entire nucleus, therefore, retains its shape. This appears to be due to the presence of a sort of crystalline or fibrous protein layer attached to the undersurface of the inner nuclear membrane. The protein layer is undisturbed by detergents, as are the pore complexes. These results indicate the separability of nuclear pore complexes from phospholipid nuclear membranes. The function of the crystalline protein layer is under intensive study at present.

### The Nucleolus

The nucleolus has no surrounding membrane, but usually is attached to a particular **nucleolar-organizing chromosome** which produces the nucleolus in a specific location along the chromosome. The chromosomal site of nucleolar formation is called the **nucleolar-organizing region,** or **NOR** (Fig. 9.6). The main function of nucleoli is production of ribosome subunits, so that nucleoli are very rich in RNA and ribonucleoprotein.

When viewed with the light microscope, the nucleolus appears to contain fibrillar as well as structureless areas. In electron micrographs of cell thin sections, however, four organizational zones can be resolved: (1) a **granular zone** containing fuzzy particles measuring 15–20 nm (or somewhat smaller than ribosomes); (2) a **fibrillar zone** with indistinct fibrils measuring 5–10 nm in diameter; (3) a structureless **matrix** in which these granular and fibrillar materials are suspended; and (4) **nucleolar chromatin** with threads of

10 nm-wide chromosomal fibers from the adjacent nucleolar-organizing chromosome. The granular and fibrillar zones are most clearly delineated in thin sections (Fig. 9.7). The relative proportions of the four nucleolar zones change according to physiological activities of the cell. Actively metabolizing cells usually have prominent granular zones in the nucleolus, undoubtedly reflecting active production of ribosome subunits.

There are several ways to determine the chemical

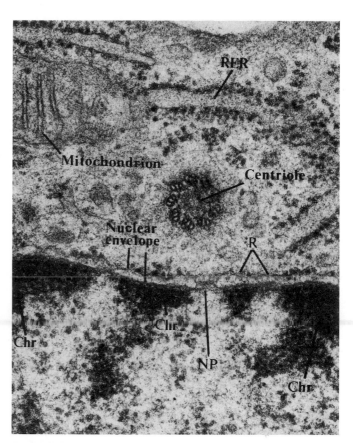

**Figure 9.2**
Part of a bone cell from the rat. Condensed chromatin (Chr) is closely associated with the inner membrane of the nuclear envelope. Ribosomes (R) are present on the outer surface of the outer membrane of the nuclear envelope, and a nuclear pore (NP) with ill-defined structure is also present. × 51,700. (Courtesy of M. Federman)

**Figure 9.1**
Electron micrograph of a cell of onion root tip, just beginning nuclear division as evident from the condensed chromosomes (Ch). There is a prominent nucleolus (Nu), and all the structures bounded by the nuclear membrane system (NM) are situated in a granular matrix of nucleoplasm. Some profiles of endoplasmic reticulum (ER) and mitochondria (M) are also present. × 12,000. (Reproduced with permission from Jensen, W. A., and R. Park, *Cell Ultrastructure*, Fig. 9-1. Wadsworth Publ. Co., Belmont, Calif., 1967.)

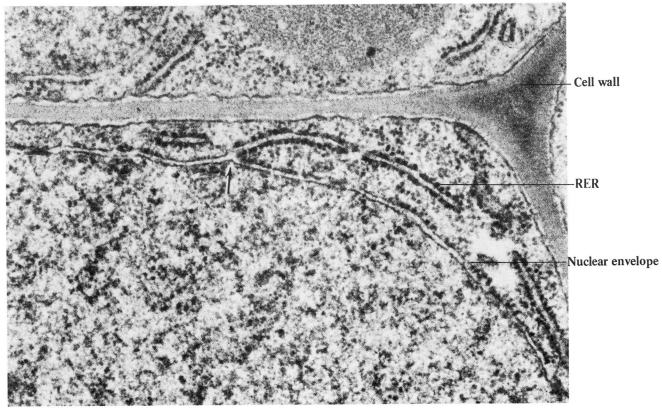

**Figure 9.3**
Thin section of root cell of *Lythrum salicaria*. There is a continuity (at arrow) between the
ribosome-studded outer membrane of the nuclear envelope and an element of rough
endoplasmic reticulum. × 46,000. (Courtesy of M. C. Ledbetter)

composition of the nucleolus including straightforward
chemical analysis of isolated nucleoli, specific staining
procedures for macromolecules, specific enzyme
digestion tests, autoradiography, and others. All
these approaches clearly show a considerable amount
of RNA and protein in nucleoli. By autoradiography it
is also possible to show that RNA is synthesized in
nucleoli, presumably by transcription of nucleolar
chromatin DNA. On adding radioactively-labeled
RNA precursors, such as $^3$H-uridine, silver grains
form in areas containing the radioactive units, and
these silver grains invariably are concentrated in
nucleoli of actively synthesizing cells (Fig. 9.8). There
is no clear evidence showing that proteins are
synthesized in nucleoli, or anywhere else within the
nucleus. Since there are few if any ribosomes functioning in nuclei there would be no machinery for such
synthesis. Nuclear proteins for the most part probably
are imported from the cytoplasm, where we know they
are made at the ribosomes.

If there is no nucleolar-organizing region in any
member of the chromosome complement, no nucleoli

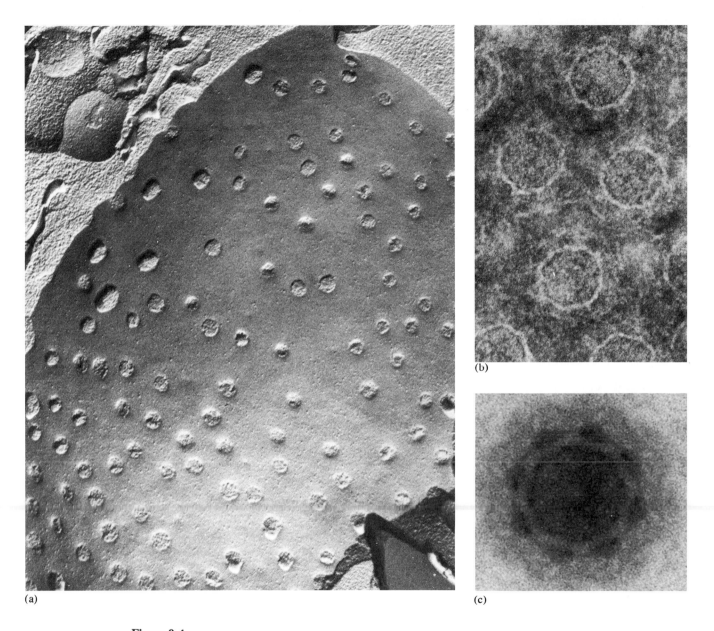

(a)

(b)

(c)

**Figure 9.4**
Nuclear pores. (a) Freeze-fracture preparation showing circular pores of the nuclear envelope. × 48,000. (Courtesy of D. Branton). (b) Negatively-stained nuclear envelope from oocyte of the newt *Triturus*. × 200,000. (c) 8-fold symmetry of the nuclear pore of the frog *Rand pipiens*, as revealed by a rotation test. × 350,000. (Courtesy of J. G. Gall, from Gall, J. G., 1967, *J. Cell Biol.* **32**:391-400, Figs. 2, 10c.)

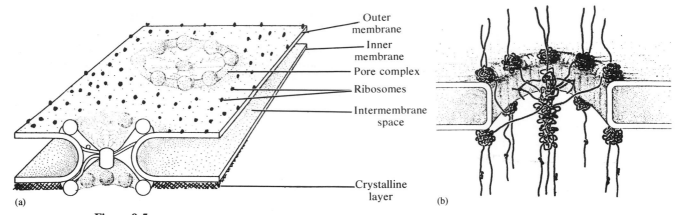

**Figure 9.5**
The nuclear pore complex. Diagrammatic concepts of (a) globular and (b) fibrillar pore complex structures as interpreted from electron micrographs of many cell types. The eight regularly spaced granules of the annulus are found at the pore margin on each surface of the nuclear envelope. Evidence for the occurrence of a crystalline layer of unknown material at the inner surface of the inner membrane has been obtained in experiments reporting the persistence of nuclear shape and nuclear pore complexes after the removal of membrane phospholipids by detergent action. (Reproduced with permission from Franke, W. W., 1970, *Z. Zellforsch.* **105**:405.)

are formed. Such a condition is lethal because ribosomes will not be made, and growing cells and organisms will be unable to survive without protein-synthesizing machinery. The NOR is unique and cannot be replaced by other chromosome regions. Its uniqueness is due to the localized presence of genes coded for ribosomal RNA (rRNA). This particular set of rRNA genes or DNA sequences is called **ribosomal DNA, or rDNA,** signifying that it is genetic information specifying ribosomal RNA. These genes are found in clusters of repeated sequences, numbering in the hundreds or thousands of repeats per cluster in the NOR of each nucleolar-organizing chromosome in a nucleus.

The rRNA genes are spaced apart within the cluster, but we are uncertain about the significance of spacer DNA. Spacer DNA sequences are not transcribed, but the rDNA sequences are transcribed into rRNA (Fig. 9.9). Numerous rRNA molecules are made at each gene sequence, since transcription goes on continuously from the beginning to the end of the DNA informational stretch in a gene. As one molecule is started and the synthesis proceeds toward the end of the gene, another molecule gets started right behind it, then another and another. We therefore see feathery collections of rRNA attached to rDNA, with graduated RNA lengths reflecting the amount of transcription that has so far occurred for each RNA being made at a single rDNA gene. These rRNA molecules then assemble together with imported ribosomal proteins, and ribosomal subunit precursors are thereby made in nucleoli. Ribosomal RNA, therefore, is never translated. It is the end product of rDNA genetic information. In fact, we can tell what part of almost any DNA molecule has rDNA by noting these feathery rRNA clusters during active synthesis.

One way of showing that rRNA is transcribed from the particular portion of nuclear DNA that we call rDNA is to conduct **molecular hybridization** tests. In these assays, nuclear DNA is purified and allowed to hybridize (by complementary base pairing) with radioactively-labeled, purified ribosomal RNA extracted from ribosomes. One of the best systems for such studies is the oocyte of the clawed toad *Xenopus laevis*. During egg cell development from the oocyte there is considerable synthesis of extra copies of rDNA, which collect in numerous individual nucleoli out in the nucleoplasm of the nucleus. The rDNA can be identified by its different density in CsCl gradients after high-speed centrifugation. By comparing the DNA from ordinary somatic cells (with the usual amount of ribosomal DNA) and from developing oocytes, one can recognize and distinguish rDNA from the bulk nuclear DNA. Using such identified, purified rDNA in molecular hybridizations with purified rRNA, rDNA-rRNA *hybrid duplex* molecules are formed (Fig. 9.10). The only basis known for such hybrid duplexes is complementary base pairing, and the only basis known for such specificity of pairing is that the rRNA must be a complementary copy of the rDNA molecules; that is, rRNA is transcribed from that part of the nuclear DNA which contains ribosomal DNA genes.

The phenomenon of additional replication of certain genes in the chromosome complement is called **gene amplification** (Fig. 9.11). It is particularly evident in amphibian oocyte development and in certain other animal oocytes, but apparently does not occur during oocyte development in plants.

Oocyte studies showed that rRNA was transcribed from rDNA that accumulated in free nucleoli in the developing amphibian nucleus. The specific localization of rDNA to the nucleolar organizing region of chromosomes had been made in 1965, in a study of *Drosophila melanogaster,* the common fruit fly of classical and modern genetic studies. Frank Ritossa and Sol Spiegelman used specially constructed strains of *Drosophila,* in which there were 1, 2, 3, or 4 NORs. The NOR is part of the X chromosome in this species, so males usually have 1 NOR (since they have only one of these sex chromosomes), while females are XX and therefore have 2 NORs. The special strains had extra NORs, since extra pieces of X chromosome were present in their nuclei. Using RNA-DNA molecular hybridization methods, Ritossa and Spiegelman showed that the proportion of RNA-DNA

hybrids formed was in exact correspondence with the number of NORs (Fig. 9.12). Since this parallel was consistent it showed that rRNA molecules must be transcripts of DNA located specifically in the NOR of chromosomes.

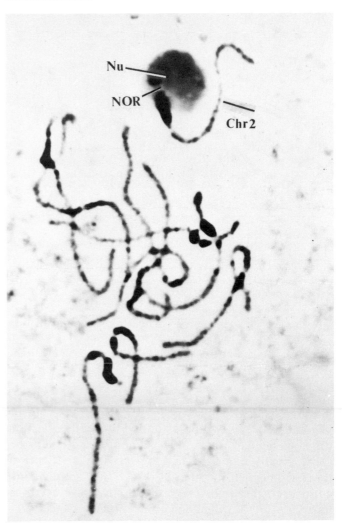

**Figure 9.6**
Pachytene stage of meiosis in castor bean (*Ricinus communis*). The main nucleolar-organizing chromosome is No. 2 in the haploid complement of ten chromosomes. The NOR is located at a secondary constriction near the end of the chromosome attached to the nucleolus (Nu). (Courtesy of G. Jelenkovic)

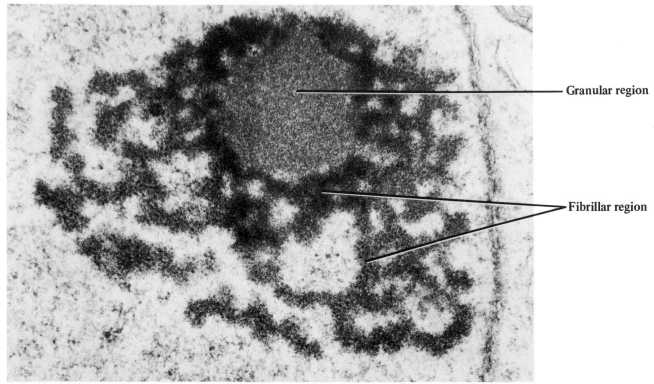

**Figure 9.7**
Thin section of nucleolus from spermatogonium of opossum testis. The granular and fibrillar regions were earlier referred to as *pars amorpha* and *nucleolonema*, respectively. × 33,000. (Courtesy of D. W. Fawcett)

## CHROMOSOMES

Chromosomes were described in the 1880s as vividly-staining, rod-shaped bodies that were always present between two poles in dividing cells. Between times of nuclear division these condensed bodies became greatly extended and appeared as a network of tangled fibers in most cells (Fig. 9.13). Permanently condensed chromosomes are known to occur in various protists, fungi, and other simple organisms.

The stainable materials of chromosomes are called **chromatin,** or **chromatin fibers.** Chromatin fibers are complexes of DNA and proteins for the most part, although some minor amount of RNA may also be present. The chromatin fiber may be considered the basic structural unit of the eukaryotic chromosome.

### Chemistry and Structure of the Chromatin Fiber

The absolute amount of DNA varies according to the length of the chromatin fiber in chromosomes. DNA from the largest chromosome in *Drosophila* may have a molecular weight of $4 \times 10^{10}$ daltons, while a typical chromosome in yeast may contain DNA that is from 1 to $8 \times 10^8$ daltons, which is a hundredfold difference. Since approximately 2 million daltons of DNA corresponds to a duplex molecular length of 1 $\mu$m, Drosophila fibers would measure 20,000 $\mu$m ($4 \times 10^{10}$ divided by $2 \times 10^6 = 2 \times 10^4$, or 20,000) from end to end! This represents 20 mm, or almost one inch of DNA duplex length, so it is considerable. Yeast duplex DNA would be about 400 $\mu$m long for a molecular weight of $8 \times 10^8$. By comparison, the

DNA molecule that contains all the genes in *E. coli* is 1300 $\mu$m, or 1.3 mm. There are 4 chromosomes in the *Drosophila* complement and 17 or 18 in a set in yeast, indicating that the total amount of nuclear DNA generally is vastly greater in eukaryotes than in prokaryotes.

Approximately 13–20 percent of the mammalian chromosome is DNA, and the remainder consists of proteins and a varying but small amount of RNA. There has been considerable interest recently in analyzing chromosomal proteins. Early studies had shown that basic, uniquely chromosomal **histone proteins** are present in regular proportion to DNA in the chromatin fiber, and that histones are synthesized (in the cytoplasm) during precisely the same time interval that DNA is synthesized in the nucleus. Other proteins are made at various times, including during DNA replication. In addition to the parallels in times of DNA and histone syntheses, there was experimental information in the 1960s that showed a relationship between transcription activity of DNA and the presence or absence of histones. When chromatin was allowed to synthesize RNA *in vitro,* it was much less active than deproteinized chromatin, that is, naked DNA. If proteins were added back to DNA, there was a significant reduction in RNA transcription, particularly when histones were added back to reconstitute chromatin.

The original interpretations led to suggestions that histone proteins were regulatory molecules that guided gene expression during development. This idea is not generally accepted today because there is too little diversity of histones to account for the diversity of genes and variability in gene expression among the cells of eukaryotic organisms and among cells at different developmental stages. Only five major kinds of histone molecules are known, and they are essentially the same in all eukaryotes that have been analyzed. Furthermore, regulatory proteins should have different specificities of interaction with different genes, but histones interact similarly with all DNAs.

**Nonhistone proteins,** on the other hand, fulfill requirements not met by histones. They exist in great variety, differ from one tissue to another in an organism and in different stages of development; and they are more abundant in actively-transcribing than in non-transcribing chromatin. Although gene specificities remain to be demonstrated, it is known that different RNAs are produced by DNA that is complexed with different sources of nonhistone proteins. Nonhistone proteins are generally acidic or neutral, in contrast with basic histones that have high arginine or lysine content.

The relationship between chromatin fiber chemistry and structural organization is the subject of

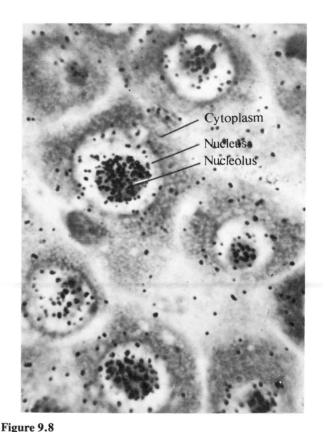

**Figure 9.8**
Autoradiograph showing localization of labeled RNA precursor in nucleoli of rat liver cells. (Courtesy of S. Koulish, from Koulish, S. and R. G. Kleinfield, 1964, *J. Cell Biol.* **23**:39-51, Fig. 8.)

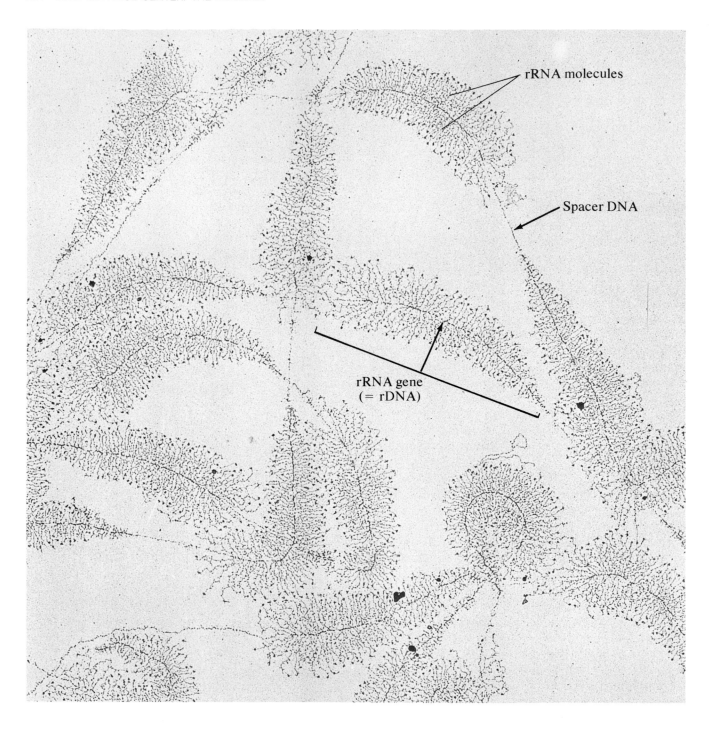

rRNA molecules

Spacer DNA

rRNA gene
(= rDNA)

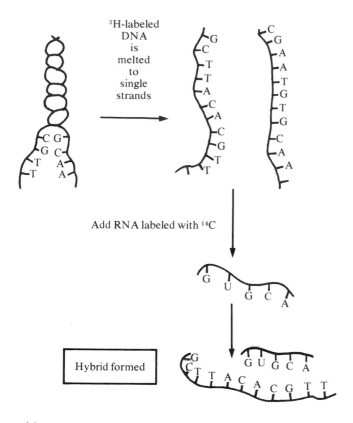

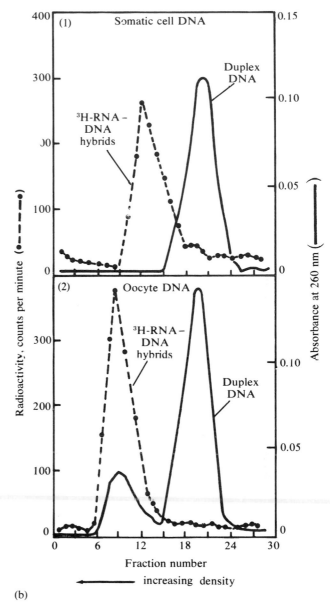

(a)

**Figure 9.10**
(a) Protocol for DNA-RNA hybridization and the detection of complementary nucleotide sequences in radioactively-labeled molecules collected after density gradient centrifugation.
(b) DNA-RNA hybridizations of ³H-labeled rRNA and rDNA from *Xenopus* cells: (1) Little or no DNA is detectable in the "satellite" region of somatic cell preparations according to light absorption at 260 nm, but rDNA must be

(b)

**Figure 9.9**
Ribosomal DNA transcribing ribosomal RNA molecules, from oocyte nucleoli of the spotted newt *Triturus viridescens*. The tandemly repeated genes are separated by nontranscribing "spacer" segments. Each rRNA gene (=rDNA) is engaged in transcribing many rRNA molecules, which present the appearance of a feathery arrowhead. The newest transcripts are shorter and closer to the initiation site for each gene. × 27,500. (Courtesy of O. L. Miller, Jr., from Miller, O. L., Jr. and B. Beatty, 1969, *J. Cell Physiol.* **74** Suppl. 1:225-232, Fig. 3.)

present there since hybrid rRNA-rDNA molecules can be located according to radioactivity measurements; (2) "satellite" rDNA is present in sufficient amount to be observed by light absorption measurements in oocyte preparations. This DNA specifically hybridizes with rRNA according to radioactivity levels. Localizations in oocyte preparations provide support for interpretations of the distribution and meaning of hybrids between rRNA and somatic cell DNA in the "satellite" region of the density gradient.

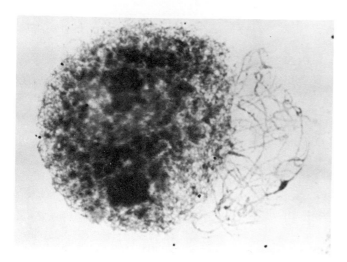

**Figure 9.11**
Light micrograph of an oocyte nucleus from the ovary of the Ditiscid beetle, *Dytiscus marginalis*. The large cap of amplified rDNA is to the left and the synapsed meiotic chromosomes are to the right. Approximately 90 percent of the nuclear DNA at this stage of meiosis (pachytene) is rDNA, which is not attached to the chromosomes at this time. × 1,200. (Courtesy of J. G. Gall, from Gall, J. G. and J.-D. Rochaix, 1974, *Proc. Natl. Acad. Sci.* **71**:1819-1823, Fig. 1.)

vigorous investigation at present. Before 1974 there were various models of chromatin fiber structure that essentially postulated a coating of protein within the narrow grooves of duplex helical DNA molecules. The protein coat would consist of basic histones bonded to DNA acidic groups of the sugar-phosphate backbone. One problem with such a model is that considerable stiffness would result, which would make it difficult to explain the known folding and packing of very long chromatin fibers into highly condensed chromosome structure. In *Drosophila*, for example, 20,000 μm of duplex DNA is packed into a condensed chromosome that may be only 1 μm long.

In 1974, Roger Kornberg presented a model of the chromatin fiber as a flexibly-jointed chain, resembling beads on a string. The model was based on biochemical analysis showing that chromatin could be broken down to units containing about 200 nucleotide

pairs and 4 histone molecules. This repeating unit has been called a **nucleosome,** and it is a bead or knot spaced at intervals along the whole chromatin fiber (Fig. 9.14). The spaces between nucleosomes consist of DNA plus at least one other histone molecule type. All the knots and spaces in between are parts of a continuous chromatin fiber that can be folded and packed into a tiny volume, very much like a flexibly jointed chain.

There is extremely little information as yet on the associations of nonhistone proteins and RNA with this

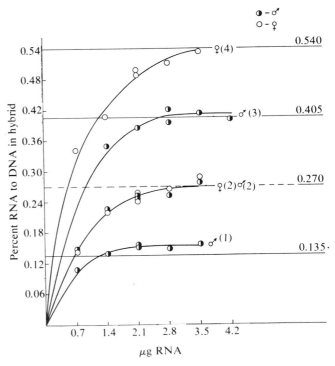

**Figure 9.12**
Saturation levels of DNA containing various dosages of nucleolar-organizing regions (indicated by the number in parentheses). The dashed horizontal line at 0.270 (percent RNA to DNA in the hybrid molecules) is the estimate for a dosage of 2 NOR, and the solid horizontal lines represent predicted plateaus for dosages of 1, 3, and 4 NOR, respectively. Numerical values of the plateaus are given on the right. (From Ritossa, F. M., and S. Spiegelman, 1965, *Proc. Natl. Acad. Sci.* **53**:742, Fig.1.)

fiber, nor do we know very much about mechanisms that explain the constant attachment and detachment of these molecules to the chromatin fiber during cellular activities. The basis for structure of the chromatin fiber is its DNA-histone protein complex. The structural *integrity* of the chromosome is due to its DNA molecule that extends from one end of the chromosome to the other. It is easily demonstrated by specific enzyme digestion tests. When protein-digesting enzymes are added to chromosomes they undergo some degradation and alteration in shape, but do not fall apart. When DNase (deoxyribonuclease) is provided, however, the chromosomes fragment into small pieces.

### Number of Chromatin Fibers per Chromosome

For many years there were controversial observations and opinions concerning the number of chromatin fibers in a chromosome. In 1973 several studies presented significant evidence showing that only one fiber was present in an unreplicated chromosome. The evidence was so convincing that there has been virtually no mention of alternatives in the succeeding years. The so-called **unineme** (one thread) model is essentially accepted as fact.

If total nuclear DNA is extracted from a known number of cells, we need use only simple arithmetic to calculate the amount of DNA in the **haploid** nucleus (a nucleus with one set of chromosomes that includes one set of genes). A little more arithmetic provides the amount of DNA per chromosome in the haploid set. For example, there are approximately $1 \times 10^{10}$ daltons of DNA per haploid nucleus in yeast cells. Since there are about 18 chromosomes in the haploid set, there would be an average of $5.5 \times 10^8$ daltons of DNA per chromosome ($1 \times 10^{10}/18 = 5.5 \times 10^8$). These estimates can be confirmed and expanded to calculate the number of DNA molecules (or chromatin fibers) per chromosome, as follows:

1.  Carefully isolated yeast nuclear DNA molecules settle in a region of the CsCl density gradient that corresponds to 50S to 130S particles, which in turn correspond to individual DNA molecules

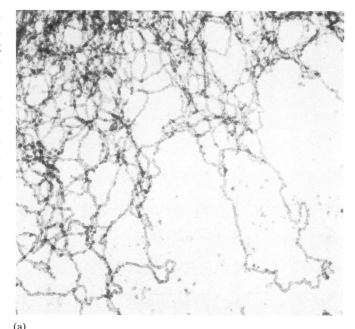

(a)

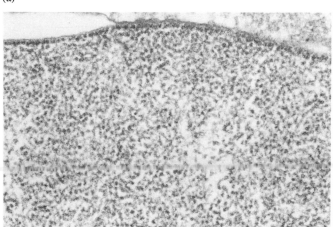

(b)

**Figure 9.13**
The chromatin fiber.
(a) Native fiber, about 20 nm thick, from frog *(Rana pipiens)* erythrocytes, spread on a buffer and air-dried. × 36,000. (b) Thin section of a nucleus from frog erythrocyte. The chromatin fibers are about 20 nm thick. (All photographs courtesy of H. Ris, from Ris, H., 1975, *The Structure and Function of Chromatin*, Elsevier Press, pp. 7-28, Figs. 2 and 3.)

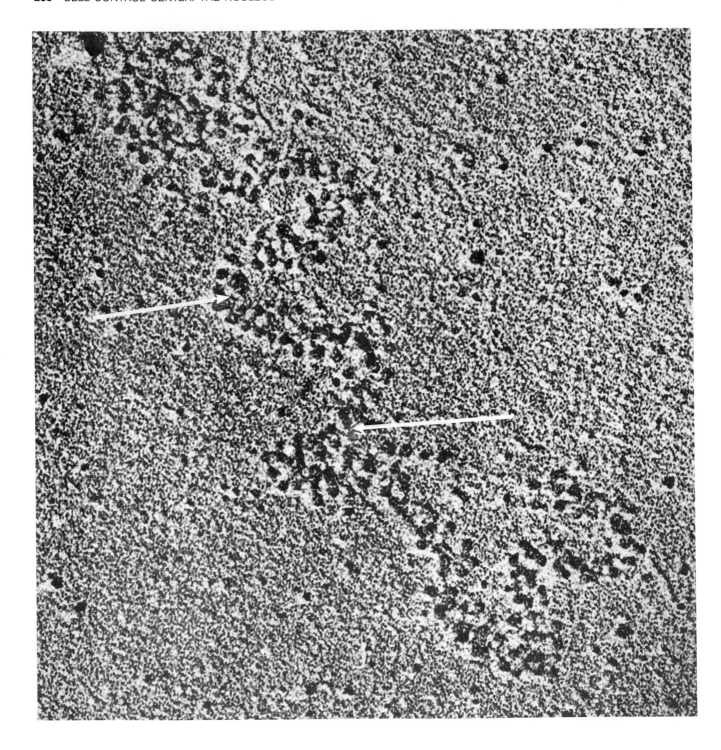

with molecular weights of $1 \times 10^8$ to $1 \times 10^9$ daltons, or an average of about $6 \times 10^8$ daltons per molecule. These data support the conclusion that each chromosome contains one continuous DNA molecule ($5.5 \times 10^8$ daltons of DNA per chromosome and $6 \times 10^8$ daltons for a single duplex DNA molecule).

2. Electron micrographs of purified yeast nuclear DNA reveal molecules that vary from 50 to 365 $\mu$m in length. At $2 \times 10^6$ daltons per $\mu$m of duplex DNA, the observed molecules would be assigned molecular weights between 1.0 and $7.3 \times 10^8$ daltons. This evidence further supported the unineme model.

Other studies utilizing still different methods provide additional support for the unineme model. Using a physical method that permitted the determination of the molecular weight of the largest molecule in a solution, based on its rate of recoil during relaxation of stretched molecules, it was shown that single duplex DNA molecules from *Drosophila* had a molecular weight as high as $4 \times 10^{10}$ daltons. Using the equivalence of $2 \times 10^6$ daltons per $\mu$m of DNA duplex length, these molecules would be 20,000 $\mu$m long. These values were further verified by independent methods.

Taking all the available information from these and other species, it is concluded that a chromosome contains a single nucleoprotein fiber which is continuous from one end of the chromosome to the other. The fiber is folded and coiled into a compact,

**Figure 9.14**
Electron micrograph of chromatin from chicken liver nuclei lysed directly on the metal grid and prepared for microscopy. The chromatin fiber appears to be a flexible chain of spherical particles, about 125 Å in diameter, connected by DNA filaments (at arrows). The spherical particle (nucleosome) contains about 200 base pairs of DNA and an equal weight of four kinds of histone proteins. The same basic structure can be reconstituted *in vitro* from DNA and these four histones (minus the lysine-rich fifth histone, Fl). $\times 252,000$. (Courtesy of P. Oudet and P. Chambon, from Oudet, P. *et al.*, 1975, *Cell* **4**:281-300, Fig. 13.)

condensed structure that varies in length during different nuclear activities, but which may be only one or a few micrometers long when it is most contracted during nuclear divisions.

### The Centromere

Chromosomes are not merely strings of chromatin which condense into highly visible bodies when seen by microscopy; they are differentiated along their length into regions that perform unique functions. These functions cannot be taken over by substitute regions located elsewhere along the chromosome, which further testifies to differentiation of chromosome structure. On yet another level, there are whole chromosomes in the set that perform unique functions for which no other chromosome in the same complement can substitute.

Two particular differentiated regions of a chromosome are the **centromere** (or **kinetochore**) and the **secondary constriction,** which is usually associated with the nucleolar-organizing region of certain chromosomes. Every chromosome has one centromere, recognized by a constriction at its place along the chromosome. Because of its importance the centromere is sometimes called the **primary constriction,** all others being secondary to it, by definition (Fig. 9.15). When stained chromosomes are examined by light microscopy there is little or no staining at the centromere. Despite its appearance, there is a continuous chromatin fiber that passes through the centromere in every chromosome. This is seen most clearly in electron micrographs of whole chromosomes (Fig. 9.16).

There is actually a **centromere region** which includes a number of structured components, and we ought to consider these components as the centromere itself. In animal cells the **centromere** is usually a disclike structure that lies close to one side of the chromosome width (Fig. 9.17). During nuclear division, particularly at metaphase, spindle fibers (microtubules) can be seen to be inserted into a three-layered centromere. The outer, moderately dense layer is called the centromere plate; the inner dense layer is formed from compacted chromatin fiber; and the mid-

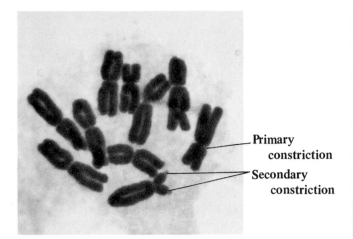

Primary
constriction
Secondary
constriction

**Figure 9.15**
Ten of the 16 metaphase chromosomes from onion root tip
cell. There is an obvious primary constriction in each
replicated chromosome (centromere region), and one
chromosome has a secondary constricton at its nucleolar-
organizing end. × 4,000.

dle layer is the most transparent of the three. The
outer centromere plate is usually present even in
animal cells that have a less defined layered centromere. In flowering plants there seems to be no
layered centromere at all; instead the spindle fibers
emanate from a tangle of chromatin fiber in a ball-
shaped centromere region (Fig. 9.18).

Regardless of the construction of the centromere
region, there must be a functional centromere if chromosomes are to move to the poles at anaphase of nuclear division. This is the specific site for spindle fiber
attachment, and if there is no attachment site there
are no spindle fibers inserted and no way for chromosomes to move directionally. The significance of
the centromere for anaphase movement of chromosomes comes from observations of **acentric** chromosomes (no centromere), and from abnormal situations in which there are two centromeres in the same
chromosome (Fig. 9.19). The two centromeres may be
oriented toward opposite poles; a chromosome bridge
forms under such conditions.

The centromere may be located anywhere along the
length of a chromosome, even at the very tip. But each

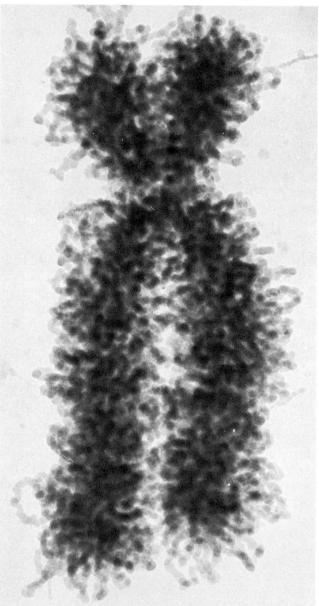

**Figure 9.16**
Electron micrograph of a whole mount of human
chromosome 12. The two chromatids are joined together at
the centromere region. × 60,400. (Courtesy of E. J. DuPraw,
from DuPraw, E. J., 1970, *DNA and Chromosomes*, Holt,
Rinehart & Winston, Fig. 9.10, p. 144.)

chromosome has its centromere located in a fixed position which is constant for that chromosome. Chromosomes in a particular set can be recognized according to their centromere location, as well as relative length, and other characteristics such as secondary constrictions.

Virtually nothing is known of the chemical composition of centromeres, or of their specific contribution and function in chromosome movement other than as a site for spindle-fiber attachment. It is unlikely that centromeres are merely holdfasts for spindle fibers, but we can do little more than speculate at the present time.

## HETEROCHROMATIN AND EUCHROMATIN

The definition of **heterochromatin** in 1928 was based on chromatin that remained in a condensed state during interphase (intervals between nuclear divisions).

Since 1928 there have been additional features assigned to heterochromatin, which distinguish it from the **euchromatin** that becomes extended rather than remaining condensed during interphase (Fig. 9.20). Both types of chromatin are parts of chromosomes, and are most easily distinguished by light microscopy according to the state of condensation seen in interphase nuclei.

An abundance of genetic studies has provided a second feature that is different, namely, that heterochromatin is relatively stable DNA. This feature is interpreted from mutation studies, which show few mutations in heterochromatic parts of chromosomes as compared with numerous mutations mapped in euchromatic regions. Some heterochromatic regions do not even transcribe RNA and in this sense are clearly noninformational DNA regions.

The third feature was noted after 1960 when autoradiographic studies of DNA synthesis in chromosomes

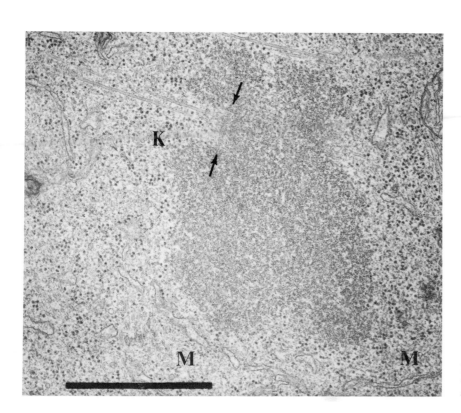

**Figure 9.17**
Electron micrograph of a thin section of Chinese hamster chromosome showing the centromere (kinetochore) region (K). A disclike three-layered centromere passes through the primary constriction (arrows) with a layer of fine fibers around it. Membrane profiles (M) are also present. × 39,000. (Courtesy of E. Stubblefield, from Stubblefield, E., 1973, *Internat. Rev. Cytol.* **35**: 1-60, Fig. 22.)

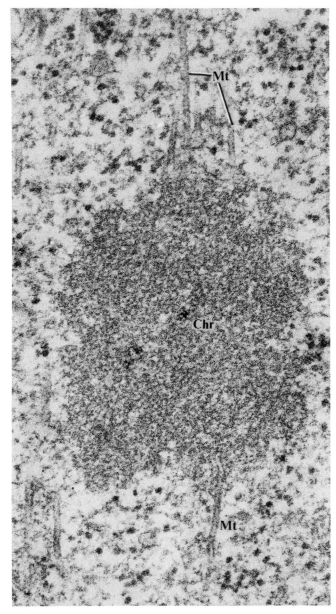

**Figure 9.18**
Mitotic cell of the African violet (*Saintpaulia ionantha*).
Microtubules (Mt) are inserted into the ill-defined
centromere region of dividing chromosomes in flowering
plants. × 81,000. (Courtesy of M. C. Ledbetter)

revealed a later time for replication of heterochromatin
than for euchromatin in the same chromosomes or
same chromosome set (Fig. 9.21). Heterochromatin is
therefore (1) condensed during interphase, which is
otherwise known as the most active metabolizing state
of the nucleus; (2) genetically stable or, at least, subject to very little mutational change and minimal gene
expression; and (c) late-replicating chromosomal material.

There are two types of heterochromatin, facultative
and constitutive. **Facultative heterochromatin** contains
active genes, but may become condensed and
genetically inactive in response to physiological and
developmental conditions, and it may revert to a

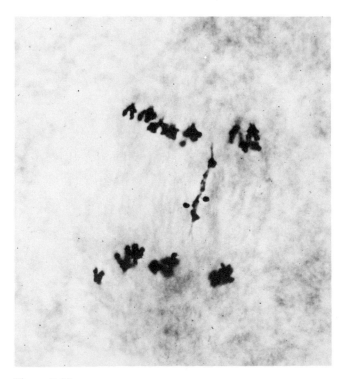

**Figure 9.19**
Anaphase bridge and two acentric chromosome fragments in
the first meiotic division of maize (*Zea mays*). (Courtesy of
M. M. Rhoades)

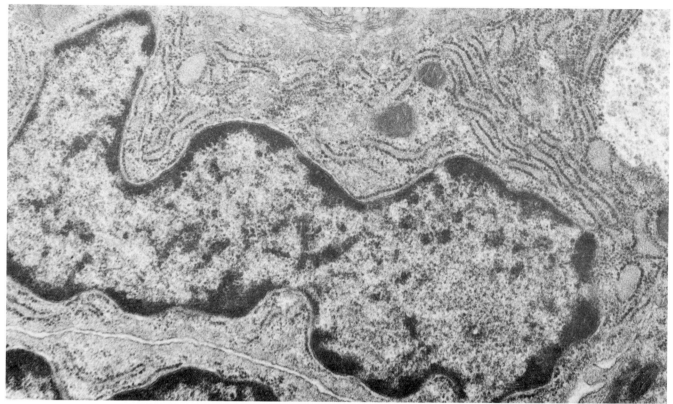

**Figure 9.20**
Thin section of rat osteoblast. The condensed heterochromatin is principally dispersed at
the nuclear envelope, and euchromatin fills the remainder of the interphase nucleus.
× 24,000. (Courtesy of M. Federman)

euchromatic state at certain times. **Constitutive heterochromatin** is permanently condensed, genetically
conservative, late-replicating material all of the time.
The most common site for constitutive heterochromatin is around the centromere region of all the chromosomes in most species studied so far. A well-known
instance of facultative heterochromatin involves the
mammalian X chromosome. There is only one X chromosome in males, and this chromosome is euchromatic along most of its length. Females have two X
chromosomes, but only one of these remains euchromatic during the life of each cell, while the second X

chromosome becomes condensed heterochromatin
very early during embryonic development.

This observation was followed through by Mary
Lyon, who used genetically-marked strains of mice
and showed that *either one* of the two X chromosomes
in female mice could become the condensed, heterochromatic unit. The heterochromatic X chromosome is seen as a dense blob in the nucleus of
nondividing cells (Fig. 9.22). This blob is called a **Barr
body,** and it has provided a simple test for sex
identification in human beings and other mammals. Females have one Barr body per nucleus, this being the

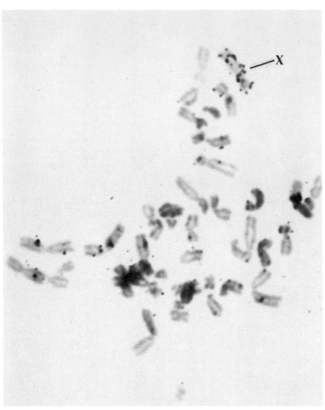

**Figure 9.21**
Autoradiograph showing silver grains over chromosome regions that replicate late in the S period of the human cell cycle. These regions are largely heterochromatic in nature. (Courtesy of L. J. Sciorra)

second X chromosome. Males have no Barr body, since they have only one X chromosome, which remains euchromatic. In some patients with clinical symptoms that include some sex-related aberration, counts of Barr bodies have provided the starting point for a more careful examination of the chromosome complement. For example, men with two X chromosomes and one Y chromosome will have one Barr body (the second, condensed X) in a nucleus. Women with one rather than two X chromosomes per cell will have no Barr body in the cell nucleus. We will discuss some of these features at greater length in Chapter 11, when we deal with medical genetics and cell biology.

## Chromosome Banding

New methods to obtain differentially stained chromosome regions were introduced in 1969. Before this it was possible to stain chromosomes and try to identify them by size, centromere position, and secondary constrictions. But many chromosomes in a complement, or complete set of chromosomes, turned out to be so similar that it was virtually impossible to identify them accurately in every case. The new staining methods now permit accurate identification of each chromosome in a set because each chromosome shows a unique banding pattern after staining, even if size and shape are identical (Fig. 9.23).

There are several types of chromosome banding procedures, but there are two main types of patterns found. The **G-banding** method and others produce a number of bands along the length of a chromosome. It has been suggested that each stained band represents

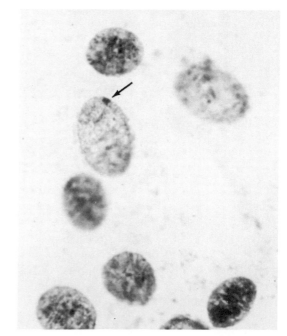

**Figure 9.22**
Barr body, or sex chromatin, in the human female. One of the two X chromosomes remains condensed at interphase, and is located at the periphery of the nucleus (arrow). (Courtesy of T. R. Tegenkamp)

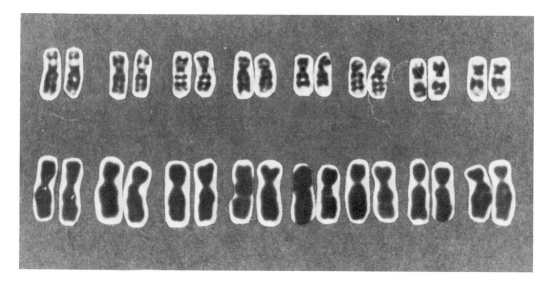

**Figure 9.23**
Photographs of the C-group human chromosomes, numbers 6 to 12 and the X chromosome (second from the left in each row). The G-banding patterns allow each of the chromosomes to be identified (top row), whereas conventionally stained chromosomes (bottom row) are very difficult to identify and recognize individually. (Courtesy of Ann M. Willey)

a heterochromatic region of the chromosome. This remains to be shown by more definitive tests, which have not yet become available. The second type of banding pattern has been called **C-banding,** and its basis is better documented and understood than G-banding.

C-bands develop around the centromere region in almost every case studied, and for virtually every chromosome in a complement. This is the region known to contain constitutive heterochromatin (the basis for naming it C-banding). The C bands apparently develop because much more stainable chromatin remains in this region after preliminary treatment to extract DNA and protein. Since there is so little DNA and protein remaining after extraction elsewhere along the chromosome, very little stain becomes absorbed and most of the chromosome looks pale. More chromosomal materials remain in the C-banded areas and therefore more stain is taken up, resulting in dark, contrasting stained bands around the centromere region. C-banding has a useful function in identifying constitutive heterochromatin regions, even when there is little other information from genetics or autoradiography of replication.

In general, banding has proven useful in routine chromosome identification and in locating constitutive

heterochromatin regions that previously went undetected. Unfortunately, we are not certain about the mechanism of G-banding and cannot make any substantial statements about the nature of heterochromatin from such information.

We do know from other lines of study that heterochromatin and euchromatin are differently folded and compacted regions of the single chromatin fiber of a chromosome. Heterochromatin is considerably more folded and collapsed, one fold against another, resulting in a condensed appearance and deep stainability because there is so much more material per unit area to absorb stains. The mechanisms that lead to different degrees of folding are under intensive study at present.

**Repetitious DNA**

There have been many studies in recent years showing that certain regions of chromosomal DNA are highly repetitious, that is, they consist of numerous repeated sequences of nucleotides in tandem arrangement (one right next to another down the line). There may be relatively few such repeats in some areas and up to 10 million sequence repeats in other regions of the chromosome. Many, and perhaps all, of these regions of repeated DNA are heterochromatic.

Repetitious DNA in any quantity can be separated

(as "satellite" fractions) from the remainder of nuclear DNA by sedimentation at high speeds using CsCl density gradients. The repetitious DNA settles in a different part of the gradient from bulk nuclear DNA, partly because of differences in molecular size and weight and partly because of differences in relative amounts of guanine and cytosine bases in the DNA. We have already encountered one of these repetitious DNAs in our discussion of rDNA at the nucleolar-organizing region. This DNA can be separated from bulk nuclear DNA (see Fig. 9.10), and there are hundreds of repeated copies of the rDNA sequence in the NOR (see Fig. 9.9). The NOR is a heterochromatic region of the chromosome, and apparently contains only the one kind of gene that specifies ribosomal RNA. It is a specialized area of DNA, with a limited genetic capacity and few, if any, known mutations in this region.

Some repetitious DNAs have many more tandem repeats, each of which may include only 7 or 8 nucleotides in a simple sequence that may be repeated 1–10 million times. One of these highly reiterated, **simple-sequence repeat DNAs** is centromere DNA, which we also know is constitutively heterochromatic. The highly repetitious, simple-sequence DNA was separable from bulk nuclear DNA in CsCl density gradients. In 1970, Mary Lou Pardue and Joseph Gall showed that a particular "satellite" repetitious DNA specifically occurred at the centromere region of every chromosome in the mouse complement. They used a method they had first developed to localize repetitious rDNA specifically to the nucleolar organizing region of the chromosome; the method is referred to as *in situ* **hybridization,** which means hybridization "in place".

In the usual situation, purified "satellite" DNA is isolated from CsCl density gradients and then heated to separate the two strands of the DNA duplex molecules. These single strands of DNA are then incubated in a system which permits the transcription (copying) of RNA as complementary strands. If the RNA precursor is labeled with a radioactive isotope such as tritium ($^3$H), it is possible to identify the copied RNA (cRNA) in later tests. Chromosome preparations are obtained on ordinary glass microscope slides, and a harsh chemical is carefully applied to slightly dena-

ture duplex DNA in the chromosome preparation. Next, the labeled cRNA is applied and time is allowed for complementary base pairing between cRNA and separated DNA strands (caused by slight denaturing) in the chromosomes on the glass slide. Excess or unbound cRNA is washed off and the material is covered with a photographic emulsion. Slides are then stored for a long enough time to allow radioactivity to act on the emulsion and produce silver grains. When the slides of mouse chromosomes were examined after *in situ* hybridization, silver grains (therefore cRNA) were present at the centromere region of each chromosome (Fig. 9.24). From these and similar autoradiography studies, centromeric DNA has been found to be one of the highly repetitious, simple-sequence repeat "satellite" DNA fractions of nuclear DNA.

In addition to repetitious rDNA and centromere DNA, at least three other particular repetitious DNAs have been identified and localized in chromosomes us-

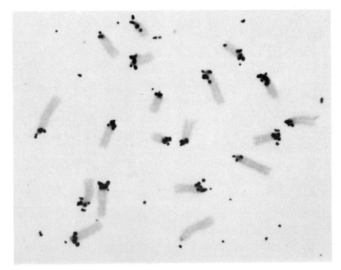

**Figure 9.24**
Mouse tissue culture chromosomes hybridized with radioactively-labeled RNA copied from mouse satellite DNA. Only centromeric heterochromatin is labeled in these acrocentric chromosomes, as seen by the distribution of dark silver grains against the lightly stained chromosomes. × 1,420. (Courtesy of M. L. Pardue and J. G. Gall)

ing the *in situ* hybridization method. These are: (1) genes for 5S RNA found in the larger ribosomal subunit; (2) genes for transfer RNAs that carry amino acids to ribosomal sites of protein synthesis; and (3) histone-specifying genes.

The precise distribution of these three types of repeated genes varies according to the species studied, sometimes being distributed in many chromosomes and sometimes in just one or two of the complement. For example, there are about 200 5S RNA genes distributed in almost all 18 chromosomes in *Xenopus laevis,* but these repeated 5S RNA genes are confined to a single region on chromosome-2 in *Drosophila*. The estimated 750 tRNA genes and about 200 to 300 histone genes are also varied in distribution, according to *in situ* hybridization studies using different species.

The five kinds of repetitious DNA are in known heterochromatic regions of chromosomes, but there must be other specific heterochromatin regions if we are to take the information from G-banding studies as indications of such materials. Except for constitutive heterochromatin at the centromeres, the other heterochromatins are most likely to be facultative in nature. They are capable of gene expression under certain physiological conditions. The rRNA, tRNA, and 5S RNA genes are unique, since their RNA products are the only ones that are never translated into proteins; each of these three kinds of RNA is an end product of gene action. The histone genes produce only five kinds of molecules, which are so similar in eukaryotes that histone from one species may be used equally well by DNA from any other species in chromosome construction tests.

Centromeric DNA cannot possibly code for anything, since 7 or 8 nucleotides are not enough for an informational sequence. In addition, we have more direct evidence of their genetic inactivity. If all the RNA is extracted from cells and presented to centromere DNA in molecular hybridizations, no hybrid duplex molecules form. Clearly, centromere DNA is not even transcribed in living cells. Its function may be strictly structural, but we really cannot say very much with certainty at this time about the way in which its function is directed.

Repetitious DNA is a trademark of eukaryotic nuclei. Prokaryotic DNA has few repeated genes; most of their genes are **unique copy** sequences (one of a kind). There is, of course, a substantial amount of unique copy genes in eukaryotes also, but not exclusively as in prokaryotes. There is reason to believe that repetitious DNA exerts a regulatory role over coordinated gene action in eukaryotes. The organization of repetitious and unique-copy DNA in chromosomes is under intensive study, in hopes of finding relationships between gene expression and chromosome structure, coordinated activities of chromosome sets, and evolutionary changes in eukaryotes as compared with their prokaryotic ancestry.

## POLYTENE CHROMOSOMES

When DNA replication occurs repeatedly and the replicated strands do not separate into individual chromosomes, the resulting multistranded chromosomes are called **polytene** (many threads). Polytene chromosomes are regularly found in certain protists, insects, and flowering plants, among others. The most familiar and distinctive polytene chromosomes are those giant chromosomes found in some larval tissues and organs of the two-winged insect group called the Diptera. Flies, mosquitoes, and midges are examples of dipteran insects. Among the best known dipteran structures are polytene chromosomes from salivary glands of the larvae. Because of this, polytene chromosomes are often called "salivary gland" chromosomes.

Polytene chromosomes from dipteran larvae may undergo as few as four or as many as 15 replications, without separation of the replicated strands. This spans a range of $2^4$ to $2^{15}$, or from 16 to 32,768 strands in a giant chromosome. The nuclei of *Drosophila* larval salivary glands contain a thousand times more DNA than an ordinary nucleus from the adult fly. This indicates that ten replications, without subsequent strand separation, produced the larval polytene chromosomes, and that each polytene chromosome consists of 1024 strands ($2^{10}$).

Because of the increased width of multistranded

chromosomes, polytene chromosomes are much thicker than ordinary chromosomes. They are also very long because they occur in the extended conformation of the interphase nucleus. An added but unusual feature is that partner chromosomes are closely paired along their entire length. This leads us to count an apparent haploid chromosome number even though these cells are diploid (have two chromosome sets). Polytene chromosomes are *banded* because of a natural difference in local DNA concentrations rather than because of any staining that can be applied. An unstained preparation clearly reveals giant, banded, identifiable chromosomes (Fig. 9.25).

### Puffing and Gene Expression

The bands of polytene chromosomes represent DNA-rich segments, while the interbands contain far less DNA per unit area or length. The chromatin fibers are continuous, however, from one end of the chromosome to the other. Different degrees of fiber folding

and compaction explain the differences in DNA per unit area of band and interband.

The chromosome band may exist in either of two alternative states: (1) compact, as wide as the interbands above and below it, with no sign of RNA transcription; or (2) looser-stranded, swollen, and forming a **puff** in which RNA and nonhistone proteins accumulate (Fig. 9.26). A puff is basically a localized decondensation of a chromosome band, but if the puffed region is unusually large and well-defined it is called a **Balbiani ring** (Fig. 9.27). The Italian biologist who is so honored was the first to describe the puffing phenomenon in 1881, but he was unaware of its nature or significance. In fact, these huge structures were not recognized to be chromosomes until 1934.

Since the 1940s a great deal of information has been obtained from these very favorable chromosome systems, which has then been applied to chromosome organization and gene action in general. It is difficult or impossible to study active interphase chromosomes

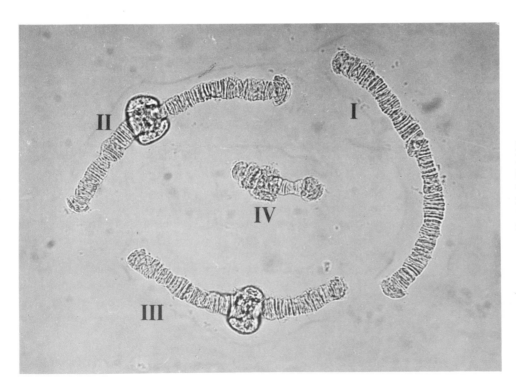

**Figure 9.25**
Phase contrast light micrograph of the chromosome complement from salivary gland nuclei of the midge *Chironomus tentans*. The unstained chromosomes are identified by the roman numerals. Each chromosome is recognizable by its morphology and banding display. The 8 polytene chromosomes are very closely paired in the diploid cell, which gives the impression of only four chromosomes. × 375. (Courtesy of B. Daneholt)

in most nuclei, but the polytene chromosomes give us an admirable interphase system for many kinds of analysis. In particular, there has been very important information gathered to show that genes are not active all the time in all the cells of an organism; rather they are turned "on" and "off" during development.

Using specific stains and autoradiographic methods to localize DNA, RNA, and proteins as well as their sites of synthesis, it was shown that DNA replication did not take place during puffing. Instead, new RNA was synthesized at the puff site, and proteins made elsewhere were then brought to these sites where they complexed with the newly-made RNA. RNA was made only at puffed sites and not elsewhere along the chromosome. If puffing was inhibited by some chemical, RNA was no longer synthesized at the site. Such evidence points to puffing as an expression of active genes, engaged in transcribing RNA; this is probably messenger RNA copied from DNA information. Wolfgang Beermann proposed a model of puffing in which he suggested that a puff arose by unfolding of DNA at the band, and that this unfolding made these DNA looped regions accessible to transcription (Fig. 9.28).

It was Beermann who also provided the first clear demonstration in 1961 of a protein product made at a

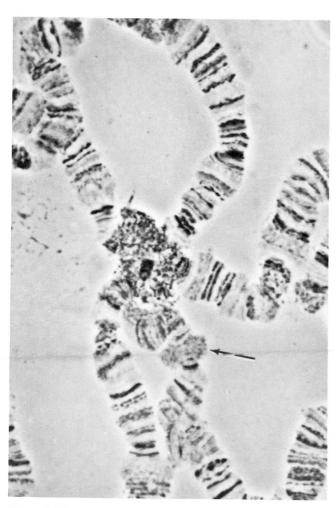

**Figure 9.26**
Light micrograph of a region of stained salivary gland chromosomes of *Drosophila virilis*. A puff is visible (arrow) in the right-hand chromosome below the center. The granular region in which all the chromosomes converge is called the *chromocenter*, and it includes most of the heterochromatin of the complement. × 1,600. (Courtesy of J. G. Gall, from Gall, J. G. *et al.*, 1971, *Chromosoma* **33**:319-344, Fig. 5.)

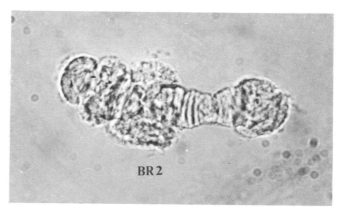

BR 2

**Figure 9.27**
Chromosome IV of *Chironomous tentans* salivary gland nuclei, showing the well-developed Balbiani ring known as BR 2. × 900. (Courtesy of B. Daneholt)

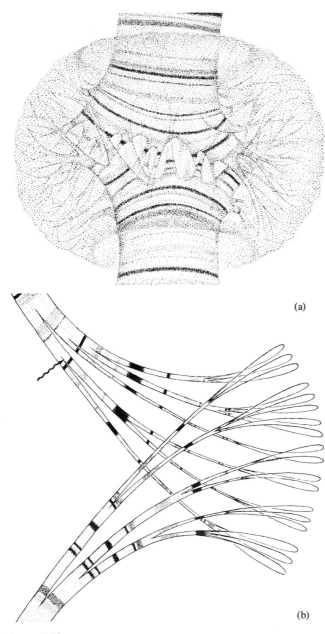

(a)

(b)

**Figure 9.28**
Diagram showing (a) the formation of a puff, as caused by (b) local unfolding of DNA at a band of the salivary gland chromosome. (From "Chromosome Puffs" by W. Beermann and U. Clever. Copyright © 1964 by Scientific American, Inc. All rights reserved.)

specific time in larval development in only a few cells; these cells being the only ones in which a puff formed at a particular band of a particular chromosome in which the gene specifying this protein was located. This demonstrated correspondence between gene action and gene product during puffing was a very important observation.

A known gene in chromosome-4 of *Chironomus,* a midge, is responsible for a protein secretion in salivary glands of larvae just about to pupate, prior to emerging as adult insects. This salivary secretion is used to "glue" the pupa to a solid surface, where it remains during metamorphosis to the adult form. There are only four cells in the entire salivary gland that produce this protein, and it is precisely and only in these four cells that a Balbiani ring forms on chromosome-4 during the period of protein secretion. If there is no puff formed, no protein is made in the cell. In a hybrid formed between two *Chironomus* species, only one of which normally synthesized the protein and developed the particular Balbiani ring, Beermann found that only one of the two partner chromosomes produced a Balbiani ring and only half as much protein was synthesized (Fig. 9.29). The correspondence among puffing, gene action, and synthesis of a specific protein product of this gene action clearly showed that puffing accompanied RNA transcription from coded DNA. It further showed that the same gene was turned "on" only in certain cells and only at certain times in development.

### Regulation of Puffing

In the early 1960s, Ulrich Clever showed that the molting hormone **ecdysone** was directly involved in puff formation in *Chironomus.* When the hormone was withheld from developing cells, the usual puffing pattern was absent and the affected cells did not differentiate. When ecdysone was injected into such cells, puffing took place in exactly the same way as during normal molting in larvae about to pupate.

In other studies Clever further showed that formation of earlier puffs depended on hormonal control, but that later-appearing puffs depended on the formation of the earlier ones and on new RNA and protein syn-

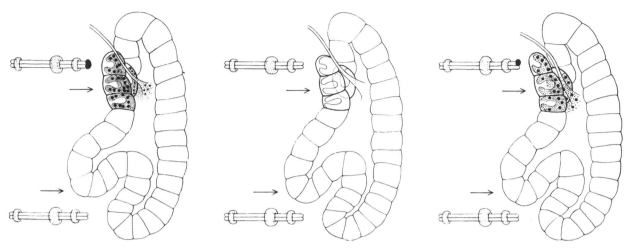

**Figure 9.29**
Diagrammatic summary of Beermann's experiments with *Chironomus* hybrids. There are four cells in the salivary gland of *Ch. pallidivittatus* (left) that produce a granular secretion. A puff is formed at one end of chromosome-4 only in these particular cells. The corresponding cells in *Ch. tentans* (center) produce a clear secretion that lacks these protein granules. No puff forms at the end of chromosome-4. In hybrids (right) between these species, half the amount of protein granules is found in the secretion as in the *Ch. pallidivittatus* parent. Only the chromosome derived from that parent forms a puff in the critical four cells of the gland. Each chromosome-4 contributed by a parent to the hybrid was identifiable by its banding pattern. (From "Chromosome Puffs" by W. Beermann and U. Clever. Copyright © 1964 by Scientific American, Inc. All rights reserved.)

thesis. This is only the beginning of an indication that many kinds of regulatory events and components interact in coordinated pathways leading to differentiation and development of complex tissue and organ systems.

The mechanism of puffing has been explained by two alternative hypotheses:

1. The primary event in the induction of gene action involves detachment of associated proteins from DNA in the chromatin fiber, making the exposed DNA accessible for transcription. Puffing is, therefore, the first event in gene action leading to transcription.
2. Puffing itself may be the result of transcriptional activity of chromosomal DNA at a particular band site.

We cannot decide between these alternatives at the present time, or know whether there may be some other hypotheses that have not been proposed but which may some day prove to be correct.

## SUMMARY

1. The nucleus is the control center of the cell and it is the trademark of eukaryotic cell organization. The nuclear envelope is a two-membrane system which participates in the dynamic exchanges of molecules and particles that lead to coordinated activities of the entire nucleocytoplasmic volume. Contained within the nuclear envelope enclosure are chromosomes, one or more nucleoli, and granular nucleoplasm. There are pore openings in the nuclear envelope formed by a local fusion between the two nuclear membranes at each nuclear pore complex. Ribosomes stud the cytoplasm-facing surface of the outer nuclear membrane.

2. The nucleolus is a secretion of the nucleolar-organizing chromosome, and is located at the nucleolar-organizing region (NOR) of this chromosome along with repeated gene sequences specifying ribosomal RNA. Ribosome subunit

precursor particles assemble within the nucleolus from rRNA produced at the NOR and from proteins shipped in from the cytoplasm. Localization of rRNA genes (= rDNA) at the NOR has been verified by several different methods, including electron microscopy, molecular hybridizations, *in situ* hybridizations, and analysis of suitable mutants.

3. During meiosis, amphibian oocytes are capable of gene amplification in which rDNA replicates while the remainder of the chromosomal DNA does not. Many copies of rDNA are made and packaged in extra nucleoli, where rRNA is transcribed and assembled into subunit precursor particles together with ribosomal proteins made at cytoplasmic ribosome sites. The huge increase in ribosomes provides a reservoir that can be used later on when the embryo develops from the fertilized egg.

4. There is one chromatin fiber in each unreplicated chromosome. Each chromatin fiber is a deoxyribonucleoprotein strand, which is continuous from one end of a chromosome to the other. Genes are housed within the duplex DNA molecules. Histone as well as numerous nonhistone proteins are complexed to the DNA molecule, which alone is responsible for structural integrity of the chromosome. The most widely accepted model of a chromatin fiber is that of a flexibly-jointed chain in which nucleosome units are spaced like beads on a string. The chromatin fiber can be folded and packed into a very small volume. Each chromosome has a differentiated centromere region, which is essential for directed movement of the chromosome to the pole at nuclear divisions.

5. Heterochromatin is condensed during interphase and euchromatin is extended in this stage between nuclear divisions. In addition, heterochromatin replicates later in the cell cycle and it is genetically less active or even inert when compared with euchromatin. Constitutive heterochromatin remains condensed at all times and is found particularly around the centromere of each chromosome. Facultative heterochromatin can revert to the euchromatic state under certain cellular conditions, as we find for any X chromosome in excess of one in a mammalian cell nucleus.

6. Staining methods that produce banded chromosomes have permitted the identification of each chromosome in a set, and provide a tentative location for heterochromatin along the length of a chromosome. Repeated sequences of DNA are characteristic of eukaryotic chromosome organization. Genes known to be present in such clusters include those specifying 45S rRNA precursor, 5S rRNA, tRNAs, and histone proteins. Simple-sequence repeated DNA around the centromere is not informational, and is not transcribed. The intermediate-sequence repeated DNAs are transcribed into nontranslated ribosomal and transfer RNAs and into messengers for histone synthesis. DNA which is not repeated, called unique-copy DNA, represents the bulk of the informational genetic material in a chromosome set.

7. When replicated strands do not separate, a polytene (multistranded) chromosome results. These huge, banded chromosomes are found characteristically in various species and cell types but are especially prominent in certain larval tissues of two-winged insects (Diptera). There may be from $2^4$ to $2^{15}$ strands per polytene chromosome, following 4 to 15 replication events. Puffs and Balbiani rings are sites of active transcription on these chromosomes, and provide evidence of differential gene action in different cells at different stages of development.

## STUDY QUESTIONS

1. What are the three structured components of the cell nucleus, and what are their major functions in the living cell? What is a nuclear pore complex, and how is it structurally organized? What is the significance of the occurrence of pore openings in the nuclear envelope? What is the nuclear envelope?

2. Where would we expect to find nucleoli in the cell nucleus? Why is it essential for a living cell to have at least one nucleolar-organizing chromosome in a chromosome set? What is the nucleolar-organizing region (NOR), and what is the evidence of the kinds of genes present at the NOR? How can molecular hybridization assays indicate the identification of certain DNA, or certain RNA, or the location of particular sequences of DNA in the set of chromosomes? What do we mean by gene amplification? What is the advantage of gene amplification of rDNA in amphibian oocyte development?

3. What is chromatin? a chromatin fiber? a chromosome? What kinds of evidence do we have showing that each unreplicated chromosome contains only one chromatin fiber? What is the role of histones in chromosome construction, and are they necessary to maintain an intact chromosome? What are nucleosomes in relation to the accepted model of a chromatin fiber? How do we explain the extensive folding and packing of a very long chromatin fiber, perhaps 20,000 $\mu$m long, into a condensed chromosome that may only be 1 $\mu$m long? How many daltons of DNA in a duplex molecule 200 $\mu$m long? How long would a duplex DNA molecule be if it had a molecular weight of 100 million

daltons? Would you consider a DNA molecule like this to be small, large, or in between in the usual chromosomes found in eukaryotic cells?

4. What is the importance of the centromere to chromosome structure? What kinds of evidence support your statement? What is a primary constriction? a secondary constriction? Are centromeres usually found in any particular region of chromosomes?

5. What are the major distinctions between heterochromatin and euchromatin? What is the distinction between constitutive and facultative heterochromatin? What is a good example of facultative heterochromatin in mammalian cells? What is a Barr body? How many Barr bodies should we find in cells from a woman? from a man? Why?

6. What is the significance of methods for staining chromosomes to produce banding patterns? What is the difference between G-banding and C-banding? Does banding allow us to identify heterochromatin with certainty? Why not?

7. What is repetitious DNA? What is the difference between simple-sequence and intermediate-sequence repeated DNA, and where should we look for examples of each type? How do we know that there is repetitious DNA at the NOR? at centromere regions? What is *in situ* hybridization? What is the advantage of this method over molecular hybridizations using isolated and purified nucleic acid preparations? What is cRNA and how is it used in locating a particular kind of DNA in microscope slide preparations?

8. What is a major difference between genes for rRNA and tRNA and genes that transcribe mRNA? Is it better to ask what is the difference among these three kinds of RNA products of gene action? Why do we think that centromeric DNA is not informational? How do we know this? How does unique-copy DNA differ from genes specifying histone proteins?

9. What is a polytene chromosome, and how does it arise? How many replications of DNA took place leading to polytene chromosomes with $2^{10}$ strands per chromosome? Does this polytene chromosome have 10 times or 1000 times the amount of DNA as an ordinary chromosome that is not polytene? What is the advantage of the bands of polytene chromosomes in biological studies of gene action and gene location?

10. What is a chromosome puff? a Balbiani ring? What do we think is taking place at a chromosome puff? How could we prove it? How did Beermann show that puffing was a visible sign of gene action? How do we think puffing is regulated?

## SUGGESTED READINGS

Beermann, W., and Clever, U. Chromosome puffs. *Scientific American* **210**:50 (Apr. 1964).

Bishop, J. O. The gene numbers game. *Cell* **2**:81 (1974).

Britten, R. J., and Kohne, D. E. Repeated segments of DNA. *Scientific American* **222**:24 (Apr. 1970).

Brown, D. D. The isolation of genes. *Scientific American* **229**:20 (Aug. 1973).

Brown, D. D., and Dawid, I. B. Specific gene amplification in oocytes. *Science* **160**:272 (1968).

Brown, S. W. Heterochromatin. *Science* **151**:417 (1966).

Cold Spring Harbor Symposia on Quantitative Biology, vol. 38, *Chromosome Structure and Function*. New York: Cold Spring Harbor Laboratory (1973).

Daneholt, B. Transcription in polytene chromosomes. *Cell* **4**:1 (1975).

DuPraw, E. J. *DNA and Chromosomes*. New York: Holt, Rinehart, and Winston (1970).

Gall, J. G. Differential synthesis of the genes for ribosomal RNA during amphibian oogenesis. *Proceedings of the National Academy of Sciences* **60**:553 (1968).

Gall, J. G., and Pardue, M. L. Formation and detection of RNA-DNA hybrid molecules in cytological preparations. *Proceedings of the National Academy of Sciences* **63**:378 (1969).

Kornberg, R. Chromatin structure: A repeating unit of histones and DNA. *Science* **184**:868 (1974).

Miller, O. L., Jr. The visualization of genes in action. *Scientific American* **228**:34 (Mar. 1973).

Oudet, P., Gross-Bellard, M., and Chambon, P. Electron microscopic and biochemical evidence that chromatin structure is a repeating unit. *Cell* **4**:281 (1975).

Pardue, M. L., and Gall, J. G. Chromosomal localization of mouse satellite DNA. *Science* **168**:1356 (1970).

Ritossa, F. M., and Spiegelman, S. Localization of DNA complementary to ribosomal RNA in the nucleolus organizer region of *Drosophila melanogaster*. *Proceedings of the National Academy of Sciences* **53**:737 (1965).

Spiegelman, S. Hybrid nucleic acids. *Scientific American* **210**:48 (May 1964).

Stein, G. S., Stein, J. S., and Kleinsmith, L. J. Chromosomal proteins and gene regulation. *Scientific American* **232**:46 (Feb. 1975).

Yunis, J. J. (ed.) *Human Chromosome Methodology*. New York: Academic Press (1974).

# Chapter 10

# Cellular Reproduction: Mitosis and Meiosis

During the nineteenth century it was firmly established that life comes from pre-existing life, and cells come from pre-existing cells. Each new generation of cells or individuals is the result of **reproduction.** Since progeny resemble their parents, there must be mechanisms that ensure faithful increase and transfer of genetic information. Increase is essential because more copies of the genetic instructions must be made if the progeny are to get all the information they need to grow up and produce their own offspring in turn. Once the genes have multiplied, they must be transferred from parent to progeny if an independent generation is to grow and develop. Both processes, *increase* and *transfer* of genes, must be accomplished with considerable fidelity, or progeny would not resemble their parents, as they do.

The increase in genetic material occurs when DNA replicates. Since we know that new DNA is faithfully copied from template DNA in parent cells, we have the basis for understanding how progeny and parents continue to be more like each other than like unrelated individuals. Transfer of the genetic information in eukaryotes is accomplished by **mitosis** or **meiosis.** These nuclear division processes include accurate systems for distributing chromosomes to progeny nuclei. If reproduction is *asexual,* mitosis is the only mechanism which ensures that daughter cells will receive equal and identical copies of chromosomes from the parent cell.

In *sexually-reproducing* systems, mitosis also takes place, but it is not primarily responsible for the unique sets of chromosomes that will be present in progeny cells. Two processes absent from asexually-reproducing forms but present in sexual life are **meiosis** and **fertilization.** The division of parental nuclei by meiosis leads to halving of the chromosome number in the **gametes.** When two gamete nuclei fuse to form a single **zygote** nucleus, fertilization has been accomplished and the chromosome complement has been returned to its former number.

Sexual cycles are punctuated by meiosis and fertilization episodes (Fig. 10.1). Asexual cycles are maintained by mitosis as the only nuclear division event, and nuclear fusions are not necessary to initiate the next generation. Sexual forms cannot initiate a new generation until nuclear fusion has produced a zygote which has information from two different sources.

Sexual reproduction is therefore often called *biparental,* while asexual reproduction is *uniparental.* These terms are accurate for species where sexes are separated into different individuals, usually males and females. Many sexual forms, however, have both kinds of sex organs so that one individual may produce both **sperm** and **eggs,** the common forms of gametes. Because of many variations on the basic themes, we should view reproduction by asexual means as a system in which one nucleus can lead to a new genera-

REPRODUCTION

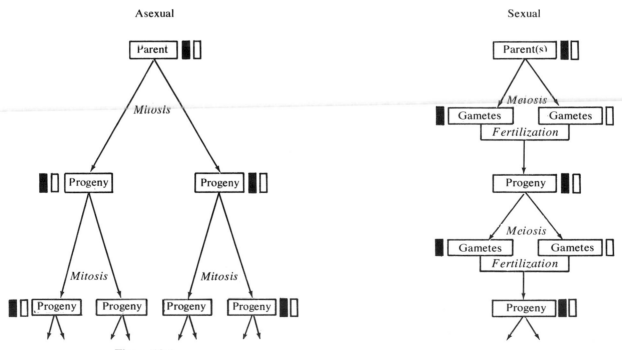

**Figure 10.1**
Schematic diagram illustrating the basic differences between asexual and sexual reproduction.

tion, in contrast with sexual generations which can only be initiated by two nuclei that fuse to form the initial product.

Asexual forms produce offspring exactly like themselves. Sexual species produce progenies that are somewhat different from their parents and from each other. The differences between sexual generations are a reflection of genetic events that lead to *new combinations of genes*. The **segregation** of genes in meiosis and the **recombination** or **reassortment** of genes at fertilization produce and maintain higher levels of variation in sexually-reproducing species. Since evolutionary changes are drawn from gene pools in populations, increased genetic variability is expected to provide increased opportunities for beneficial evolutionary developments. The significance of sexual reproduction in evolution cannot be overemphasized. The focus of this chapter is on the nuclear division processes that lead to new cell generations and on other cell activities that accompany the reproductive events.

### THE CELL CYCLE

We have known about the visible events of mitosis for 100 years. The gradual condensation of chromosomes, their separation to opposite poles of the cell, and their reorganization into new daughter nuclei can be observed by ordinary light microscopy. These dramatic events were a focus for many cell reproduction studies until the early 1950s. At this time the concept of DNA as the genetic material was being established, and new methods were developed and applied to analyze growth and reproduction at the cellular level.

In particular, methods to measure the amount of DNA in nuclei stained with the Feulgen reagents were developed. It had long been known that this staining method revealed DNA by its magenta coloring, but it was now possible to determine how much DNA was present by measuring the amount of stain that was bound to the preparation. Another important innovation in the 1950s was the development of autoradiography. The time and place of DNA replication could be monitored by examining silver grain distribu-

tions, which indicated newly-synthesized DNA containing incorporated radioactive precursors. Both experimental methods clearly showed that DNA replicated hours before there was any visible sign of mitosis.

With this new information, attention was directed toward the seemingly quiet **interphase** stage between mitotic divisions. DNA replicated during interphase. Other studies soon showed that interphase was also the time in which synthesis of proteins and RNA took place. In fact, the most active and vital time in cells was between divisions and not during mitosis itself. As more information was obtained about the **cell cycle,** a gap in time was discovered between the end of mitosis and the beginning of DNA replication. Another gap of time separating the end of replication from the onset of mitosis was also noted (Fig. 10.2).

Following a convention first suggested in 1953, the phases of the cell cycle have been labeled in the following way: The two time gaps are called $G_1$ and $G_2$; the period of DNA replication is called *S;* and the time devoted to mitosis is called the *M* phase. These four phases constitute the nuclear portion of the cell cycle.

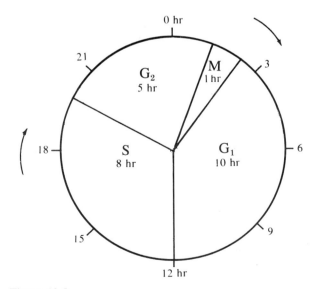

**Figure 10.2**
Stages of the cell cycle and their proportional duration in an average 24-hour mammalian cell cycle.

In most eukaryotic systems, there is an accompanying division of the cell when the new nuclei are enclosed in their own separate cell boundaries. Cell division, or **cytokinesis,** usually begins during the telophase stage of mitosis. While it is not an integral part of mitosis itself, cytokinesis usually is synchronized with mitotic events. We will discuss cell division in another section of this chapter.

### Variations in Cell Cycle Phases

Autoradiographic studies showed that DNA replicated in the $S$ phase and also provided information on the length of the $S$ phase. Since mitosis can also be visualized by microscopy, the time spent in each of the four phases could be measured. $S$ and $M$ phases were measured directly, and $G_1$ and $G_2$ phase durations were inferred by calculating the amount of time which elapsed between $M$ and $S$ ($G_1$ phase) or $S$ and $M$ ($G_2$ phase).

Mammalian cells in culture are favorite materials for cell cycle studies. These cells have relatively similar cell cycles in general, but cycles vary somewhat from one cell type to another. The whole cycle takes 18–24 hours. In a typical adult mammalian cell cultured from human tissue, $G_1$ lasts 8 hours, DNA is synthesized for 6 hours in the $S$ phase, $G_2$ continues for about 4.5 hours, and mitosis is completed within 1 hour. The largest variation in cell cycles of similar cells is usually found in the duration of $G_1$. The length of the $S$ period also varies, but the combined time for $S$ and $G_2$ shows the least change in response to external conditions.

Cell cycle measurements have also been made for cells from higher plants, where a similar 10–30 hour cycle characterizes mitotic cells. Mitosis generally takes more than 1 hour in plant cells, and $S$ is also slightly different in duration in plant cells in comparison with animal cells.

Embryonic cells and many lower organisms show variations of the cell cycle that provide important insights into the general nature of the sequence of events. Many animal embryos undergo rapid division, with smaller and smaller cells produced after each of the divisions. The rate of DNA synthesis is about 100 times faster in these nongrowing embryonic cells than it is in adult cells from the same species, and there is usually no $G_1$ phase. DNA synthesis begins therefore during, or immediately after, mitosis is completed.

Adult *Xenopus* (South African clawed toad) cells have a long $S$ period lasting about 20 hours, while embryonic *Xenopus* cells carry out DNA synthesis for most of the 25 minutes of their cell cycle. In embryonic *Xenopus* cells, there is no $G_1$ phase, $G_2$ is very brief, and the $S$ phase begins before mitosis has been completed. In sea urchin embryos, a 70-minute cell cycle is divided into $M$ lasting about 40 minutes, less than 15 minutes of $S$ (which begins during telophase of mitosis), and 20 minutes of $G_2$. These examples are fairly typical of vertebrate and invertebrate embryo cell cycles.

Many protozoa, fungi, and other lower organisms have no $G_1$ phase in their cell cycle. Like embryonic cells, the cycle lasts a relatively brief time. Other simple organisms, of course, may have a typical $G_1$ phase. The occurrence of $G_1$ seems to be correlated with the amount of growth and biosynthesis going on in the cell. If growth is minimal, as in animal embryos, or very rapid, as in some lower organisms grown under optimum conditions, then $G_1$ is brief or absent altogether. Since DNA synthesis may even begin before mitosis has been completed, the requirement for $G_1$ seems the most dispensable in a cell cycle. There usually is no continued cycling without $S$ and $M$ phases, however, and these rarely occur without an intervening $G_2$.

These observations point out that mitosis takes place after DNA synthesis has occurred. If DNA synthesis stops, the cell will not undergo mitosis; it becomes noncycling, rather than being "stuck" at some point in the cycle. Mitosis cannot begin immediately after DNA replication since there is always a $G_2$ phase, generally of relatively short duration. Other preparations for mitosis must be made, in addition to synthesis of DNA. If protein synthesis is inhibited during $G_2$, the cell will not divide. Very little is known about the particular proteins needed by the cell to enter mitosis, but they probably include structural proteins as well as enzymes. Histones are known to be synthesized at the same time as DNA, during the *S* phase, but nonhistone proteins are made at various times during interphase.

Turning to another part of the cycle, some cells are not ready to begin DNA synthesis when a previous mitosis has ended. These cells have a $G_1$ phase. Other cells have everything they need and dispense entirely with $G_1$. The preparations required for DNA synthesis are not known, but various proteins must be needed, including structural, catalytic, and regulatory proteins for DNA replication and chromosome formation.

The critical transitions in the cell cycle are from the $G_1$ phase to the *S* phase when replication begins and from $G_2$ to *M* when mitosis begins and chromosomes will be distributed to daughter cells. There is a point of "readiness" in cells that enter the *S* phase. When *S*-phase cells are fused with $G_1$-phase cells in specially-treated cultures, the two nuclei of a fusion cell remain separate and each can be identified by its chromosomes. The $G_1$ nucleus begins to synthesize DNA much earlier than it normally would. This experiment shows that something is present in *S*-phase cells which triggers or allows DNA replication to take place.

Other experiments involving fusions between cells in different phases of the cell cycle have been done to test the nature of the second crucial transition from $G_2$ to *M*. In these cases the *M*-phase nucleus causes condensation of chromosomes in $G_1$, *S,* or $G_2$ nuclei in the same fusion cell. $G_1$ chromosomes condense even though they have not yet replicated, and they appear as single chromosomes. $G_2$ chromosomes have already replicated, so they have the conventional organization as they enter mitosis. When *S*-phase and *M* cells are fused, however, small fragments of condensed chromosomes appear from the *S* nuclei. Some feature of *M* cells, therefore, leads to chromosome condensation, even in unprepared nuclei. Unprepared nuclei do not proceed normally through a mitotic division, however, if they are prematurely condensed in a fusion cell. We will review some aspects of DNA replication next.

### Replication of DNA

*E. coli* cells growing at a maximum rate will double their numbers every 30 minutes. DNA is continually synthesized in such cultures. Autoradiographic studies

by John Cairns showed that *E. coli* cells synthesized their new duplex DNA at a rate of about 40 $\mu$m per minute. At this rate, only 30 minutes would be needed to replicate the entire 1300 $\mu$m-long DNA molecule. This prokaryotic DNA has only one point of origin at which each cycle of replication begins (Fig. 10.3). New semiconserved duplexes are formed as replication proceeds in both directions away from the point of origin. The resulting image is similar to the Greek letter *theta* ($\theta$), so molecules like this are referred to as replicating $\theta$-forms.

A new replication cycle can begin before the previous one is finished; this can be seen in autoradiographs which show more than one **replication fork.** When the amount of radioactivity at each replication fork is determined, it is clear that the forks did not begin to form at the same time. Since almost all the cell activities are taken up with DNA synthesis, while other syntheses are going on simultaneously, many investigators do not think that prokaryotes have sequentially-ordered cell cycle activities.

Eukaryotic chromosomal DNA also replicates bidirectionally. But, electron micrographs of replicating DNA isolated from chromosomes show multiple points of origin (Fig. 10.4). Each of these "bubbles" along the duplex DNA represents a bidirectionally-replicating loop, according to autoradiographic analysis.

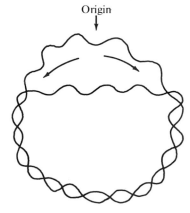

Origin

**Figure 10.3**
Bidirectional replication of the circular DNA molecule of *E. coli*. Newly-synthesized strands are gray.

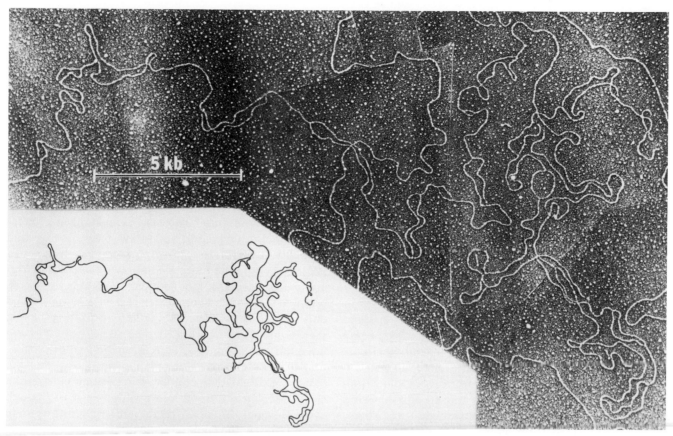

**Figure 10.4**
Electron micrograph and interpretive drawing of replicating chromosomal DNA from
nuclei of cleaving fertilized eggs of *Drosophila melanogaster*. The portion of the molecule
shown here is 119,000 base-pairs long and contains 23 "eye" forms. A kilobase (kb) is a
unit of length equal to 1000 bases or base pairs in single- or double-stranded nucleic acids,
respectively. (Courtesy of H. J. Kriegstein and D. S. Hogness, from Kriegstein, H. J., and
D. S. Hogness, 1974, *Proc. Natl. Acad. Sci.* **71**:135-139, Fig. 1.)

DNA is synthesized at a slower rate in eukaryotes
than in bacteria. The estimated rate for eukaryotic
chromosomal DNA synthesis is about 1–2 $\mu$m per
minute in adult cells in culture. Since the average $S$
phase lasts about 6–8 hours in these vertebrate cell
types, only 720 to 960 $\mu$m of DNA would be replicated
at a rate of 2 $\mu$m per minute (2 $\mu$m × 480
minutes = 960 $\mu$m in 8 hours). There may be 100–200
times this amount of DNA in a chromosome complement. To explain replication of such a large amount of
DNA during the $S$ phase, it has been suggested that
100–200 different points of origin for replication must
exist among the chromosomes. These replicating segments, called **replicons,** can be seen directly in
electron micrographs as regions with "bubbles" or
loops.

The triggering events that initiate DNA synthesis in
the cell are not known. Certain proteins are obviously
needed if replication is to begin and continue, but
external factors are also important in controlling

internal events. For example, some kinds of mammalian cells grown in culture form only one layer of cells attached to the glass dish. These cells become noncycling once the confluent monolayer of cells has covered the glass surface of the dish. If some cells in the monolayer are scraped away, cycling begins again in the cells adjoining the opened space in the dish until the space is refilled. After this, the cell cycle shuts down again.

The shutdown of cell multiplication in confluent cell monolayers has been called **density-dependent** or **postconfluence inhibition.** This phenomenon is separable and probably distinct from contact inhibition of cell locomotion in many such monolayer cultures, but the two kinds of inhibition have often been confused when the single term "contact inhibition" is applied to both phenomena (see Chapter 7). A number of specific signals must be involved in cell cycling because many kinds of cultured mammalian cells continue to grow in masses and do not shut down after a single-layer sheet is formed. Hormone-like agents, ions, and small molecules in the system also exert controlling effects on different kinds of cells under different conditions.

In some cases there appear to be communication sites in adjoining membranes of those cells that stop cycling after forming a confluent single-layer sheet. Cells that go on to form multi-layered masses do not have these **junctions** between cells. The nature of the junction ultrastructure and other features of plasma membranes make it very likely that signals can be communicated among cells of a population in some cases (see Chapter 4).

Cancer cells grow and divide without restraints. They may be cells which are noncycling in normal development but which have reinstituted their cell cycle after transformation to neoplastic growth. There are many variations involved in tumor cell phenomena, since different cells act and respond in different ways to internal and external conditions. At present it seems that in cancer cells the cell cycle restraints are lost. The restraints cannot be restored at this time. The best therapy that can be applied in cancer situations, at present, is to destroy or remove the aberrant cells as selectively as possible. If viruses are found to be the primary causative agent of cancer induction, then perhaps more specific therapies, in the form of antisera and vaccines, can be applied. The basic research problem, however, will continue to be focused on the nature of the cell cycle controls and the ways in which these controls are overridden or swamped by external and internal influences.

## MITOSIS

The continuous sequence of events in mitosis is divided, by convention, into arbitrary stages (Fig. 10.5): **prophase** (*pro:* before), **prometaphase, metaphase** (*meta:* between), **anaphase** (*ana:* back), and **telophase** (*telo:* end). The morphological features of mitotic division are quite well known, but the underlying molecular and biochemical mechanisms which produce the visible events of mitosis are being studied only now. One point, which is known from cell cycle studies mentioned earlier, is that nuclei normally will not enter mitosis unless there has been a prior replication of chromosomal DNA. It is also known that the cycle of chromosome condensation during mitosis can be induced in nuclei from any other part of the cell cycle, since $G_1$, $S$, or $G_2$ nuclei undergo chromosome contraction if they are present in a fusion cell along with an $M$-phase nucleus.

The capacity of cells to undergo DNA replication and mitosis varies across a broad spectrum. Mature muscle and nerve cells do not divide; bone cells stop dividing in the adult; cell division in skin and blood-forming tissues takes place throughout the life of the organism, but is modulated to compensate for losses of old cells; and other tissues divide only in response to some external stimulus such as wounding. Higher plants that live for many years have **meristematic** tissues that grow and divide as long as the plant lives. In these divisions, some cell products remain meristematic and others go on to differentiate into the mature tissues of the roots and stems. Onion or broad bean root-tip mitosis is studied in many introductory biology classes. The dividing cells are part of the root-tip meristem that produces epidermis, vascular

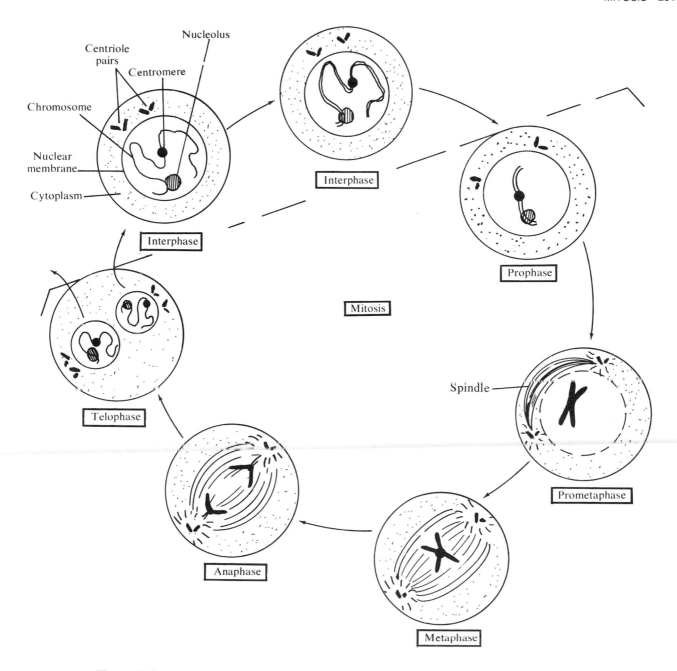

Centriole
pairs
Nucleolus
Centromere
Chromosome
Nuclear
membrane
Cytoplasm
Interphase
Interphase
Prophase
Mitosis
Spindle
Prometaphase
Telophase
Anaphase
Metaphase

**Figure 10.5**
Stages of mitosis. Chromosome replication takes place during interphase, and chromosome
distribution to daughter nuclei is accomplished during mitosis. The chromosome number in
daughter nuclei is the same as it was in the original parent nucleus.

tissues, and other parts of the root at the same time that a reserve of meristem cells is retained.

DNA replication produces the new genetic material for the next cell generation. *Mitosis is a mechanism for distributing this material to the daughter nuclei.* It is a remarkably accurate process that works just as well for a few chromosomes as for a few hundred.

### Stages of Mitosis

The first obvious signs of mitosis, signaling the beginning of **prophase,** are condensations of the chromosomes. As prophase continues, the chromosomes become shorter and thicker. Their morphological characteristics are clearer and individual chromosomes can be recognized. They can clearly be seen as double structures by about mid-prophase (Fig. 10.6). Toward the end of prophase the nucleolus and nuclear envelope disappear, their components becom-

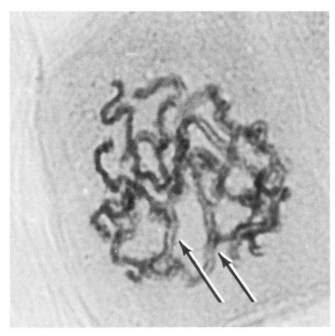

**Figure 10.6**
The dyad structure of mitotic prophase chromosomes from broad bean (*Vicia faba*) indicates that these are replicated chromosomes.

ing dispersed and generally indistinguishable from other parts of the cytosol.

Centrioles, which doubled in number in the preceding interphase, now separate with one pair migrating to each pole of the cell. The spindle microtubules become evident next to the nuclear envelope. After the nuclear envelope disappears, the spindle occupies a more central location, between the two pairs of centrioles. Cells that normally have no centrioles will still have a spindle that fills the position which had been occupied by the intact nucleus.

During **prometaphase,** the contracted chromosomes move toward the equatorial plane of the spindle. Movement is somewhat erratic, as seen in time-lapse photography; the chromosomes do not move unwaveringly toward their ultimate positions. Some individual chromosomes streak across the plane of the spindle while other chromosomes stay put or even jiggle about aimlessly. Prometaphase movement of chromosomes is poorly understood, since spindle fibers are not yet attached. Finally, as though a signal has been given, all the chromosomes line up by their **centromeres** at the equatorial plane of the spindle figure, and metaphase begins.

Each **metaphase** chromosome is a replicated structure, made up of two **chromatids.** These chromosomes are aligned so that the centromere of each chromatid faces the opposite pole of the cell. By this time the chromosome-to-pole fibers are inserted at the centromere of each chromatid, so it is even easier to see that sister centromeres face opposite poles at metaphase.

The relatively brief metaphase is followed by **anaphase,** which begins when chromatids of each chromosome separate (Fig. 10.7). Sister chromatids move to opposite poles of the cell, since their centromeres were aligned in this way at metaphase and their fibrous attachments point toward only one pole. When chromatids separate from each other, each becomes a full-fledged and independent chromosome, no longer acting together with its sister.

Once the chromosomes have arrived at their respective poles, **telophase** begins. During this final stage of mitosis, the condensed chromosomes begin to

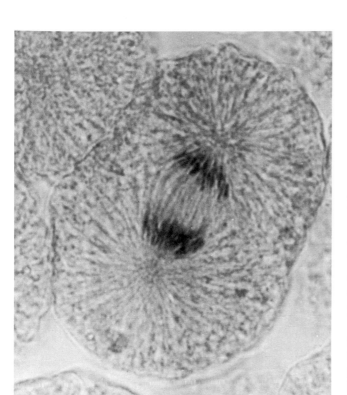

**Figure 10.7**
Mitotic anaphase in a blastula cell of whitefish.

assured of receiving a complete set of genetic instructions, since DNA was duplicated in the $S$ phase prior to mitosis. The *same distribution mechanism* operates successfully for cells that are haploid, diploid, or polyploid; or that are normal or aberrant in chromosome number. This mechanism ensures identical chromosome complements in continuing generations. The fidelity of distribution is responsible for genetic constancy in mitotic generations. Variations can be introduced by mutations or other random modifications of chromosomes, but these variations in turn will be transmitted faithfully to all descendants of the modified cell. Asexually reproducing species are therefore genetically rather uniform. Variability exists between populations rather than within populations in asexual species, depending on chance mutations that occurred and were incorporated into population gene pools.

Each body (somatic) cell in a multicellular organism arises by mitosis, whereas gametes or spores sooner or later will form after meiotic divisions in sexual species. The billions of somatic cells in a human being have the same genetic content, all derived originally from the fertilized egg and all its cell lineages produced by mitosis during development and differentiation. The differences in cell appearance, function, and activity are due to systems of regulation that control gene action, and to other internal as well as external influencing factors. Mitosis delivers the chromosomes, but other processes then direct the expression of genetic potential into the variety of cells, tissues, and organs of the individual.

unfold and gradually assume the appearance they had during interphase. Nuclear reorganization takes place, including organization of new nucleoli and nuclear envelope. Nucleoli form specifically at nucleolar-organizing regions of particular chromosomes in the complement and nowhere else. The new nuclear envelope first appears around individual chromosomes. These membranous elements eventually coalesce to form the continuous nuclear envelope surrounding all the contents of the nucleus. Interphase has then begun.

### Consequences of Mitosis

At the end of mitosis each daughter nucleus contains an identical set of chromosomes, which is also identical to that of the original parent nucleus. By the processes of mitotic distribution, each cell is

### Modifications of Mitosis

As we should expect for biological systems, mitosis in some cells and organisms may be accomplished in a somewhat different way from the average format we just described. Only a few general examples will be described.

One of the commonest variations is *intranuclear* mitosis, which is mainly encountered in some of the lower eukaryotes. The nuclear envelope remains intact throughout mitosis, so that the chromosomes are always compartmented. Apart from this feature, there are other nuclear differences, depending on the species. For example, centrioles may or may not play an active role in spindle formation or in any mitotic events.

The location of spindle microtubules varies. Sometimes they are formed within the intact nucleus, while at other times they are not. In the latter case, there often is a connection between the spindle fiber in the cytoplasm and the intranuclear chromosome. The connection is generally made at the centromere, which is embedded in the nuclear envelope (Fig. 10.8). Even though the nuclear envelope persists, the nucleoli may or may not be retained during mitosis. Chromosomes may be permanently condensed, as in *Euglena* and other species, or they may undergo the usual cycle of unfolding and folding that is typical of most mitotic systems.

Ciliated protozoa such as *Paramecium* and *Tetrahymena* have two kinds of nuclei in the cell. The smaller micronucleus does not undergo mitosis whereas the larger macronucleus does (Fig. 10.9). The macronucleus is a compound structure that contains many copies of the chromosome set. It is the metabolically active component in ciliates. The micronucleus may be lost without causing immediate damage.

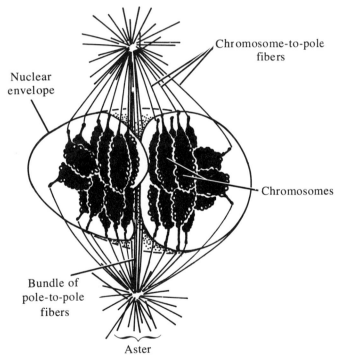

**Figure 10.8**
Drawing of an anaphase nucleus in the protozoan *Barbulanympha* showing the pole-to-pole fibers located outside the nuclear envelope and the chromosome-to-pole fibers attached to the chromosomes within the nuclear envelope where the centromeres are attached. (Redrawn from Cleveland, L. R., 1953, *J. Morphol.* **93**:371-403, Plate 59, Fig. 6. With permission of The Wistar Press, Philadelphia.)

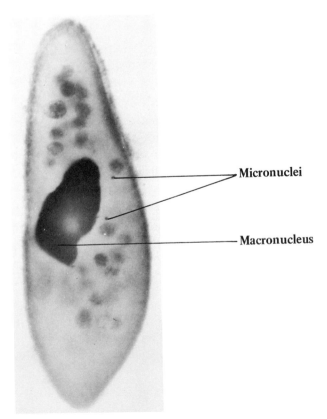

**Figure 10.9**
There are two tiny micronuclei near the large macronucleus in this species of *Paramecium*. × 700. (Courtesy of R. V. Dippell)

It is the component which undergoes meiosis. An amicronucleate cell is therefore asexual. New cells are produced by fissions, but a gradual depletion of the macronuclear material occurs, and such amicronucleate strains cannot be maintained indefinitely.

The polytene chromosomes in larval cells of dipteran insects and in some other species furnish striking examples of permanent interphase systems. DNA replication takes place in repeated rounds, but there is no subsequent separation of the strands in a mitotic division because the *M* phase is missing in this cell cycle.

In other kinds of cells there are rounds of DNA replication, but the new chromosome strands separate from each other. Since the separated chromosomes remain within the confines of the same nuclear envelope, the process leads to an increase in chromosome sets (Fig. 10.10). Cells of some tissues, such as mammalian liver, commonly have nuclei with 4 or 8 sets of chromosomes, rather than the usual 2 sets present in diploid cells in other tissues of the organism.

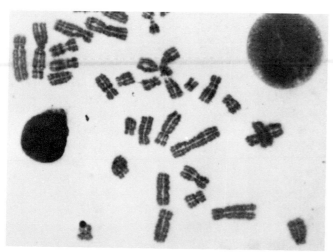

**Figure 10.10**
There have been two rounds of replication without separation of the chromatids (endoreplication). Such aggregates are called *diplochromosomes*. Chromosomes have been stained using a G-banding method that produces a unique band pattern for each of the human chromosomes in the haploid complement. (Courtesy of C. Hux)

The number of chromosome sets can be increased at will by applying colchicine to cells. Since spindle formation is prevented, the replicated chromosomes separate but are not carried to opposite poles of the cell. When nuclear reorganization occurs, the new nucleus has double the usual number of chromosomes.

## CYTOKINESIS

Depending on inherent properties of the species and cell type, division of the cell contents may or may not be synchronized with mitosis. Cell division, or **cytokinesis,** may not take place at all in some organisms. Slime molds have multinucleated protoplasm without a single partition present. Other lower eukaryotes may have one, two, several, or many nuclei per cell, sometimes in regular patterns and sometimes erratically. In the higher eukaryotes cytokinesis and mitosis are usually synchronized, and uninucleate cells are generally produced.

Animal cells typically divide by **furrowing,** also called **cleavage** (Fig. 10.11). This pattern of division is also found in a few kinds of plant cells and among some groups of protists. The first sign of cell division is a constriction or furrowing, in a line with the midpoint of the spindle figure. The furrow becomes progressively deeper, until the cell assumes an exag-

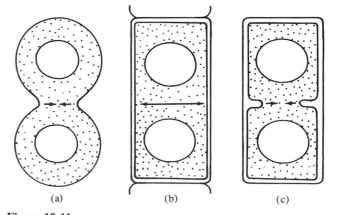

| (a) | (b) | (c) |

**Figure 10.11**
Drawings showing the differences between (a) furrowing and (b) cell-plate formation in higher eukaryotes and (c) new wall formation as it occurs in many algae, fungi, and bacteria.

gerated dumbbell shape. Finally, the cell pinches into two individual and separate daughter cells. Depending on the position of the spindle in the cell, the daughter cells may or may not be of equal size.

Most higher plant cells divide by **cell plate formation** rather than furrowing. The cell plate begins to form in the center of the spindle midpoint, and material is added at both ends until it stretches completely across the width of the cell. Furrowing begins at the cell periphery and moves inward to the cell center; cell plate formation moves outward toward the edges of the cell (see Fig. 6.13). Cell plate materials are provided in part by secretion vesicles made at the Golgi apparatus and partly by membranous elements of the endoplasmic reticulum. After these have been incorporated into the developing cell plate, other substances are added to make up the new cell wall. Cell wall carbohydrates are deposited only on the outside surface of the new plasma membranes which separate the two daughter cells.

Algae and fungi generally divide by a process of cytokinesis that resembles bacterial cell division. In these eukaryotes, the plasma membrane usually invaginates around the cell midline, and new cell wall forms alongside. As the plasma membrane pinches in closer toward the center of the cell, the new wall is extended until the two growing ends of the wall join at the center and completely separate the daughter cells. Since the separation proceeds from the outside inward, this process is similar to animal cell furrowing. The two mechanisms, however, are entirely different in their basic features.

## MEIOSIS

Almost every kind of eukaryotic cell can carry out mitosis at least sometime in its existence, but meiosis is a nuclear-division process that is highly restricted both as to cell type and time of occurrence in a cell lineage. Only cells of sexually reproducing species have the capacity to undergo meiosis, and only special cells in the multicellular individual switch from mitosis to meiosis at specified times in a life cycle (Fig. 10.12).

Since meiosis is a **reduction division,** whose end products have half the number of chromosomes as the original mother nucleus, we should suspect that meiosis is related to **gamete** production. In most animals, some lower plants, and various protist and fungus groups, meiosis leads directly to formation of gamete nuclei. These nuclei can fuse with other

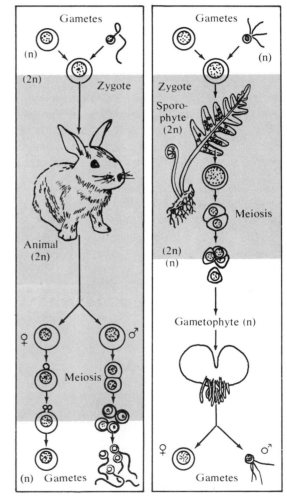

**Figure 10.12**
Comparison of animal and plant life cycles showing the differences in the time of meiosis in sexual reproduction.

gamete nuclei to produce the **zygote,** which is the initial cell representing the new generation. The fusion nucleus has double the number of chromosomes, since one set from each of the two gamete nuclei has merged into a single zygote nucleus.

Higher plants regularly produce **spores** as direct products of meiosis. These spores undergo different degrees of development, depending on the group of plants, and they eventually develop into some gamete-producing system or structure. From such systems, and only from such systems, will gametes then be produced after ordinary mitosis. Plants, therefore, have an intermediate stage in the life cycle, but gametes are eventually formed only from cells that can originally be traced back to a meiotic division. Many variations on the life cycle theme have been described in algae, fungi, and protists.

Every sexual species is recognized by the two decisive events of its life cycle: **meiosis** and **nuclear fusion** to produce a zygote. The kind of fusion called **fertilization** usually refers to union between a larger egg and a smaller sperm. The term has been used in a more general sense, however, and we will use it synonymously with other descriptions of sexual nuclear fusion.

Organs or other structures in which gametes are produced are called **gonads.** The female gonad is an **ovary** or **oogonium** in which eggs are produced. The male equivalent is a **testis** or **spermatogonium** where sperm are formed. The **oocyte** is a specialized female reproductive cell which undergoes meiosis to produce one functional egg and three abortive cells, while the **spermatocyte** is a specialized male reproductive cell which produces four functional sperm after meiosis is completed (Fig. 10.13). In plants, meiosis takes place in **sporocytes** in a **sporogonium,** and spores are the meiotic products (Fig. 10.14). Megasporocytes and microsporocytes (equivalents of oocytes and spermatocytes) produce megaspores and microspores, respectively. Megaspores ultimately develop into egg-producing systems, while microspore development leads to sperm production. Because oocytes, spermatocytes, and sporocytes have the unique property of undergoing meiosis, all can be considered as a variety of **meiocyte.** It is very convenient in general discussion to avoid the specific term and use only the more general description of a cell as a meiocyte. The sequence of events during meiosis is basically the same in all sexual species and in most meiocytes. There are always some variations, however, as might be expected.

The significant features of meiosis are: (1) reduction of the chromosome number by one-half in cells produced by meiosis, and (2) formation of genetically different gametes from a meiocyte or population of meiocytes. The mechanics of meiosis clearly show how chromosome reduction is accomplished, and we will discuss these features first. Afterward we will consider the processes which lead to gene distribution into different meiotic products. All these events contribute to potentially high genetic variability in sexually reproducing species, and to long-term evolutionary advantages.

**The Two Divisions of Meiosis**

DNA replication takes place during premeiotic interphase, in the $S$ phase of the cell cycle. Cells which are haploid normally do not undergo meiosis, whereas cells with two or more sets of chromosomes have meiocyte potential. In a typical diploid species, a diploid cell has meiocyte potential. This diploid or 2X amount of DNA doubles to 4X during $S$ phase. There is no further substantial increase in the amount of DNA during the two divisions that make up meiosis. The two divisions may proceed in immediate sequence or after a period of delay in the first of the two divisions. For example, some meiocytes may remain suspended in prophase of the first division and not continue through to the end of the entire process for months or years. In these cases a stimulus is required for the process to resume. For oocytes to resume meiosis, penetration by the sperm stimulates continuation of the oocyte division. After the egg nucleus has formed at the end of meiosis, it fuses with the sperm nucleus that had entered earlier.

Since there is a 4X amount of DNA in the repli-

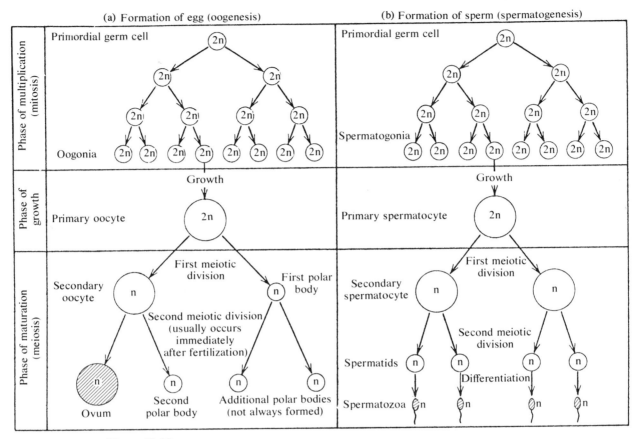

**Figure 10.13**
In higher animals (a) oogenesis leads to one functional egg per oocyte, whereas (b) spermatogenesis typically results in four functional sperm per spermatocyte at the conclusion of meiosis.

cated meiocyte nucleus, the meiotic products would have 2X DNA if division stopped after one round. These would still be diploid cells, and they would be equivalent to any body cell in DNA content. By going through a second division, the meiotic products are reduced to the 1X DNA level and are truly haploid cell products. When two such 1X cells fuse later on, they restore the conventional 2X state of diploid cell DNA. The DNA content of the species remains constant generation after generation because of the cycle of meiosis and fertilization. If meiosis did not precede fertilization, the chromosome number would double endlessly, and this does not happen. Even before meiosis was described in the 1880s, it had already been predicted that a compensating reduction process would be found to explain the constancy of chromosome number in sexual species.

THE FIRST MEIOTIC DIVISION (MEIOSIS I). Meiocyte nuclei proceed through a sequence of prophase, metaphase, anaphase, and telophase intervals during the continuing series of events (Fig. 10.15). We arbitrarily designate stages because of convenience in analysis and discussion. Because the prophase of Meiosis I is the most complex, protracted, and significant interval, it has been further subdivided. The

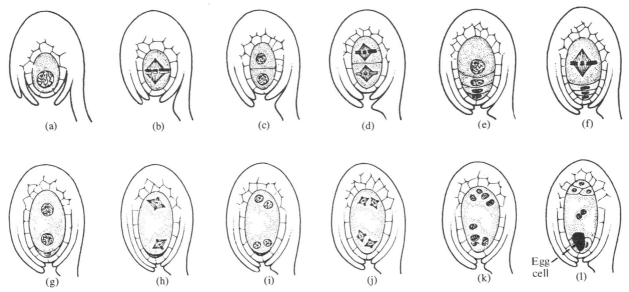

**Figure 10.14**
Development of the egg cell in flowering plants: (a-e) Meiosis occurs in the megasporocyte within the female structures of the flower; (f-k) three successive mitotic divisions take place in the one functional meiotic cell product, leading to eight nuclei; and (l) one of these eight nuclei is enclosed in the egg cell, while the others may or may not proceed to develop when fertilization has been accomplished. Once the chromosome number has been reduced by one half during meiosis, there is no further change in the haploid number until the egg is fertilized and the diploid zygote has been produced.

substages of leptotene, zygotene, pachytene, diplotene, and diakinesis are generally recognized as parts of prophase I.[1]

**Leptotene.** This earliest substage of prophase I is difficult to isolate. The nuclei are highly hydrated and usually suffer some alteration during preparation of the material for microscopy. Where it has been possible to fix and stain leptotene nuclei, the chromatin threads are seen as a jumble of tangles. It is almost impossible to follow any single thread for even a micrometer of its length without losing it in a maze of other threads. The replicated nature of the chromosomes is not apparent by light microscopy.

**Zygotene.** The continuation of prophase I into zygotene is signalled by close pairing of homologous chromosomes. This specific associating process is called **synapsis.** Both sets of chromosomes in a diploid meiocyte have the same gene complements, except for the X and Y or other **sex chromosomes** present. The nonsex chromosomes (**autosomes**) and any kind of sex chromosome present in pairs will begin to synapse at various places along the chromosome length. Synapsis does not seem to begin at a particular part of a chromosome, nor do certain chromosomes pair before the others.

As synapsis takes place a complex structure becomes organized in the space between paired chro-

[1]The suffix *-tene* designates an adjective, whereas *-nema* is used in noun forms of a word. There is some preference for leptonema, zygonema, pachynema, diplonema, and diakinesis among biologists because these are more correct grammatically in references to the names of the stages. They retain leptotene and other adjectival terms to describe structures or activities of each stage. For example, "*leptonema* is the first substage of prophase I and *leptotene* chromosomes are finely dispersed threads of chromatin." I think it will be less confusing to use the *-tene* form throughout, even if it is less correct from a grammatical standpoint.

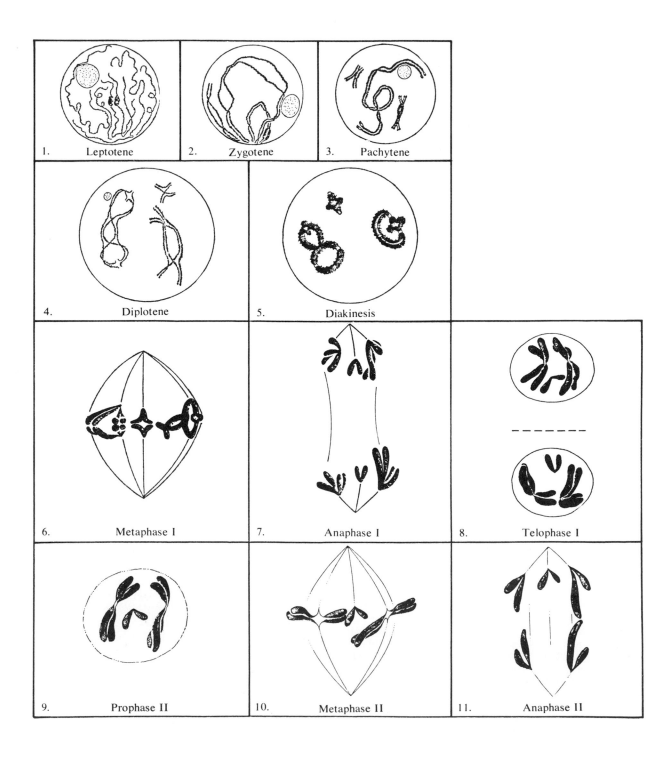

1. Leptotene
2. Zygotene
3. Pachytene
4. Diplotene
5. Diakinesis
6. Metaphase I
7. Anaphase I
8. Telophase I
9. Prophase II
10. Metaphase II
11. Anaphase II

mosomes. This structure is called the **synaptinemal complex,** and it is found uniquely in meiotic cells during early prophase I (Fig. 10.16). When pairing is complete and the synaptinemal complex is fully developed, zygotene is also ended. The paired chromosomes still appear as two single structures, since microscopy does not resolve the doubleness of the strands.

**Pachytene.** This substage is recognized by completely paired homologous chromosomes. The chromosomes have condensed to a greater degree, but do not yet show their replicated condition at this stage. A chromosome pair is called a **bivalent,** and each chromosome of this pair is made up of two replicated halves called **chromatids.** These two terms are very useful in describing the significant events of diplotene and later stages of meiosis.

As pachytene continues the bivalents condense further, but they remain as closely synapsed pairs (Fig. 10.17). In favorable materials the pachytene chromosomes can even be individually identified according to relative length and other features. Far more important, there is evidence that exchanges between homolgous chromosome segments, a process called **crossing over,** takes place during pachytene. We will consider crossing over and related events in a later part of this chapter.

**Diplotene.** This part of prophase I begins with *opening-out* of synapsed chromosomes at various places along their length. The homologues remain associated only at certain places called **chiasmata** (sing.: **chiasma**), which are considered to be sites where previous crossovers had occurred (Fig. 10.18). In favorable materials such as insect spermatocytes, it is clear that *each* chiasma involves only two of the four chromatids at that site of the bivalent. Other chiasmata may involve the same or any other two of the four chromatids, so the total effect is of exchanges involving all parts of the bivalents. Only the individual chiasma is traced to a two-chromatid exchange event.

The diplotene stage may last for weeks, months, or even years in some meiocyte types. Large oocytes from amphibians have an especially striking diplotene chromosome complement (Fig. 10.19). These **lampbrush chromosomes** were so named because of their superficial resemblance to the lampbrush that was a common utensil in the 19th century. The chromatin fibers loop out from a central axis. They are matched pairs of actively transcribing genetic regions. The pattern of loops is constant for each chromosome of a species and serves as one basis for chromosome mapping.

**Diakinesis.** Chromosomes continue to condense during diplotene and are at their most contracted length by the time of diakinesis. Bivalents are short and thick, chiasmata are evident, and the individual bivalents are relatively well spaced within the nucleus. They are very easily counted at this stage (Fig. 10.20). As diakinesis reaches its conclusion, the nucleoli and nuclear envelope disappear and spindle microtubules are visible within the chromosome area.

**Prometaphase.** Chromosomes move somewhat erratically, but congression at the midpoint of the spindle eventually takes place. By the conclusion of this preliminary phase, the bivalents are aligned along the equatorial plane of the spindle, and metaphase I has begun.

**Metaphase.** In meiotic metaphase the ends of the bivalent chromosome arms are positioned on the spindle equator, and the centromeres of the homologous chromosomes are as far apart as physically possible (Fig. 10.21). This situation is the opposite of chromosome alignment during a mitotic metaphase, when centromeres are aligned at the spindle midpoint, and the chromosome arms wave about in all directions on either side of this zone.

Each chromosome of the bivalent is made of two chromatids, and the centromeres of each sister chromatid pair remain closely associated on the same side of the spindle midpoint. They face the same pole, while the sister chromatids of the homologous chromosome of the same bivalent both face the opposite

**Figure 10.15**
The stages of meiosis outlined in diagram form for a meiocyte containing three pairs of chromosomes. (Adapted from Lewis, K. R., and B. John, 1963, *Chromosome Marker*. With permission of J. & A. Churchill, Ltd., London.)

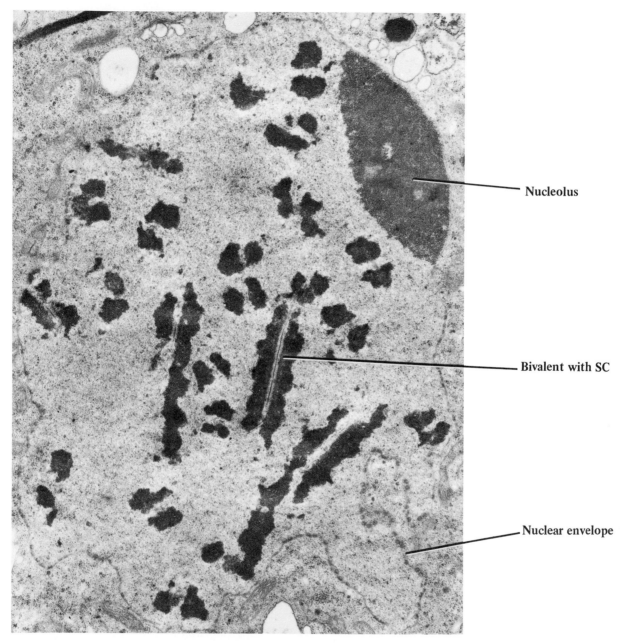

Nucleolus

Bivalent with SC

Nuclear envelope

**Figure 10.16**
Electron micrograph of thin section of a meiotic cell of the ascomycetous fungus
*Neottiella*. One of the pachytene chromosome pairs has been sectioned favorably to
display a synaptinemal complex (SC) in the space between the paired chromosomes of the
bivalent. × 16,000. (Courtesy of D. von Wettstein, from Westergaard, M., and D. von
Wettstein, 1970, *Compt. Rend. Lab. Carlsberg* **37**:239-268, Fig. 1.)

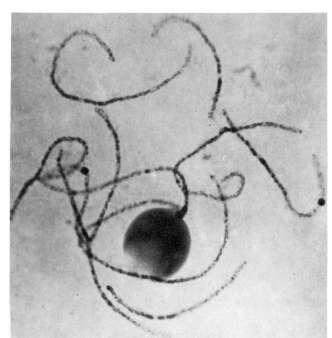

**Figure 10.17**
Pachytene stage of meiosis in maize. The ten pairs of
chromosomes can be identified in such favorable material.
(Courtesy of M. M. Rhoades)

pole. When anaphase begins, the whole duplicated
chromosome separates from its homologue. The duplicated chromosome is called a **dyad,** which describes
its double nature and indicates it is composed of two
chromatids.

**Anaphase.** Anaphase begins when homologous
chromosomes separate and move toward opposite
poles. Each chromosome has the 2X amount of DNA,
so reduction to haploidy is not yet achieved if we view
it from the standpoint of DNA content. On the basis of
chromosome number, however, half the number of
chromsomes are present at each pole after anaphase as
were in the original meiocyte nucleus (Fig. 10.22).
Completely haploid nuclei, in terms of DNA content
as well as chromosome number, will not be formed
until the second meiotic division when individual chromatids of each dyad are separated into individual nuclei. In some species the anaphase I nuclei may enter
almost immediately into the second meiotic division.
In other species there is a telophase stage beforehand.

**Telophase.** Chromosome reorganization takes place
if this stage occurs in a meiotic sequence. Chromosomes unfold to greater lengths, nucleoli and nuclear envelopes reappear, and daugher nuclei are then
evident. Telophase is abbreviated in many species,
and nuclei may go directly to prophase or even
metaphase of the second meiotic division. In other
cases there may be an extended interphase before the
reorganized nuclei divide in Meiosis II. An accompanying cytokinesis may or may not occur at the
end of Meiosis I. In most cases, cytokinesis is delayed
until both divisions are finished, at which time *four
separate cells are formed to enclose the four nuclear
products of meiosis.*

THE SECOND MEIOTIC DIVISION (MEIOSIS II). There is
nothing particularly striking or unusual about Meiosis

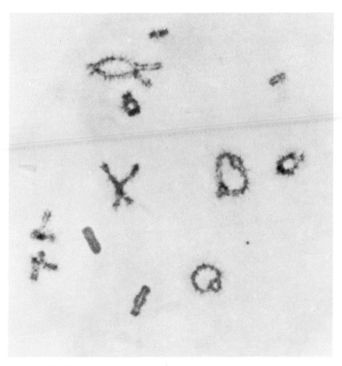

**Figure 10.18**
Diplotene stage of meiosis in grasshopper spermatocyte is
signaled by opening-out of the paired homologous
chromosomes, except at chiasmata. (Courtesy of N. V.
Rothwell)

II. Nuclei proceed through conventional prophase, prometaphase, metaphase, anaphase, and telophase, or at least the last few stages of this division. There is considerable variation among species in their M I-M II transition. Meiosis II is usually described as a "mitotic" division because its stages are similar to conventional chromosome distribution events of a mitosis. It is not mitosis, however, but it is the second set of events in a total meiotic division cycle. The main result of Meiosis II is separation of the dyads into individual chromatids, at which time these become full-fledged chromosomes in their own nuclei. At the end of Meiosis II, the DNA content of each nucleus is reduced to the 1X level which matches the haploid number of chromosomes also present. At this time, each meiotic product has half the DNA content and half the chromosome number of the original meiocyte

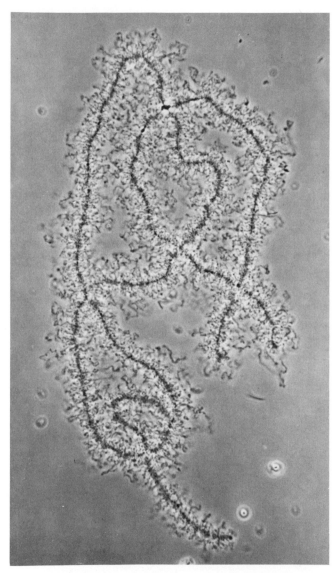

**Figure 10.19**
Phase contrast photograph of unfixed lampbrush chromosome (diplotene bivalent) from the newt *Triturus viridescens*. Three chiasmata hold the paired chromosomes together. × 440. (Courtesy of J. G. Gall)

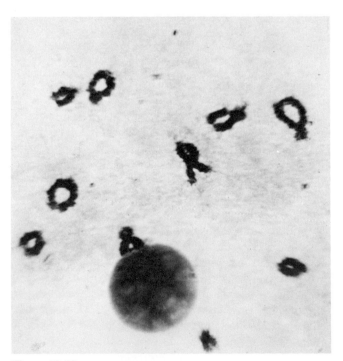

**Figure 10.20**
Diakinesis in maize meiocyte. Note the nucleolar-organizing chromosome (No. 6 in the complement of ten) associated with the nucleolus. (Courtesy of M. M. Rhoades)

**Figure 10.21**
Metaphase I in maize. The centromere is oriented closest to the pole for each homologous chromosome of a pair (note spindle fiber attachments). (Courtesy of M. M. Rhoades)

nucleus. Reduction division is complete only when this point is reached.

Once telophase and cytokinesis are completed, each meiotic product can make its unique contribution to the life cycle. In sexual cycles, a gamete must first fuse with another gamete of a compatible type before a new individual can develop. Spores, on the other hand, can develop directly into new individuals without prior fusion. In fact, the major distinction between a spore and a gamete is the ability of the spore to develop by itself into a new individual.

### Meiosis as a Source of Variability

Variability is the hallmark of living systems. New genetic information arises by **mutation** of existing

genes, and each gene can therefore occur in alternative forms called **alleles.** The particular combination of alleles in an individual is called its **genotype,** and almost every individual in some populations may have a different genotype.

All the progeny of an asexual generation are identical, since the same genotype has been transmitted by mitosis from parent to offspring. The progenies of sexually reproducing species have different genotypes because there are mechanisms that regularly lead to new combinations of genes. The principles of **segregation** and **reassortment** of alleles were first postulated by Mendel in 1866 to explain the results of breeding studies. If progenies are different from those of their parents, the inheritance factors must first be segregated and then recombined in different sets (Fig. 10.23). We know that meiosis is the time that homolo-

**Figure 10.22**
The dyad (replicated) structure of each chromosome is evident during homologue separation at anaphase I of meiosis in maize. (Courtesy of M. M. Rhoades)

gous chromosomes are segregated into different nuclei. If there were different alleles on one or more pairs of homologous chromosomes, alleles would be segregated into different haploid nuclei at meiosis.

When gametes carrying different alleles fuse to form the new generation, **reassortment** of alleles takes place, and the progeny may have different combinations of alleles (genotypes) than were present in their parents. The Mendelian rule of **independent assortment** of alleles states that the proportion of each genotype in the progeny depends on random recombinations. This rule has been verified in every case for alleles of genes *carried on different chromosomes*. As the chromosomes assort at random, so do the alleles they carry. Depending on the numbers of different pairs of alleles and chromosomes, considerable diversity may be produced in each generation in sexually reproducing species, since alleles are segregated regularly at each meiosis and reassorted regularly at gamete fusion.

There may be 5,000 to 10,000 different genes in a species, but there are usually fewer than 50 different chromosomes in the haploid set. How do alleles on the same chromosome become separated and recombined into new genotypes? We know that they do appear in combinations that are different from the parental combinations. Not only do **linked** genes (those on the same chromosome) recombine, but they do so with predictable frequency. Some mechanism must exist that first separates and then recombines alleles originally present within the same chromosome. This mechanism must be a regular feature of the meiotic nucleus, since meiosis is the time that alleles are segregated into haploid nuclei.

We know that the mechanism responsible for recombining linked genes is **crossing over,** which involves exchange of homologous chromosome segments. Recombination of genes on different chromosomes takes place by independent assortment, but recombination between linked genes depends on crossing over.

## CROSSING OVER

Based on studies from cytology (microscopy of cell structures) and genetics, we believe that **crossing over**

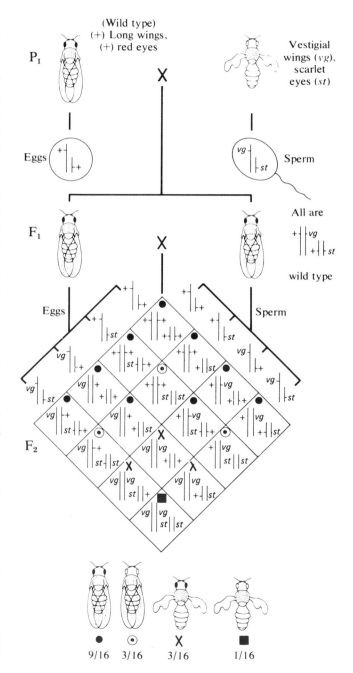

involves a physical exchange between homologous (partner) chromosome segments. The exchange takes place after DNA replication, when each chromosome is composed of two chromatid halves. Two of the four chromatids in a homologous chromosome pair undergo a **breakage and reunion** event, which leads to chromatid exchange (Fig. 10.24). After the broken ends have rejoined, the chromatids involved in a crossover event may have a different combination of alleles. When gametes containing exchanged chromosomes fuse at fertilization, gene recombination is evident in the new kinds of genetic progeny.

Breakage and reunion is the mechanism of crossing over leading to gene recombination in progenies. The breaks and rejoinings are caused by specific enzymes that act on duplex DNA; these enzymes also direct DNA synthesis in general (Fig. 10.25). The visible signs for previous crossovers are the **chiasmata,** which can be seen by microscopy in meiotic chromosomes at late prophase I and metaphase I stages (see Fig. 10.18). A chiasma is the site of a previous crossover or exchange event and often serves as the major evidence of crossing over and gene recombination, especially where genetic analysis is difficult to conduct, as in natural populations of many species.

Crossing over is a regular occurrence in meiosis, but occurs only erratically and infrequently in mitotic nuclei. The basis for this important distinction lies in activities that take place during the earliest phases of meiotic prophase I. At zygotene, homologous chromosomes begin to pair precisely, and they are completely paired when pachytene begins. **Synapsis,** or pairing of homologous chromosomes, brings partner chromosomes together on a regular basis in every meiosis, and provides the opportunities for chromatids to exchange parts during pachytene. The forces re-

sponsible for synapsis are not yet known. But in 1956 Montrose Moses provided the first electron microscopic evidence for a newly-formed structure that helps explain some of the special features of meiosis and crossing over. During pachytene we can see a three-layered **synaptinemal complex** between paired homologous chromosomes (see Fig. 10.16). It is the synaptinemal complex which *stabilizes* paired chromosomes, providing the time and continued access required for crossing over during pachytene.

Each chromosome synthesizes a **lateral element** during leptotene. At zygotene, when synapsis begins, homologous chromosomes begin to pair, and at these paired sites the two lateral elements are held in register by a developing **central region.** The two lateral elements and the shared central region constitute the synaptinemal complex of each bivalent (Fig. 10.26). The complex is about 100 nm wide, and in ways we still do not understand DNA strands of the paired chromosomes meet across this wide space and undergo exchanges during crossing over.

When pachytene is completed and the paired chromosomes *open out* at diplotene, the synaptinemal complex is shed everywhere along the chromosomes except at the chiasmata. These remnants of the complex at chiasmata are also shed sometime before metaphase I. The *regularity of synapsis* and the *stabilization of paired chromosomes* by their synaptinemal complex provide the basis for understanding the *regularity of crossing over* at meiosis. Synaptinemal complexes do not form in mitotic nuclei, so any crossover would result only from chance contact between homologous chromosome parts and the occasional exchanges that these random meetings permit.

There are, of course, other features of synapsis and crossing over that remain to be explained. For example, how do chromosomes *align* before pairing takes place? Since the synaptinemal complex forms *after* pairing, it cannot be responsible for alignment of matched chromosomes and chromosome parts. How do homologous chromosomes "find" each other? We have tentative answers to these questions, but more rigorous evidence is needed to complete the story to everyone's satisfaction.

**Figure 10.23**
Independent assortment. Mendelian inheritance of two different genes, each located on a different chromosome, leads to the familiar 9:3:3:1 ratio of $F_2$ progeny phenotypes when mutant alleles act as recessives to their wild-type alternatives.

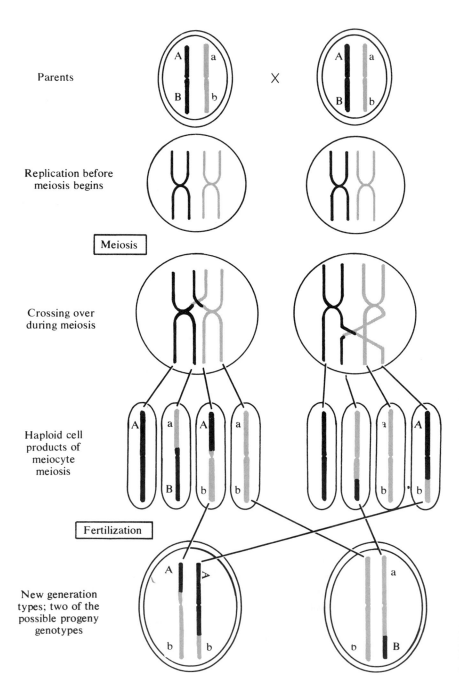

Parents

Replication before
meiosis begins

Meiosis

Crossing over
during meiosis

Haploid cell
products of
meiocyte
meiosis

Fertilization

New generation
types; two of the
possible progeny
genotypes

**Figure 10.24**
Diagram illustrating recombination of
lined genes as the result of crossing over
during meiosis.

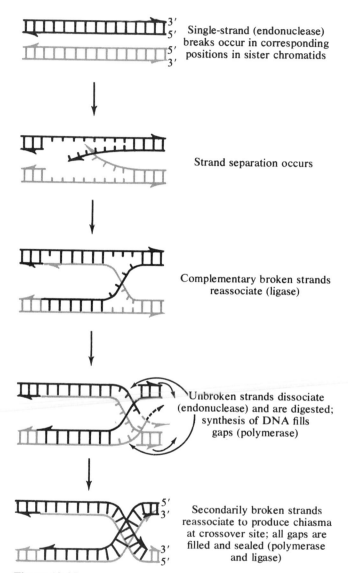

**Figure 10.25**
Molecular model for crossing over and chiasma formation in eukaryotic bivalent chromosomes, according to R. Holliday. Enzymes involved in breakage and subsequent rejoining by DNA-repair synthesis are indicated in parentheses.

The steps shown in the figure are labeled:

Single-strand (endonuclease) breaks occur in corresponding positions in sister chromatids

Strand separation occurs

Complementary broken strands reassociate (ligase)

Unbroken strands dissociate (endonuclease) and are digested; synthesis of DNA fills gaps (polymerase)

Secondarily broken strands reassociate to produce chiasma at crossover site; all gaps are filled and sealed (polymerase and ligase)

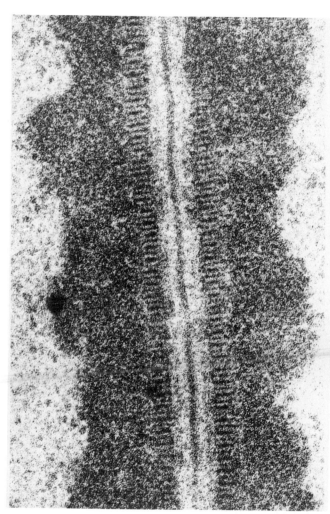

**Figure 10.26**
Longitudinal section through bivalent with synaptinemal complex at pachytene of meiosis in *Neottiella*. Sandwiched between the chromatin of the two homologous chromosomes is a banded lateral element associated with each chromosome and a lighter central region in which a dense central element is situated. × 90,000. (Courtesy of D. von Wettstein, from von Wettstein, D., 1971, *Proc. Natl. Acad. Sci.* **68**:851-855, Fig. 1.)

## SUMMARY

1. The cell cycle, which characterizes populations capable of nuclear division, is subdivided into $G_1$, $S$, and $G_2$ phases of the interphase stage between mitoses, and the $M$ phase of mitosis itself. The $G_1$ phase before DNA replication in $S$ is the most variable, and may even be lacking in some rapidly growing cell systems. The $G_2$ phase, after DNA replication, is less variable in duration. Besides being the time for DNA replication, interphase is the time of most active metabolism in general. Semiconservative replication of DNA usually proceeds in both directions away from the point of origin in a replication unit (replicon). There are many replicons per chromosomal DNA molecule in eukaryotes, but only one replicon in most bacterial DNA.

2. Mitosis is a mechanism for distributing replicated chromosomes into daughter nuclei with great accuracy regardless of the number of chromosomes present. The continuum of mitosis is arbitrarily divided into the stages of prophase (when chromosomes condense), prometaphase (when chromosomes begin to move toward the equatorial plane of the spindle), metaphase (when chromosomes are aligned at the spindle equator), anaphase (when sister chromosomes move toward opposite poles), and telophase (when nuclear reorganization takes place). The new nuclei are identical to each other and to the original parent nucleus from which they arose. Mitosis ensures that a full set of genes is distributed in every cell generation.

3. Various modifications of the mitotic theme have been described, such as intranuclear mitosis in which the nuclear envelope does not break down, and permanent interphase systems such as those leading to formation of polytene chromosomes. Cell division or cytokinesis is often synchronized with nuclear division, leading to uninucleate cells. Furrowing typifies animal cell division while cell plate formation is characteristic of plant cell division. Other modes of cytokinesis have been described for algae and fungi.

4. Meiosis consists of two sequential nuclear divisions which take place only in specific cells at specific times during development. The reduction division process leads to product nuclei with half the chromosome number and half the DNA content of the parental meiocyte nucleus. The chromosome number is reduced by one-half in the first division, and the DNA content of the replicated nucleus is reduced to the 1X amount by the end of the second meiotic division.

5. The prophase stage of Meiosis I is extended, and we recognize a sequence of leptotene, zygotene, pachytene, diplotene, and diakinesis. Homologous pairs of chromosomes move toward the spindle equator in prometaphase and align

there in metaphase I. Partner chromosomes separate at anaphase I, leading to half the number of chromosomes in each resulting nucleus. The 1X amount of DNA in a haploid nucleus is achieved by the end of Meiosis II when sister replicates of each chromosome separate at anaphase II.

6. The haploid products of meiosis directly or eventually give rise to gametes. When two haploid gametes fuse in a sexual event, the resulting zygote cell has the diploid chromosome number. Each sexual cycle is punctuated by meiosis and gamete fusion (fertilization), which accounts for the constancy of chromosome number in a species.

7. New combinations of alleles arise regularly in sexual species in every generation, since alleles are segregated into separate cells at meiosis and come together in various combinations when gametes fuse. Alleles of genes on different chromosomes reassort independently during randomized gamete fusions. Alleles of genes on the same chromosome (linked genes) recombine into new combinations if there is an exchange of homologous chromosome segments by breakage-and-rejoining during crossing over.

8. The regularity of synapsis during zygotene of Meiosis I and stabilizing of the paired homologous chromosomes by synaptinemal complex formation in pachytene provide the basis for understanding the regularity of crossing over during meiosis and production of recombinations in sexual species. The synaptinemal complex is a three-layered structure that forms between partner chromosomes after pairing has begun, and it is shed from the paired chromosomes after crossing over has been completed, before the end of prophase I. Chiasmata are visible signs of previous crossover events and are easily recognized in microscope slide preparations of meiotic nuclei in diplotene, diakinesis, or metaphase I stages of division.

## STUDY QUESTIONS

1. What is the basis for progeny resembling their parents? for progeny receiving a complete set of genes from each parent generation after generation? How do asexual and sexual reproduction differ? How are they similar? What are the two critical events in every sexual reproductive cycle? Why are they critical? When do alleles of a gene segregate and when do they reassort or recombine in sexual species?

2. What are the recognized phases of a typical cell cycle, and what events take place during each of these phases? What part(s) of the cell cycle occur during interphase? What phase of the cell cycle is missing in various rapidly-growing cell populations? What phase is missing in cells that have polytene chromosomes? What is a noncycling cell? What is

the evidence showing that there is some feature of cells in *M* phase which triggers chromosome condensation?

3. What is the physical appearance of a duplex DNA molecule with one point of origin for DNA replication? with many replicons? What is the difference between density-dependent inhibition and contact inhibition phenomena of cells in culture? How are these observations believed to be related to uncontrolled growth and multiplication in cancer cells? Why can't we develop better means of cancer therapy using this knowledge?

4. What are the major stages of mitotic division of the nucleus? How would you recognize a particular mitotic stage from observations of dividing cells seen through the light microscope? What is the major significance of mitosis? What is the most likely basis for the remarkable accuracy of mitotic divisions of the nucleus? What is a chromatid, and how is it different from a chromosome? Why does mitosis permit eukaryotic species to have any number of chromosomes, from two to two hundred or more, and to maintain the particular chromosome number generation after generation without many mistakes? If progeny cells produced by mitosis are genetically identical, how can we explain the presence of so many kinds of cells in our own bodies, when all these cells arose by mitosis from an original fertilized egg?

5. Describe at least three kinds of modifications of mitosis, explaining why they are considered to be modifications on a basic theme. Why do cells of higher organisms tend to be uninucleate? How do cell divisions differ in plants and animals? What is the possible contribution of the Golgi apparatus to cell-plate formation in plants?

6. Why do we call mitosis an "equational" division and meiosis a "reductional" division? What is the immediate or eventual outcome of meiosis in terms of cell function? What happens to plant spores produced by meiosis? What is a gamete? a zygote? a gonad? an oocyte? a spermatocyte? a sporocyte? a meiocyte?

7. What are two genetically significant results of meiosis? Why are they significant? When does DNA replicate in the meiocyte? What is the DNA content of the meiocyte after replication? What is the chromosome number of the meiocyte before replication? after DNA replication? During which of the two meiotic divisions is the chromosome number reduced by one-half? When does the nucleus contain the 1X amount of DNA if the meiocyte contains the 4X amount of DNA at the onset of meiosis? Why is it necessary for both divisions to be completed before meiosis reaches its conclusion?

8. Describe the prophase of Meiosis I in terms of appearance of chromosomes, pairing of chromosomes, and occurrence of chiasmata. What is the consequence of anaphase I of meiosis? What sorts of differences do we find in various species between anaphase I and metaphase II of their meiotic divisions? Are these important differences?

9. Discuss meiosis in terms of segregation of alleles of genes. Why is there a steady production of variability in sexually-reproducing species which is not found in asexual species? Why is this an important distinction among living organisms?

10. What is crossing over? How does crossing over take place? When does it take place during meiosis? What are the visible signs of previous crossover events in meiotic cells? What is the significance of chromosome synapsis? What is the function and significance of synaptinemal complex formation? Why does crossing over take place regularly during meiosis and not during mitosis in cells of the same organism? What is the significance of crossing over in sexual species?

## SUGGESTED READINGS

Hanawalt, P. C., and Haynes, R. H. The repair of DNA. *Scientific American* **216**:36 (Feb. 1967).

Hartwell, L. H. Genetic control of the cell division cycle in yeast. *Science* **183**:46 (1974).

Mazia, D. The cell cycle. *Scientific American* **230**:54 (Jan. 1974).

Meselson, M., and Radding, C. M. A general model for genetic recombination. *Proceedings of the National Academy of Sciences* **72**:358 (1975).

Meselson, M., and Weigle, J. J. Chromosome breakage accompanying genetic recombination in bacteriophage. *Proceedings of the National Academy of Sciences* **47**:857 (1961).

Mitchison, J. M. *The Biology of the Cell Cycle.* Cambridge: Cambridge Univ. Press (1971).

Moens, P. B. The structure and function of synaptinemal complexes in *Lilium longiflorum* sporocytes. *Chromosoma* **23**:418 (1968).

Nicklas, R. B. Chromosome segregation mechanisms. *Genetics* **78**:205 (1974).

Potter, H., and Dressler, D. On the mechanism of genetic recombination: Electron microscopic observation of recombination intermediates. *Proceedings of the National Academy of Sciences* **73**:3000 (1976).

Stern, H., and Hotta, Y. Biochemical controls of meiosis. *Annual Reviews of Genetics* **7**:37 (1973).

Westergaard, M., and von Wettstein, D. The synaptinemal complex. *Annual Reviews of Genetics* **6**:71 (1972).

# Chapter 11

# Cell Biology in Medicine

In recent years there has been a considerable infusion of basic cell biology into medical research and medical practice. Some of the more important contributions have come from basic and applied studies involving human chromosome analysis, inherited diseases that involve some altered cell structure or vital molecule, problems of cellular recognition and communication between cells through cell surface properties and activities, and fundamental information about the immune responses in the body's defense against disease. We will look at some of these studies.

## HUMAN CHROMOSOME STUDIES

In 1956 it was first shown reliably that there were 46 chromosomes in human cells. Before this milestone accomplishment, based on improved methods for preparing and staining human cells, there were several erroneous reports. The first reported counts of human chromosomes were made in 1912, when it was stated that men had 47 chromosomes, including one sex (or X) chromosome, while women had 48 chromosomes of which two were X chromosomes. This inaccurate report was contradicted in the early 1920s when the very small Y chromosome was detected in males. The count was therefore revised so that both men and women had 48 chromosomes, men being XY and women being XX. This 1923 published account was retained for the next 33 years, but was finally shown to be wrong in 1956 when XY males and XX females were each shown to have a total of 46 chromosomes (Fig. 11.1).

### The Human Chromosome Complement

Human cells can be collected directly for study or can be grown in culture and used in various experimental situations. The most commonly studied cells

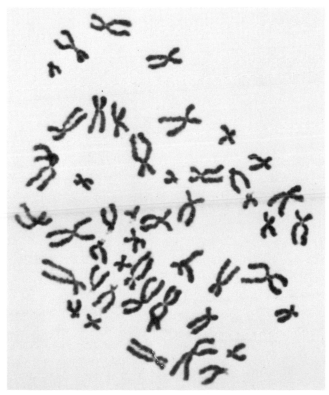

**Figure 11.1**
The 46 chromosomes of the human diploid nucleus. The two chromatids of each replicated chromosome are joined at the centromere region. (Courtesy of L. J. Sciorra)

are **lymphocytes** (a type of white blood cell) and **fibroblasts** (a type of connective-tissue cell). Whether taken directly from an individual in a blood sample, or from cells grown in culture, chromosomes are best seen in mitotic metaphase. The cells are stimulated to divide, and then arrested in metaphase upon addition of colchicine. Suitable metaphase chromosome sets are located by light microscopy and photographed. The individual chromosomes are cut out of the photograph and arranged in a conventional way to construct the **karyotype** of the individual (Fig. 11.2). The chromosomes are arranged with the largest first, proceeding through intermediate-sized, with the smallest positioned at the end of the series. Before new methods were developed to stain chromosomes having banded patterns, it was rarely possible to distinguish one chromosome from another in a group of similar size and shape. For this reason chromosomes were first grouped conventionally into categories from A to G. Since we can now recognize each chromosome by its banding pattern, we refer more often to a specific chromosome number. On occasion, however, it is still convenient to talk about a group of chromosomes, such as the D group, G group, and so on (Fig. 11.3).

Using photographed metaphase chromosome sets directly, or after they have been arranged into a karyotype, it has been possible to correlate certain clinical symptoms with variations in chromosome number or chromosome structural alterations. Because of the relatively short time since the correct number of 46 chromosomes was announced, there have been studies mainly of the more commonly occurring chromosome anomalies. New discoveries continue to be made as rare conditions are occasionally found in population studies or in cells from patients and other sources.

### Sex Chromosome Anomalies

We were unaware of the basis for sex determination in human beings and other mammals until the 1960s. Before this it had been assumed that our sex determination system was the same as that worked out in *Drosophila* and other insects in the earlier years of combined studies of gene and chromosome behavior, or **cytogenetics.** The various combinations of **sex chromosomes** and nonsex chromosomes, or **autosomes,** showed that *Drosophila* females had a ratio of 1.0 for X chromosomes to sets of autosomal chromosomes (X:A ratio). Males, on the other hand, developed when the X:A ratio was 0.5, that is, one X chromosome per two sets of autosomes (Table 11.1). The Y chromosome was irrelevant to sex determination in *Drosophila;* males developed in XY as well as in XO

(pronounced ex-oh) individuals who lack a second sex chromosome altogether. In fact, many insects of the grasshopper and other groups lack a Y chromosome altogether, and their males are always XO while females are XX. Any ratio other than 0.5 or 1.0 led to aberrant individuals of some sort, according to this **Sex Balance Theory** of sex determination.

When it became possible to study human and other mammalian chromosomes, it was soon realized that the mammalian Y chromosome was a crucial com-

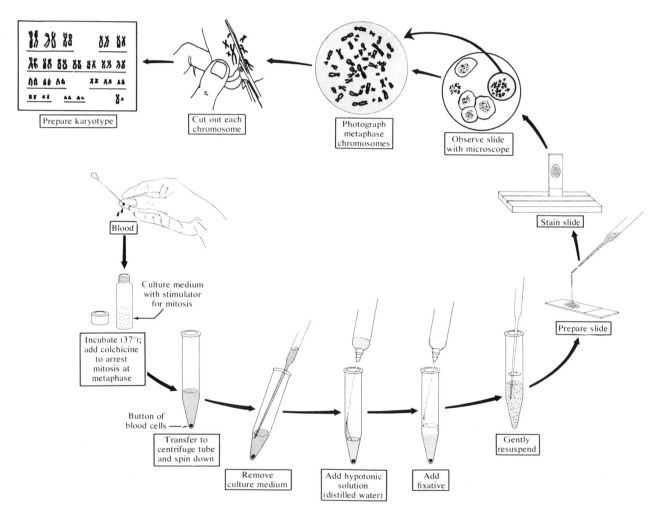

**Figure 11.2**
Flow diagram illustrating the procedures for karyotype preparation.

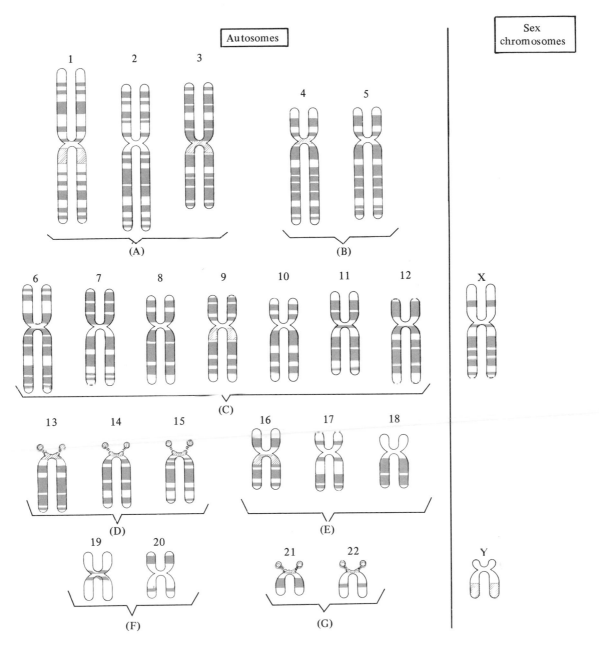

**Figure 11.3**
The twenty-two autosomes and two sex chromosomes of the human chromosome
complement, arranged according to size and centromere location into seven groups (A to
G) and numbered individually. The satellited chromosomes 13, 14, 15, 21, and 22 are all
nucleolar-organizing.

**TABLE 11.1** **Chromosomal basis for sex determination in** *Drosophila*

| Sex Chromosomes Present | Number of Autosome Sets | Ratio of X : A | Sex of Individual |
|---|---|---|---|
| XX | 2 | 1.00 | Female |
| XY | 2 | 0.50 | Male |
| XXX | 2 | 1.50 | Metafemale (sterile) |
| XXY | 2 | 1.00 | Female |
| XX | 3 | 0.67 | Intersex |
| X | 2 | 0.50 | Male (sterile) |

ponent in sex determination (Fig. 11.4). If a Y chromosome is present, even when there is an abnormal number of X chromosomes also present, the individual is male (Table 11.2). Males normally are XY. If there is no Y chromosome present, regardless of the number of X chromosomes, from one to four or five, the indi-

vidual is female. Females normally are XX. Two particular sex chromosome combinations show the difference between the Sex Balance system in *Drosophila* and the system in mammals: the XO and the XXY situations. In mammals, XO individuals are females whereas XO flies are males. In mammals, XXY individuals are males whereas XXY flies are females.

There is no known case in human beings or other mammals of an individual entirely lacking in X chromosomes; at least one is required for life. The Y chromosome obviously is not essential for life, since females never have a Y chromosome. The sex-deter-

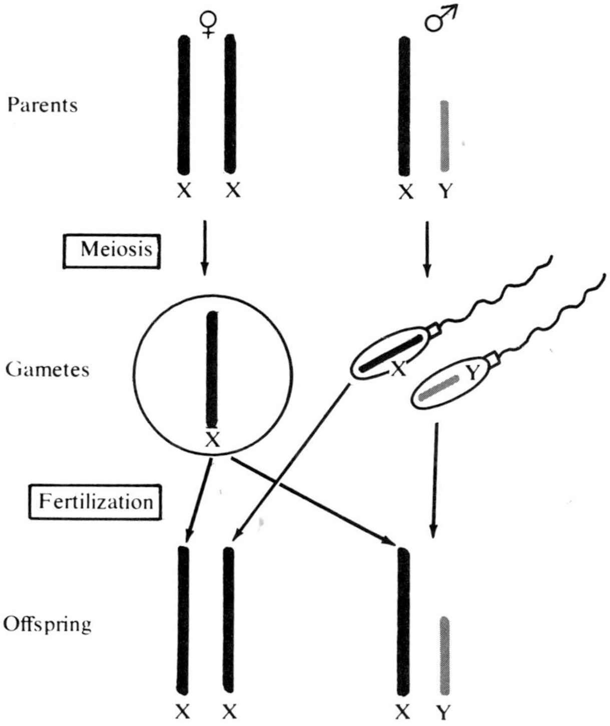

**Figure 11.4**
The chromosomal basis for sex determination in human beings and other mammalian species.

**TABLE 11.2 Anomalies involving sex chromosomes in humans**

| Individual Designation | Chromosome Constitution* | Sex |
|---|---|---|
| Normal male | 46, XY | Male |
| Normal female | 46, XX | Female |
| Turner syndrome | 45, X | Female |
| Triplo-X | 47, XXX | Female |
| Tetra-X | 48, XXXX | Female |
| Penta-X | 49, XXXXX | Female |
| Klinefelter syndrome | 47, XXY | Male |
| Klinefelter syndrome | 48, XXXY | Male |
| Klinefelter syndrome | 49, XXXXY | Male |
| Klinefelter syndrome | 48, XXYY | Male |
| Klinefelter syndrome | 49, XXXYY | Male |
| XYY-male | 47, XYY | Male |

*The two-digit number indicates the total number of chromosomes in diploid cells, followed by the exact number and kinds of sex chromosomes in this complement.

mining genes on the X and Y chromosomes are believed to act only during the first months of human embryo development. By about 6 weeks, the embryo begins to produce male hormones if a Y chromosome is present. These hormones influence differentiation of hitherto unspecified gonad tissues which develop into testes. External genitalia differentiate in the fetus at 3 months. If development continues normally in the fetus and after birth, male reproductive tract and other secondary sex characteristics will appear under the directing influence of male sex hormones produced in the testes. Apart from these sex-determining genes there is no unambiguous evidence that other genes are present on the Y chromosome in mammals.

If the embryo has no Y chromosome, its gonads begin to differentiate into ovaries by about the twelfth week. Internal reproductive structures differentiate, as do the external genitalia, and both **primary** (gonad) and **secondary** (all other) **sex characteristics** continue to develop under hormonal influences and gene direction. Many genes on the X chromosome are totally unrelated to sex or sexual development. There is ample evidence from inheritance studies for **sex-linked genes,** that is, genes on the X chromosome. We will discuss these again shortly.

There are two main types of sex chromosome anomalies in human beings. One type is characterized by *45,X* chromosome constitution (total number of 45 chromosomes, with only one sex chromosome present), and clinical symptoms in these females that are usually referred to as **Turner syndrome** (Fig. 11.5). The second frequent sex chromosome anomaly is called **Klinefelter syndrome,** and it is expressed in men with two or more X chromosomes in addition to the Y chromosome. The commonest of these is the *47,XXY* male. About 1 in every 500 live male births every year is a Klinefelter male; about 1 in every 1200 live female births is a Turner female. Most *45,X* conceptions result in miscarriages, but there seems to be a low miscarriage rate for women carrying a *47,XXY* fetus.

It is a simple matter to determine whether an individual male or female has too few or too many X chromosomes. By taking a sample of cells scraped from the lining of the cheek inside the mouth, standard in-

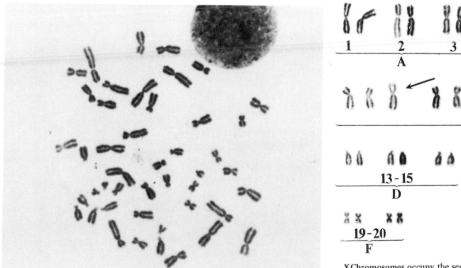
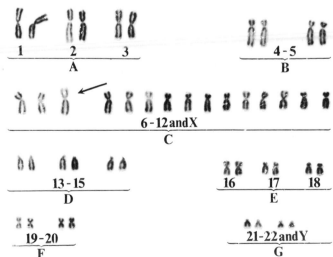

X Chromosomes occupy the second position in the C group

**Figure 11.5**
Chromosomes of a *45,X* female with Turner syndrome: (a) metaphase chromosome spread; (b) karyotype prepared from the metaphase chromosomes photographed in a. (Courtesy of T. R. Tegenkamp)

terphase nuclei can be stained and examined routinely by light microscopy. We know that one X chromosome always remains in the extended euchromatic configuration, and that all other X chromosomes become condensed, heterochromatic blobs. These condensed X chromosomes are called **Barr bodies** (Fig. 11.6). Any male with one or more Barr bodies must have one or more extra X chromosomes; any female lacking a Barr body must be XO. Females with more than two X chromosomes have more than the usual one Barr body. Barr body counts, together with other kinds of information, can therefore prove useful in preliminary screening and in subsequent medical decisions about each case.

Until 1965 there were only a handful of known cases of *47,XYY* males. The current statistical studies indicate that about 1 in 650 live male births is this chromosome type. There seems to be no general effect on fertility or intelligence, and there is probably little or no correlation between the *47,XYY* chromosome constitution and "antisocial" behavior, as had been suggested in earlier and less careful studies. Women with Turner syndrome usually have some physical symptoms, in addition to their being sterile because the internal reproductive structures do not differentiate. There is no particular correlation, however, between the XO condition and intelligence. Klinefelter males, on the other hand, are usually sterile and show increasingly high mental retardation with increasing numbers of extra X chromosomes. Many *47,XXY* males have normal intelligence. Women who are *47,XXX* have been found in random samplings in populations. They are outwardly and intellectually the same as *46,XX* females. Women who are *48,XXXX* are mentally retarded, however.

### Autosomal Anomalies

The first report of a human chromosome anomaly was made in 1959, and described a case of 47 chromosomes in a patient with **Down syndrome** (formerly called "mongolism"). The extra chromosome is chromosome-21 (Fig. 11.7). The condition is also referred to as **trisomy-21.** By trisomy we mean that there are three (*tri*) of a particular chromosome (rather than the

usual two), and the number that follows tells which chromosome is in excess. There are a few other trisomies that have been reported, but trisomy-21 accounts for 96 percent of all these autosomal chromosome anomalies. The frequency of trisomy-21 live births is 1 in 650.

Individuals with Down syndrome may be male or female, since autosomes are identical in both sexes. All these individuals have some degree of mental retardation, often severe, and an assortment of physical defects that lead to a greatly reduced life expectancy.

There is an increasing risk of giving birth to a child with Down syndrome in older mothers (Fig. 11.8). According to genetic studies, the main problem seems to result from aging of the oocytes from which the eggs develop. Human females are born with a full complement of oocytes, suspended in an early stage of meiotic prophase. One oocyte is released each month between puberty and menopause, and an oocyte completes the first division of meiosis after its release from the ovary during a menstrual cycle. An oocyte which

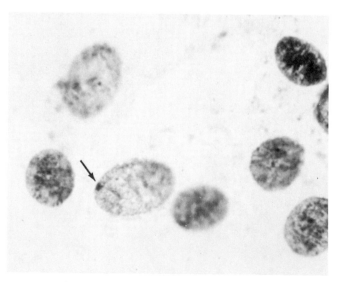

**Figure 11.6**
Barr body, or sex chromatin, in the human female. One of the two X chromosomes remains condensed at interphase, and is located at the periphery of the nucleus (arrow). (Courtesy of T. R. Tegenkamp)

has been stored for 40 years before its release and activation may be more likely to undergo a faulty meiotic division, particularly during separation of the paired chromosomes at anaphase I of meiosis. If paired chromosomes do not disjoin and both go to the same pole of the cell at anaphase, a phenomenon called **nondisjunction,** one extra chromosome is incorporated into those gametes. Gametes formed from nuclei that received less than the usual chromosome number may also be formed, although they probably are less viable. Nondisjunction also takes place during mitotic divi-

sions in the developing embryo, and may lead to trisomies either of some autosome or of the sex chromosomes (Fig. 11.9).

## PRENATAL DIAGNOSIS

There are at least 2000 known inherited diseases, and more are discovered each year. Most of these diseases are relatively rare, sometimes being known to occur in only one family group. Others, however, occur more often in the population. In a number of

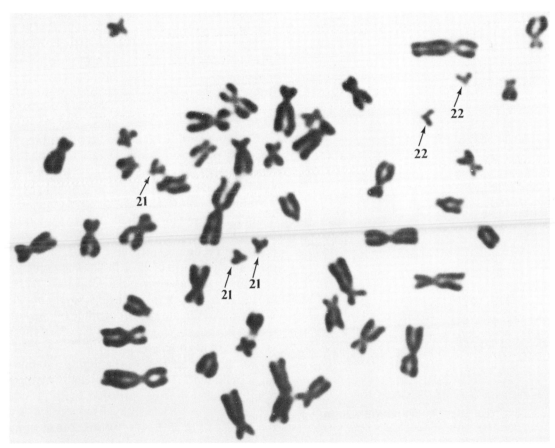

**Figure 11.7**
Chromosome complement of a female patient with Down syndrome, or trisomy-21. There are forty-seven chromosomes, five of which are from the small 21-22 (G) group. The extra chromosome is easily detected and identified in such a well-spread chromosome preparation. (Courtesy of C. Hux and T. R. Tegenkamp)

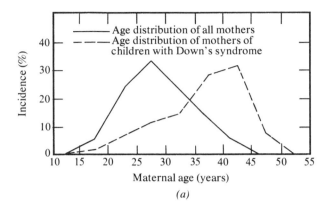

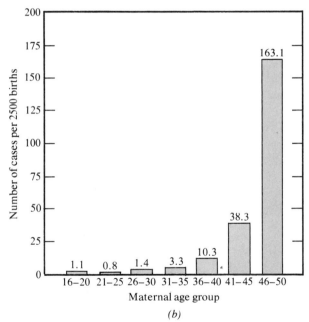

**Figure 11.8**
There is an increasing frequency of children with Down syndrome born to mothers in the older childbearing years. (a) Although affected children are born to mothers of various ages, the greatest percentage occurs for mothers who are over 30 years of age. (b) The absolute frequency of children with Down syndrome increases with the increasing age of the mother and is dramatically higher for women in the 46-50 age group.

instances it may be possible to detect a genetic defect in the fetus, or to find that one or both parents are **carriers** of a mutant allele and therefore run a particular risk of producing a genetically defective child. Similarly, it is often helpful to the family and physician to know the chromosome constitution of the fetus; for example, to determine whether it may be a trisomy-21 situation in the case of an expectant mother over the age of 40. In cases of suspected gene or chromosome defects, methods are available to examine cells of the fetus and provide the family with more specific information than only the percentage risk of the fetus having a defect. Women over the age of 40–45 have a risk of 1:50, or 2 percent, of having a child with Down syndrome. If the expectant family can be told that the fetus definitely is or is not carrying an extra chromosome-21, they can plan more confidently for the future event, or perhaps decide to abort the pregnancy.

**Amniocentesis**

The human fetus floats in liquid within the **amnion,** a sac formed from embryonic membrane. During its development, the fetus regularly sheds cells into the **amniotic fluid** which fills the amnion cavity. Using the method of **amniocentesis,** a physician withdraws a sample of amniotic fluid by hypodermic syringe through the mother's abdomen (Fig. 11.10). The fetal cells are predominantly fibroblasts, and these can be grown in a culture dish or flask. The cells may later be examined to determine the fetal karyotype or chromosome count, or to measure biochemical activities which may indicate the occurrence of a genetic disease due to some defective gene even when the karyotype is perfectly normal.

As mentioned above, the high probability of bearing a child with Down syndrome may be the basis for a family's decision to proceed with amniocentesis. When a trisomy-21 fetus is discovered it is then up to the family to decide on a future course of action, that is, whether to have the child or to abort the pregnancy. In some cases it is helpful to know whether the fetus is male or female, because some inherited diseases are sex-linked, that is, due to a gene on the X chromosome. Sex-linked inheritance leads to a higher percentage *expression* in males than in females because males have only one X chromosome whereas females have two. If a recessive allele of the gene is present on the only X chromosome in a male, disease expression will result. Because many of these recessive alleles are rare, it is highly unlikely that a female will have received an X chromosome carrying the recessive allele from each of her parents. In most cases one of the X chromosomes carries the dominant allele for normal expression, which masks a recessive allele that may be present on the second X chromosome (Fig. 11.11). If there is a family history of some sex-linked genetic problem, the family may thus wish to know if the expected child will be a boy or a girl. If it is a boy they can be told what the chances are for its having the disease or being disease-free. If it is a girl, the risk is often zero, depending on the genes carried by both parents.

In addition to advising families once the pregnancy has been initiated, it is possible in certain situations to make a biochemical determination of the parents themselves and see if one or both of them are carriers of a recessive allele. For a recessive condition based on a single gene, there is a 25 percent risk in each birth that the child will develop the disease if both parents are carriers (Fig. 11.12). The actual risk depends on the nature of the inheritance pattern, the genetic constitution of the parents, and other lesser factors.

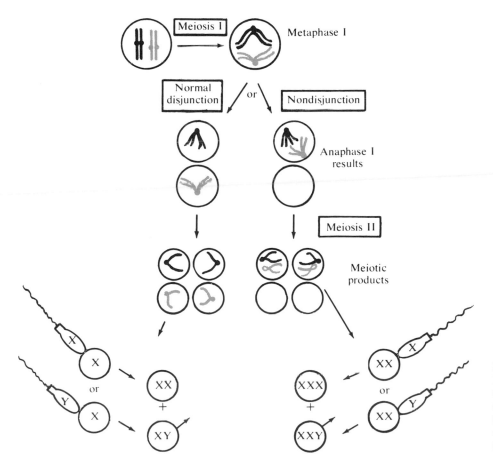

**Figure 11.9**
Flow diagram illustrating the origin of cells or individuals with aberrant chromosome number as the result of nondisjunction at meiosis (right), as compared with normal disjunction (left).

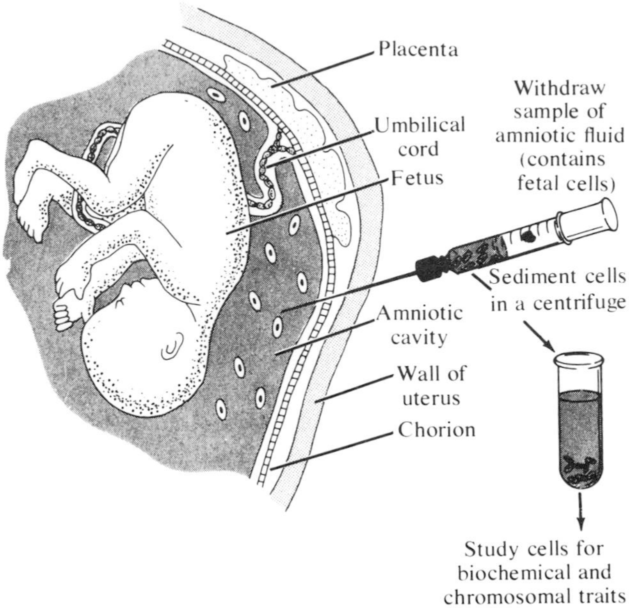

**Figure 11.10**
Amniocentesis. A sample of amniotic fluid is withdrawn and processed for cytological and biochemical analysis of fetal traits.

agent damages chromosomes, causing mutation or cancerous cell development in some instances. The method can therefore be used to test a broad spectrum of chemicals and physical agents (such as x rays) that are suspected to cause tissue damage or diseases or to shorten the life expectancy of people.

The usual staining methods for mitotic metaphase chromosomes do not produce any differences or contrast between the two sister chromatids of the replicated chromosome seen at this stage of mitosis (see Fig. 11.1). In newly developed procedures, however, one can clearly distinguish each of the two sister chromatids (Fig. 11.14). These stained metaphase chromosomes are examined for **sister chromatid exchanges** that have occurred, by breakages and rejoinings, to determine whether an experimentally-treated cell shows a different "harlequin" staining pattern from untreated control cells. Any increase in sister chromatid

Often the particular recessive allele occurs more commonly in some population groups than in others. If this is the case, members of such groups may be able to take advantage of a simple blood test to find out if they are carriers, since carriers often show some trace of a difference even though they are normal, functioning individuals. Examples of such situations include the more frequent occurrence of Tay-Sachs disease among descendants of Eastern European Jews, and of sickle cell anemia among blacks (Fig. 11.13). Family planning becomes more directed once the prospective parents are aware of the risk, if any, in each pregnancy.

### Detection of Chromosome Damage

A new staining method was developed during the 1970s which permitted the detection of breakages and rejoinings in mitotic chromosomes. There is tremendous theoretical importance to this simple method because it can provide evidence that some chemical

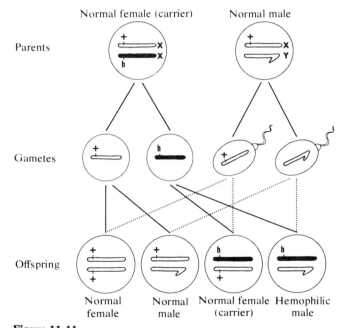

**Figure 11.11**
Sex-linked recessive afflictions such as hemophilia are expressed more frequently in males than in females because males have only one X chromosome and therefore have no masking allele, as XX females do.

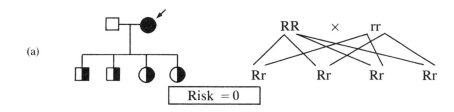

(a)

RR  ×  rr

Rr    Rr    Rr    Rr

Risk = 0

*Recessive inheritance*
Each of progeny receives one dominant allele from one parent and one recessive allele from other parent. All progeny are carriers.

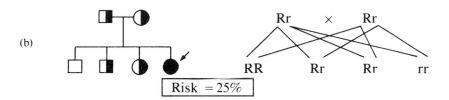

(b)

Rr  ×  Rr

RR    Rr    Rr    rr

Risk = 25%

When each parent is a carrier, there is a 25% chance in each birth that r + r will give rise to rr

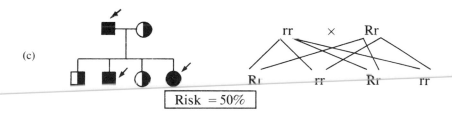

(c)

rr  ×  Rr

Rr    rr    Rr    rr

Risk = 50%

When one parent expresses the trait and the other is a carrier, there is a 50% chance in each birth of rr occurring.

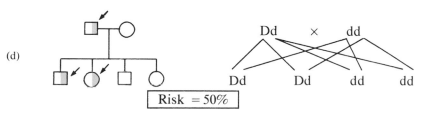

(d)

Dd  ×  dd

Dd    Dd    dd    dd

Risk = 50%

*Dominant inheritance*
There is a 50% chance in each birth that the parent with the trait will transmit the dominant allele (and a 50% chance of transmitting the recessive allele).

**Figure 11.12**
Risk of birth of a child with an inherited defective trait varies according to the gene constitution of the parents and the nature of the inheritance pattern involving two alleles of one gene. In recessive inheritance (a - c), both recessive alleles are required to be present before the trait is expressed, but in dominant inheritance (d) only one allele is necessary for expression. Arrows indicate persons expressing the trait in pedigrees shown at the left. All possible combinations of alleles are shown in diagrams at the right of each pedigree, these combinations being the theoretical basis for prediction of risk in a particular pedigree.

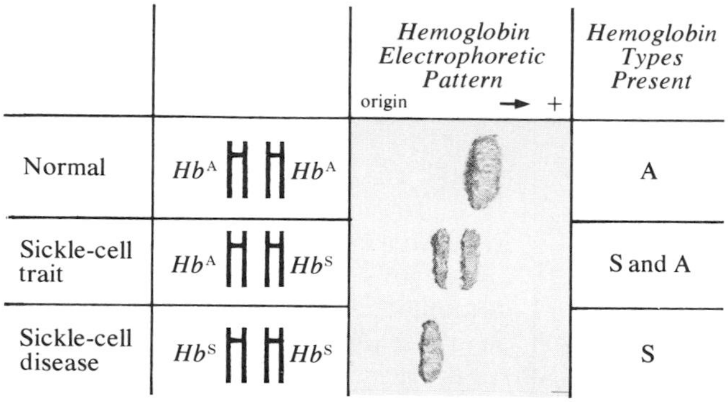

| | | Hemoglobin Electrophoretic Pattern | Hemoglobin Types Present |
|---|---|---|---|
| Normal | $Hb^A$ ⊓⊓ $Hb^A$ | | A |
| Sickle-cell trait | $Hb^A$ ⊓⊓ $Hb^S$ | | S and A |
| Sickle-cell disease | $Hb^S$ ⊓⊓ $Hb^S$ | | S |

**Figure 11.13**
People who are carriers of the sickle-cell allele ($Hb^S$) have both A and S hemoglobin proteins, which can be detected by exposing a blood sample to an electrical field. The proteins migrate to different end positions in such a field. Only A and S protein is present in the homozygous $Hb^AHb^A$ or $Hb^SHb^S$ individuals, respectively.

exchanges, when compared with control cells, is an indication of damage done to chromosomes by the agent being tested.

In addition to its usefulness as a general screening method for potentially harmful agents in the environment or foods, it can also be applied to individuals who may have been exposed to some dangerous situation. If their exposure has led to chromosome damage, which may later lead to disease, malformation, or early death, the individual can be so advised. If an expectant mother has been exposed to some potentially harmful agent or situation, analysis of sister chromatid exchanges in mitotic metaphase chromosomes taken during amniocentesis can then be used to assess any damage that may have been done to the fetus as well as to the mother.

### Genetic Counseling

An increasingly important area of health care involves the genetic counseling team that specializes in medical genetics. The team includes the physician, whose province is diagnosis and rehabilitative treatment, and others such as human-genetics professionals, laboratory personnel who perform and may interpret tests and chromosome analyses, and counsel-

(a)

(b)

**Figure 11.14**
Metaphase chromosomes from Chinese hamster cells prepared and stained to reveal sister chromatid exchanges (SCEs) in chromosomes whose two chromatids are differentially stained: (a) Example of an untreated control cell showing 12 SCEs among the 20 chromosomes, and (b) chromosomes from a cell exposed to nitrogen mustard, a mutagenic chemical agent, showing approximately a tenfold increase in SCEs relative to the example from the controls. (Reproduced with permission from Perry, P., and Evans, H. J., *Nature* **258**:121-125, Figs. 1 and 4. 1975.)

ing associates who contribute in various ways to the effective functioning of the team.

Families with a history of genetic disease or with recent experience in having a child with a genetic defect may obtain counsel from the team. Predictions of risk can be estimated, family pedigrees can be constructed for more accurate knowledge of the risks a couple may face, and appropriate identification and treatment can be advised once the condition is understood. Genetics clinics, usually affiliated with medical schools and medical centers, are relatively few in number at the present time, but there is every hope of increasing the emphasis on genetic counseling as a branch of preventive medicine along with more widely used rehabilitative programs.

One important problem is the inadequacy of training in medical genetics now available to medical students, and another problem is that of totally inadequate governmental funding for research and treatment of birth defects and prenatal care. Perhaps priorities will be shifted toward medical problems of the young, rather than maintaining the present emphasis on diseases of middle and old age.

## CELL SURFACE STUDIES

During the past twenty to thirty years there has been a considerable amount of information obtained about the body's defenses against infection and cancer. Important observations in these areas led directly to our current concepts of cell-membrane organization and a growing appreciation of the contributions made by the cell-surface membrane to systems involved in cell-cell recognition and interactions. These cellular properties underlie the immune response reactions involved in warding off infection and restraining uncontrolled cell growth.

### Organization of the Cell Surface

You may wish to look over parts of Chapter 4 in which membrane structure was discussed, but a brief restatement here may be helpful. We broadly accept the model of cell membranes, including the plasmalemma, as a mosaic of proteins in and on a fluid bilayer of phospholipid molecules. Some **integral** proteins are partially embedded within the lipid phase while some completely span the membrane thickness and protrude on both surfaces of the bilayer. Integral proteins are an essential component, together with lipids, for membrane structural intactness. If integral proteins or lipids are removed or dissolved away, the membrane falls apart. Other proteins **(peripheral)** interact through ionic bonds, at the surfaces of the membrane, with structural membrane components. Peripheral proteins are involved in various membrane functions other than membrane structural integrity, however, since removal of peripheral proteins influences cell activities but does not lead to destruction or dissolution of the membrane itself.

According to this **fluid mosaic model** of membrane organization, there are two important implications for membrane function. First, membranes may be organized asymmetrically, which allows certain components to be localized mainly in the outer, exterior-facing membrane surface or at the inner, interior-facing membrane surface. For example, membrane glycoproteins and glycolipids (conjugated compounds involving some carbohydrate residue bonded to protein or lipid, respectively) are so arranged that their carbohydrate residues are exposed only at the outer face of the plasma membrane. These compounds function at the cell surface as specific receptors for antibodies, hormones, viruses, and other agents. Second, since components can diffuse *laterally* within the plane of the membrane, a mechanism is available for rapid and reversible changes in the distribution of certain membrane constituents. For example, molecular distribution patterns may provide a basis for different and specific cell-surface patterns involved in cell contact and cell recognition phenomena, or for providing information on cell positioning within a tissue. In addition, redistribution of cell-surface components can occur in response to various environmental stimuli, providing a mechanism for rapid and reversible modulation of cell-surface properties after surface interactions with other cells or with such agents as antibodies, hormones, and viruses.

## The Cytoskeletal Framework

Based on numerous studies using various drugs, and microscopical and chemical analyses, it is suggested that a "skeletal" framework of microtubules and microfilaments is linked to protein components across the membrane. It is further suggested that the cytoskeletal fibrous framework is part of a control system which modulates movement of certain cell-surface components. If structures on the inner side of the membrane can regulate mobility of components in the membrane and on the opposite surface of the membrane, then it is implied that information must pass *across* the membrane thickness and influence molecule movement. Such a **trans-membrane control system,** in which the cytoskeletal framework regulates mobility of membrane components, presumably influences integral and peripheral protein mobilities *simultaneously* through linkages between proteins and cytoplasmic fibers (Fig. 11.15).

Microtubules and microfilaments are underlying components in various cellular movements, as we discussed in Chapter 7. We may now extend the basic concept to include movements of molecules on or within the mobile cell-surface membrane, particularly to mobility leading to redistributions of cell-surface **receptor** molecules. As has been suggested at the levels of cell and cell compartment movements, microtubules and microfilaments may play opposing roles in molecule movement. While only a hypothesis at present, it is thought that microtubules may function as a system of *anchoring elements* while microfilaments may function as part of a *contractile system.* The anchoring microtubules would restrain or limit the mobility of certain receptors, while microfilaments would act in redistributing clusters of receptor molecules unless held in restraint by microtubule anchorages. The means of interconnecting the elements of the cytoskeletal framework and protein receptors of the membrane are unclear at present. Microtubules have a diameter of about 25 nm and microfilaments are about 6–10 nm wide, whereas molecules usually have smaller dimensions. There are several problems in geometry which must be solved, as well as problems in the nature of the postulated interactions in mobility control.

## THE IMMUNE SYSTEMS

Two different but interacting systems of immunity are responsible for protecting the body from hazards of infection and cancer; these are known as the **humoral response system** and the **cell-mediated response system.** When any foreign material, called an **antigen** by definition, is introduced into the body, the humoral immune system responds by synthesizing and secreting **antibodies,** while the second system responds by making and releasing cells specifically sensitized to the invading antigen. Each system interacts in exquisitely specific ways with antigens, like the fit between a lock and key. The immune systems, however, are so specifically discriminating that they can respond precisely and differentially to millions of different kinds of antigens. The humoral response system is effective against bacterial infections and viral reinfections. The cell-mediated response system combats fungus and virus infections and also initiates the rejection of tumors and such foreign tissues as transplanted organs. The cell-surface membrane is the site of the primary immune reaction involving the two distinct but cooperating response systems.

**Figure 11.15**
Cell-surface membrane components and the cytoskeletal framework of myosin molecules (MM), microfilaments (MF), and microtubules (MT), which may control molecule mobility in the overlying plasma membrane (PM). One microvillus (MV) protrusion is shown. (a) The usual proportions of proteins, carbohydrate residues (branched projections of the membrane proteins), and cytoskeletal components are shown in the hypothesized arrangement. (Reproduced with permission from Nicolson, G. L., 1976, *Biochim. Biophys. Acta* **457**:57, Fig. 13.). (b) A more accurate representation of the proportionate differences in component sizes, showing how large the MT, MF, and MM really are in relation to molecules of the plasma membrane. A tobacco mosaic virus (TMV) particle provides the basis for size comparisons. If the cytoskeleton is so disproportionate relative to membrane molecules, the microfilament—microtubule control over membrane molecule movements seems more difficult to visualize. (Reproduced with permission from Loor, F., 1976, *Nature* **264**:272, Fig. 1.)

(a)

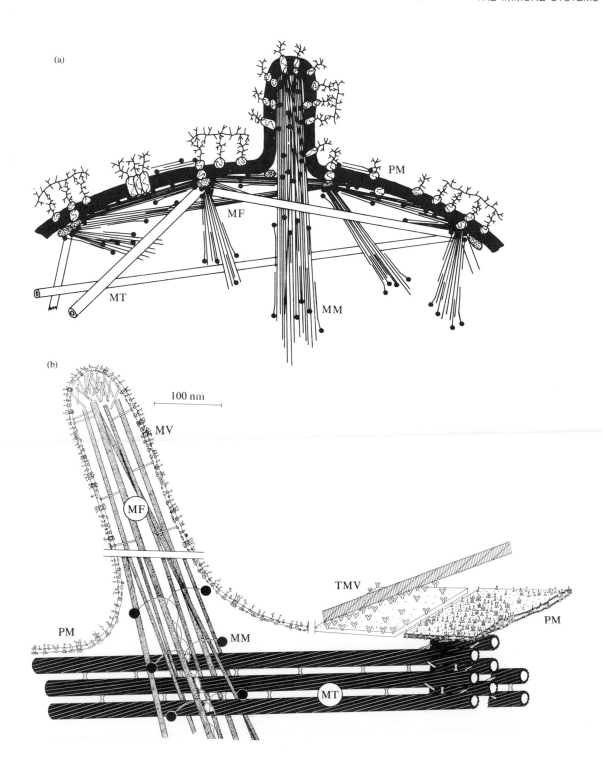

(b)

### T Cells and B Cells

It was established in the 1950s that the cells responsible for the immune reactions were particular kinds of white blood cells called **lymphocytes.** Until then the function of lymphocytes was unknown. It gradually became clear in the 1960s that there were two classes of lymphocytes: (1) *B* lymphocytes which become the *B* cells that synthesize and secrete antibodies in response to introduced antigen, and (2) *T* lymphocytes which become the *T* cells that are sensitized to the specific introduced antigen, and then direct the cell-mediated immune response.

Both kinds of lymphocytes differentiate from the same type of precursor **stem cell** which occurs in the bone marrow (Fig. 11.16). Within the bone marrow some stem cells give rise to bone-marrow lymphocytes, which migrate to the peripheral lymphoid tissues and there become *B* lymphocytes. On stimulation by introduced antigen, *B* lymphocytes enlarge and proliferate as activated *B* cells (also called plasma cells). It is these activated *B* cells which synthesize antibody that is specific for the introduced antigen, and secrete these antibodies into the circulating blood stream. Other stem cells migrate from the

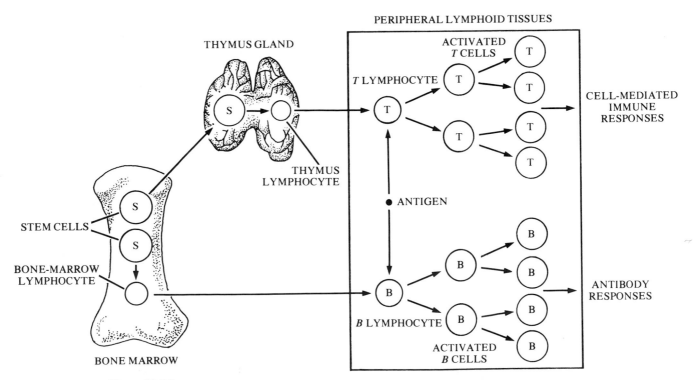

**Figure 11.16**
Stem cells give rise to lymphocytes, which are the cells responsible for immune reactions. Some stem cells migrate from the bone marrow to the thymus gland and develop into thymus lymphocytes, which in turn migrate to peripheral lymphoid tissues and become *T* lymphocytes. Other stem cells develop into bone-marrow lymphocytes which become *B* lymphocytes after migration to peripheral lymphoid tissues. *T* cells are activated to produce cell-mediated immune responses and *B* cells are activated to produce antibodies of the humoral immune response system. (From ''Cell-Surface Immunology'' by M. C. Raff. Copyright © 1976 by Scientific American, Inc. All rights reserved.)

bone marrow to the thymus gland where they differentiate into thymus lymphocytes, which in turn migrate to peripheral lymphoid tissues and there become *T* lymphocytes. When a specific antigen is encountered, *T* lymphocytes enlarge and proliferate as activated *T* cells which direct the cell-mediated immune response reactions. Both *T* and *B* lymphocytes are also found in the blood and in other parts of the mammalian body, such as liver and spleen. Bone-marrow stem cells give rise to other blood-cell types as well as to lymphocytes (Fig. 11.17).

Differentiation into activated *T* cells and *B* cells requires interaction with antigen, so it was expected that the cell-surface receptor proteins would be antibodies or some equivalent class of molecules. It was found that the receptors were indeed antibodies, which are members of a class of proteins called **immunoglobulins.** The immunoglobulins of *B* cells are molecules made up of two pairs of polypeptide chains, or globins; one pair of "light" and one pair of "heavy" chains which differ in amino acid composition and molecular weight and length (Fig. 11.18). *T* cells also have im-

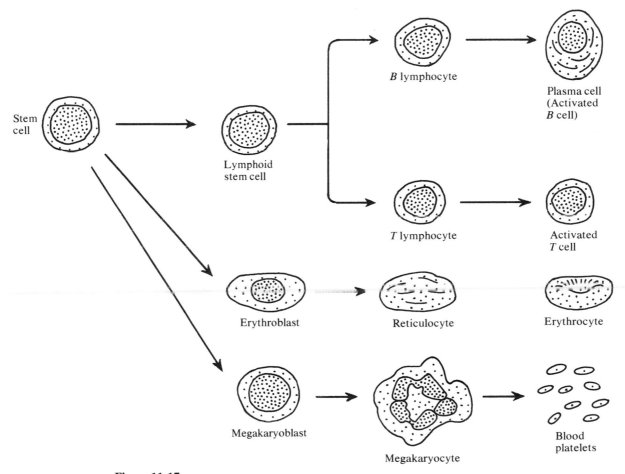

**Figure 11.17**
Different blood cell components, of which some are shown, all originate from differentiation of the same kind of precursor stem cell.

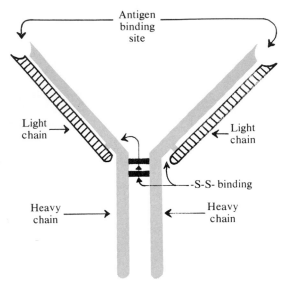

**Figure 11.18**
An immunoglobulin molecule, or antibody, consists of four polypeptide chains. The two longer chains are designated as "heavy" and the two smaller ones are "light" chains. Disulfide bridges (-S-S-) hold the molecule together. There are two sites on each immunoglobulin molecule where antigen can bind; that is, an antibody is divalent.

munoglobulin receptors, but of a somewhat different type from those in the surface membrane of *B* cells. The *B* cell immunoglobulins have two sites that specifically recognize and interact with parts of antigen molecules that are known as **antigenic determinant sites.** Since antigens have a number of determinant sites, antibodies can cross-link with antigens and produce lattices of molecules (Fig. 11.19). Each antibody interacts only with its specific antigen, so that discrimination among millions of different antigens depends on specificity of fit and interaction between corresponding sites of antigen and antibody molecules.

When some antigen is encountered by the immune systems, the antigen interacts only with *T* and *B* lymphocytes already displaying the specific antibody receptors for this antigen. The interaction apparently signals these lymphocytes to undergo enlargement and cell division, that is, to become activated *T* and *B* cells. Since little or no antigen is found within the *T* and *B*

cells, the activation appears to be a cell-surface phenomenon. This has been verified repeatedly by studies using labeled antigens, which are found clustered at the cell surfaces. In addition, each *T* or *B* cell is committed to *only a single kind of antigen*. An introduced antigen apparently stimulates *T* and *B* cells *already committed* to that antigen to undergo cell divisions, giving rise to a **clone** or family of cells that are also specifically committed to that same antigen (Fig. 11.20).

This **theory of clonal selection** had been proposed in the 1950s by Sir Macfarlane Burnet, and has been verified by many investigators in recent years. According to this idea, lymphocytes of many kinds exist in the body, each of which is specific for one kind of antigen molecule. On encountering this same antigen later, the particular committed lymphocytes proliferate as activated *T* or *B* cells and thereby produce

**Figure 11.19**
The divalent antibodies bind to antigens at antigenic determinant sites, forming a network or lattice that leads to inactivation of the invading antigen and to its ultimate removal from the system through this immune reaction.

large numbers of identical cells each of which is sensitized to this same antigen. The remaining lymphocytes do not respond to this antigen, but do respond to some other antigen to which each is committed. Infection by some invading agent is therefore met by a large number of *T* cells or of antibody-secreting *B* cells that will interact only with the invading antigenic agent. In a way, each invading agent signals its own destruction since it stimulates immune cell responses directed specifically at itself. Similarly, we may view the body as containing a huge reservoir of millions of kinds of specified lymphocytes, any of which may be stimulated to respond to its particular antigenic correspondent. Triggering of different lymphocytes by different antigens provides a broad screen of defense against a host of potential invader proteins and agents.

When *T* cells or antibodies in the bloodstream meet their specified antigen, antigen-antibody interactions take place which immobilize the antigens and lead to their removal or destruction. Antigen-antibody complexes or lattices will precipitate out of solution and be eliminated from the body. *T* cells bind antigen and may either engulf the antibody-antigen complexes by endocytosis, destroying them within the cell, or these complexes may be shed from the cell and enter the blood stream, from which they also are eventually eliminated (Fig. 11.21). All these events are primarily cell-surface reactions and require a membrane construction in which protein mobility is possible. The fluid mosaic membrane admirably suits this requirement. Control over mobility of membrane protein receptors may in turn be based on the cytoskeletal framework of microtubules and microfilaments.

### Interferon

In 1957 another kind of protein was found to act in immune responses to virus infection. This protein type, called **interferon,** acts as an inhibitor of virus multiplication through effects on the host cell itself and not on the virus directly. There are various kinds of interferon which regulate the immune response, but we know very little at present about the specific roles for specific interferons in the response system. It is believed, however, that the virus-type interferon acts by blocking translation of viral messenger RNA in the host cell.

Interferon acts on the cell-surface membrane and not within the cell directly. This surface reaction, however, induces antiviral activity within the host cell. Cells that are stimulated to produce interferon must first externalize this protein, and afterward the host cell is induced to manufacture antiviral protein that prevents virus multiplication within the cell. The cell-surface receptors that interact with interferon are believed to be glycoproteins, including various gangliosides. Once interferon is bound to cell-surface

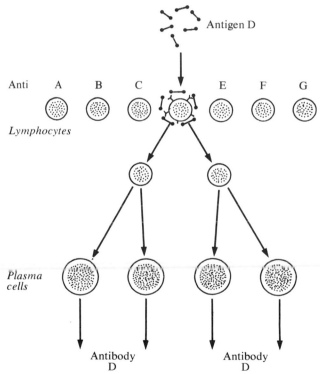

**Figure 11.20**
Diagram illustrating the theory of clonal selection. Various sensitized lymphocytes occur in the body, from previous exposure to specific antigens. When some antigen, such as D, again invades the body, sensitized *B* lymphocytes of the anti-D variety are stimulated to become activated plasma cells (*B* cells), which proliferate and secrete D-antibodies that combat the invading D-antigens.

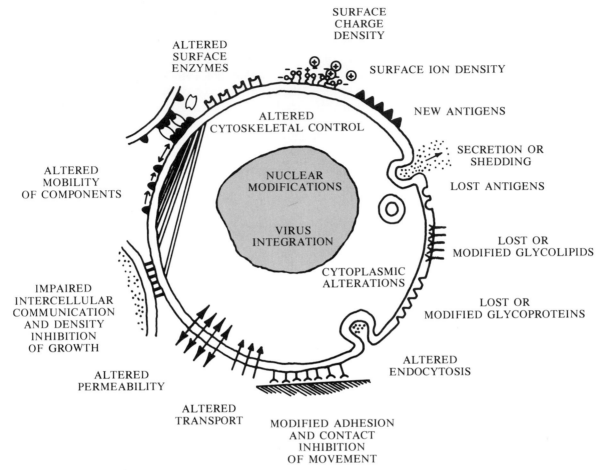

SURFACE CHARGE DENSITY

SURFACE ION DENSITY

NEW ANTIGENS

ALTERED SURFACE ENZYMES

ALTERED CYTOSKELETAL CONTROL

SECRETION OR SHEDDING

LOST ANTIGENS

ALTERED MOBILITY OF COMPONENTS

NUCLEAR MODIFICATIONS

VIRUS INTEGRATION

CYTOPLASMIC ALTERATIONS

LOST OR MODIFIED GLYCOLIPIDS

LOST OR MODIFIED GLYCOPROTEINS

IMPAIRED INTERCELLULAR COMMUNICATION AND DENSITY INHIBITION OF GROWTH

ALTERED PERMEABILITY

ALTERED TRANSPORT

ALTERED ENDOCYTOSIS

MODIFIED ADHESION AND CONTACT INHIBITION OF MOVEMENT

**Figure 11.21**
Some cell-surface alterations found after transformation to cancerous cell growth.
(Reproduced with permission from Nicolson, G. L., 1976, *Biochim. Biophys. Acta* **458**:1.)

receptor, a signal is generated by this interaction to the appropriate host-cell gene(s) to begin transcription of messenger RNA responsible for production of the **antiviral protein** (Fig. 11.22). While only hypothetical at present, it has been suggested that the mechanism of transmission of the interferon effect from the cell membrane to the nucleus may be by migration of a surface-membrane protein to the chromosomes in the nucleus. This mechanism has been shown to be true for effects of certain hormones on cells, and a similar

situation may also explain the signal for antiviral protein manufacture in virus-infected host cells.

The two known proteins, to date, of the interferon system are interferon and antiviral protein. Interferon production requires some inducing substance, such as a virus or a nucleic acid, but various factors influence and regulate the whole system. Some information has been obtained on the genetics of the interferon system in human beings, but genetic control seems to be rather complex according to current studies. At least

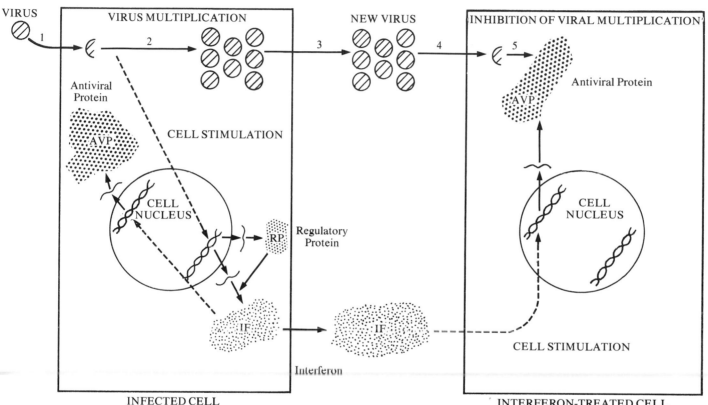

**Figure 11.22**
Cellular events of the induction and action of interferon (IF): (1) virus comes into contact
with cell and penetrates cell membrane; (2) replication of virus occurs, directed by genetic
material released by the virus; (3) new viruses leave cell and enter fluids around the cell; (4)
some of the new viruses infect other cells; and (5) release of virus genetic material takes
place. During the early stages of infection of the first cell, the host cell is stimulated to make
interferon (perhaps under regulatory protein, RP, control) according to nuclear gene
information. Some IF released from the first cell enters another cell and stimulates it to
produce a messenger RNA-specifying antiviral protein (AVP). AVP inhibits virus
multiplication in the host cell by inhibiting host synthesis of virus-specified molecules. The
first cell may also be stimulated to make AVP, perhaps leading to lowered production of
viruses in the infection. (From DHEW Publ. No. (NIH) 75-700.)

one genetic factor coding for antiviral protein has been localized to chromosome 21, and genes for production of human interferon have been assigned to chromosomes 2 and 5. Additional genetic complexities are now under investigation.

## THE CANCER CELL: CELL-SURFACE MODIFICATIONS

There are three essential features of tumor cells that distinguish them from their normal counterparts: (1) proliferation in an uncontrolled fashion, (2) invasion of normal tissues, and (3) spreading to distant sites (metastasis). Changes in surface properties of tumor cells as compared with normal cells probably are important in determining each of these major behavioral characteristics.

Within the tumor, alterations in surface properties of tumor cells contribute to their escape from many of the restraints to which normal cells are subject. Proliferation of tumor cells is no longer effectively regulated by cell-cell contact interactions, and the cells become increasingly unresponsive to growth regulation by hormones, serum factors, and other agents that exert their effects after binding to the cell surface. Tumor cells can therefore achieve varying degrees of independence from normal growth restraints, and this ability is reflected in uncontrolled growth and progressive enlargement of the primary tumor.

Altered surface properties of tumor cells may also result in aberrant cell-cell recognition, allowing tumor cells to escape from the control mechanisms that are responsible for maintaining proper cell position. In malignant tumors these processes are aggravated by metastasis, in which the surface properties of tumor cells contribute to invasion of surrounding tissue, to initial separation of cells from the primary tumor, and to determination of the subsequent pattern of cell distribution and establishment of metastases (growths at sites distant from the cell source). In addition, the outcome of the interaction of tumor cells with parts of the host's immune apparatus is influenced very largely by the surface properties of tumor cell populations.

There is now a large catalogue of differences between normal and tumor cell surfaces, but relatively little insight into mechanisms leading to these differences, or how these differences are maintained, or how the observed cell-surface alterations contribute to tumor cell proliferation, invasiveness, and metastasis. One general observation, however, is that tumor cells have modified glycolipid content and molecular structure. Specifically, there is a significant decrease in the more complex glycolipids, and significant loss of the terminal sugar residues of glycolipid molecules in the cell-surface membrane. Other modifications in cell membrane glycolipids have also been found.

Since cancer cells contain novel tumor-associated antigens which are foreign to the host organism, it is expected that the host's immune apparatus would detect and destroy such cells. In many cases, of course, this is precisely what happens. In other cases, however, cancer cells escape the immune system and establish themselves in the host. The means by which cancer cells escape destruction probably depend on their first escaping detection by the immune apparatus. How is this achieved? There is good reason to believe that tumor-associated antigens undergo **redistribution** at the cell surface, rather than some other escape mechanism such as unbound antigens being shed from the cell or being internalized by endocytosis. Using test systems, it has been shown that "patches" and "caps" of bound antigen-antibody complexes appear at the cell surface *after* tumor antigens bind with antibodies (Fig. 11.23). Such complexes are then shed from the cell surface so that tumor cells escape destruction, since they have disposed of their telltale antigens and thereby effectively neutralized the cell-mediated response system.

Changes in the trans-membrane control system probably underlie certain behavioral characteristics of tumor cells such as altered properties of contact inhibition of cell movement and the ability to be properly positioned within the tissue. Such changes point to modifications of microtubule and microfilament components, which are considered to control membrane protein mobilities and resulting normal surface interactions between cells. Disruption or alteration of normal displays of surface components may

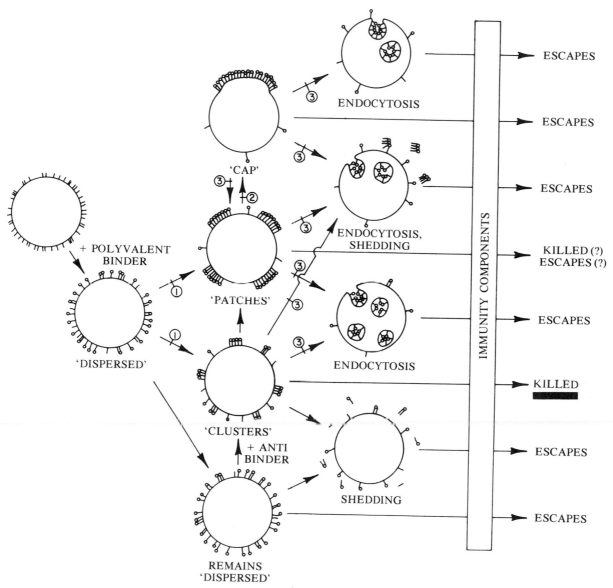

**Figure 11.23**
Some possible mechanisms for tumor-cell escape from surveillance by the immune system, depending on events involving cell-surface antigens of these tumor cells. (Reproduced with permission from Nicolson, G. L., 1976, *Biochem. Biophys. Acta* **458**:1.)

also contribute to loss of "signals" for responses to other cells, and to invasiveness and metastasis once positional controls are lost from the cell surface.

Avenues of investigation are now available to analyze the major contribution to the malignant state of tumor cells made by changes in mobility and distribution of plasma membrane components. One line of study has pointed out that increased **protease** (protein-digesting enzyme) activity occurs in tumor cells as compared with normal cell types. More of such enzyme activity leads to loss of surface components and perhaps to greater predilection for initiating the tumor state. But we do not yet know what the particular changes are or how they are caused. Each new advance in distinguishing between normal and cancer cells, and in learning more about normal cell activities, holds the hope of bringing us closer to the solution of the cancer problem. But cancer is a general term for any cell gone "wild", and there may be more than one signal, agent or mechanism that induces the deranged cell state. In some instances we know that viruses are inducers of cancers, in other cases we know there is a genetic basis to cancer development. In the large majority of cancers, however, we have yet to learn the causative factors and therefore are still not in a position to produce remedies or cures for this threat to the quality of living and to human life.

## SUMMARY

1. Human beings have 46 chromosomes in their body cells, two of which are sex chromosomes. Females are XX and males are XY, and the remaining 44 autosomes are the same in both sexes. Human chromosomes can be studied from karyotypes constructed by organizing each chromosome from a photographed metaphase nucleus into a conventional arrangement beginning with the largest and ending with the smallest chromosome in the set. Location of the centromere provides another basis for identification and karyotype arrangement of chromosomes.

2. The Sex Balance Theory of sex determination states that a female is determined if the ratio of X chromosomes to autosome sets is 1.0, but a male is determined if the X:A ratio is 0.5. This system is applicable to insects but not to human beings and other mammals, where the Y chromosome is strongly male-determining. Human males may have more than one X chromosome but there is always a Y, whereas human females may be determined even if there is only one X chromosome, as long as there is no Y chromosome present. Among the more common sex chromosome anomalies are XO females $(45,X)$ and Klinefelter or XXY $(47,XXY)$ males. Counts of Barr bodies in interphase nuclei correspond to the number of X chromosomes minus one, and provide a simple and rapid method to identify a person's sex chromosome content.

3. Autosomal chromosome anomalies of various kinds have been found since 1959 when the first case of Down syndrome was identified as trisomy-21, in which there are three copies of chromosome 21 to give a total chromosome count of 47 instead of 46. Aberrant chromosome numbers probably arise initially because of failure in accurate separation of sister or homologous chromosomes during nuclear divisions.

4. Prenatal diagnosis can provide information about the chromosomes and metabolic activities of the fetus using a sample of fetal cells collected from amniotic fluid during amniocentesis. Prospective parents can be screened to determine whether they are carriers of recessive alleles, and can then be informed of the relative risk in each pregnancy of producing a child with the inherited recessive condition. Damage to chromosomes by harmful agents can be assessed from stained chromosomes in which sister chromatid exchanges have occurred. In general, genetic counseling services form a part of health care delivery systems and aid in family planning in situations where risk of some inherited defect has been shown to exist according to pedigree analysis.

5. Proteins and lipids, some of which are conjugated with carbohydrate residues, can move laterally within the lipid phase of the fluid mosaic cell-surface membrane. Movements are influenced by a transmembrane control system composed of a contractile microfilamentous component and an anchoring microtubular component, which act coordinately during molecule redistribution within and on the membrane. These dynamic features of the cell surface underlie host cell responses to infection and cancer development.

6. Antigens are foreign materials that (a) stimulate antibody secretion by activated $B$ cells, which mediate the humoral immune response system, and (b) stimulate the proliferation of activiated $T$ cells, which are sensitized to the invading antigen and interact through a cell-mediated immune response system. Both $B$ and $T$ cells develop from the same kind of precursor stem cells which are found in the

bone marrow, and which initially differentiate into *B* and *T* lymphocytes that can become activated immune response components. Interactions between antigens and immunoglobulins, which are cell-surface receptor antibody proteins, can lead to rejection of the invading agent. According to the clonal selection theory, invading antigens are met by a large number of *B* or *T* cells which are already committed to respond to specific antigens. Immune responses are primarily cell surface phenomena.

7. Interferon is a protein inhibitor of virus multiplication, made in the host cell and active on the host rather than on the virus directly. After interferon has moved to the cell surface, its action there leads to synthesis of antiviral protein within the host cell. Antiviral protein then inhibits virus multiplication in the host.

8. Tumor cells may escape the restraints to which normal cells are subject and proceed to proliferate uncontrollably, invade normal tissues, and spread to distant sites. Such cancer cells often escape destruction by the body's immune response systems and become established at one or more sites. Tumor antigens should provide signals to the host immune response systems, but may be shed after binding to cell-surface antibody receptors and therefore neutralize the cell-mediated immune response system, since the telltale antigens are no longer present at the cell surface. Cell-surface modifications leading to altered cell-cell recognition, altered contact inhibition, and other altered properties also characterize cancer cell development.

### STUDY QUESTIONS

1. How many chromosomes in a human cell? Which sex is XX? Which sex is XY? When we examine a human karyotype, what are we looking at? How can such material be used to explain a basis for an aberration, or a correspondence between a karyotype variation and a particular disease in human beings? How do we construct a karyotype? Is it worthwhile, or could we simply examine microscope slide preparations showing well spread out chromosomes and gain the same amount of specific information?

2. What is the basis for sex determination in insects such as *Drosophila*? What is the basis for sex determination in mammals? What is the sex of XO flies and of XO human beings? What is the sex of XXY flies and of XXY human beings? If the Y chromosome is not essential for life either in *Drosophila* or in human beings, what is its importance in development of an organism? What happens in a human embryo that is XY as compared with one that is XX? What is

Turner syndrome? Klinefelter syndrome? How many Barr bodies in people who are XX? XY? XO? XXXXY? XXXX? What is a Barr body?

3. What is trisomy? What is trisomy-21? Why do we find roughly equal numbers of males and females who are trisomy-21 but only women who are XO and only men who are XXY? How do trisomies arise?

4. What is a genetic carrier? Why is there a 25% risk of producing a child with an inherited defect when both parents are carriers, but either a zero or 50% risk when one parent is a carrier of the recessive allele? What is the risk in each pregnancy for the child to inherit a dominant trait when one parent shows this trait and the other does not? If both parents are carriers of a recessive allele and their first three children are normal, what is the chance that their fourth child will be normal? Why? In what situations would it be desirable to undergo amniocentesis? Why can't amniocentesis provide information in every case about a genetic or chromosomal defect in the fetus?

5. What use can we make of the method of staining sister chromatids differentially? What is a sister chromatid exchange? How can we recognize it in stained chromosome preparations? How does genetic counseling contribute to preventive medicine and to health care delivery systems? What sorts of families would benefit most from genetic counseling? What is genetic counseling? Why aren't more genetic counseling services available in the United States today? Should there be more? Why?

6. How has it been proposed that microfilaments and microtubules interact with membrane proteins? Why does a fluid mosaic membrane fit particularly well into our general concept of cell-surface activities as dynamic events? How can we reconcile the vast differences in real size of microtubules and microfilaments in relation to size of membrane protein and lipid molecules? If these problems can be set aside for the moment, what do we propose to be the function of microfilaments and microtubules in the transmembrane control system?

7. What are an antigen, an antibody, and an immune reaction? What is the difference between the humoral and cell-mediated response systems of immunity? What are lymphocytes, stem cells, *T* and *B* cells, and plasma cells? How do *T* and *B* cells arise in the body? Where should we expect to find activated *T* and *B* cells? What is the relationship of immunoglobulins to immunity? What is the theory of clonal selection? How does it propose to explain the rapid response of the body's immune systems to an invading antigen? Why can we say that each invading antigen

essentially signals its own destruction on entering the body? How does the cell surface contribute to immune responses?

8. What is interferon? How does interferon lead to inhibition of virus multiplication in an infected host cell? in nearby host cells? What is antiviral protein?

9. What are three essential differences between normal cells and tumor cells? In what ways do cell-surface alterations contribute to particular properties of tumor cells? How can tumor cells escape immune surveillance and avoid being destroyed? What is cancer? Why haven't we found a cure for cancer even though we know how to describe it?

## SUGGESTED READINGS

Cairns, J. The cancer problem. *Scientific American* **233**:64 (Nov. 1975).

Capra, J. D., and Edmundson, A. B. The antibody-combining site. *Scientific American* **236**:50 (Jan. 1977).

Cerami, A., and Peterson, C. M. Cyanate and sickle-cell disease. *Scientific American* **232**:44 (Apr. 1975).

Cold Spring Harbor Symposia on Quantitative Biology, vol. 41. *Origins of Lymphocyte Diversity.* New York: Cold Spring Harbor Laboratory (1976).

Cooper, M. D., and Lawton, A. R. III. The development of the immune system. *Scientific American* **231**:58 (Nov. 1974).

Drets, M. E., and Shaw, M. W. Specific banding patterns of human chromosomes. *Proceedings of the National Academy of Sciences* **68**:2073 (1971).

Edelman, G. M. Surface modulation in cell recognition and cell growth. *Science* **192**:218 (1976).

Friedmann, T. Prenatal diagnosis of genetic disease. *Scientific American* **225**:34 (Nov. 1971).

Jerne, H. K. The immune system. *Scientific American* **229**:52 (July 1973).

Lerner, R. A., and Dixon, F. J. The human lymphocyte as an experimental animal. *Scientific American* **228**:82 (June 1973).

Mayer, M. M. The complement system. *Scientific American* **229**:54 (Nov. 1973).

Mittwoch, U. Sex differences in cells. *Scientific American* **209**:54 (July 1963).

Munro, A., and Bright, S. Products of the major histocompatibility complex and their relationship to the immune response. *Nature* **264**:145 (1976).

Nicolson, G. L., and Poste, G. The cancer cell: Dynamic aspects and modifications in cell-surface organization. *New England Journal of Medicine* **295**:197, 253 (1976).

Perry, P., and Evans, H. J. Cytological detection of mutagen-carcinogen exposure by sister chromatid exchange. *Nature* **258**:121 (1975).

Perutz, M. F. Fundamental research in molecular biology: relevance to medicine. *Nature* **262**:449 (1976).

Raff, M. C. Cell-surface immunology. *Scientific American* **234**:30 (May 1976).

Rothwell, N. V. *Human Genetics.* Englewood Cliffs, N.J.: Prentice-Hall (1977).

Ruddle, F. H., and Kucherlapati, R. S. Hybrid cells and human genes. *Scientific American* **228**:82 (July 1974).

Silvers, W. K., and Wachtel, S. S. H-Y antigen: Behavior and function. *Science* **195**:956 (1977).

Tijo, J. H., and Levan, A. The chromosome number of man. *Hereditas* **42**:1 (1956).

# Chapter 12

# Cellular Evolution: From Chemicals to Life

A few hundred million years after a solar dust cloud began to develop into our sun and planets, the Earth and other planets had solidified to approximately their present form. The oldest dated rocks on Earth are only 4 billion years old (in Greenland), but from datings of meteorites as representative solid samples of our solar system there have been consistent age determinations of 4.6 billion years.

Conditions during sun and planet formation could not support life as we know it, but from fossils in the record of our planetary history we think that life on Earth may have originated about 3.5 to 4 billion years ago. How did that life arise? What were the profound events that ultimately produced living systems from a nonliving, primeval mix of chemicals? How did cells come into existence and, once in existence, what evolutionary changes gave rise to eukaryotic life from prokaryotic ancestors?

## ORIGIN OF LIFE

No one is certain about the exact time or conditions that sponsored life from nonlife, but there are certain distinctions between the two which allow us to put together a logical sequence of events. In addition, we have some knowledge of the kinds of chemistry that exist in the world today and those which probably existed during primeval times. Laboratory experiments have shown that organic molecules could have arisen **abiogenically;** that is, in the absence of life at the time. The first clues to abiogenic synthesis of biologically significant organic molecules were presented in 1953 by Stanley Miller.

### Abiogenic Synthesis

Using a simple apparatus that mimicked the primeval air and watery phases of the ancient Earth, Miller showed that amino acids were formed from methane, hydrogen, and other simple gases (Fig. 12.1). We have good reason to believe that there was water on Earth from the earliest days, and that the atmosphere was *reducing* rather than oxidizing as it is now. For example, carbon existed in reduced form as methane rather than in oxidized form as carbon dioxide; nitrogen existed in ammonia form; hydrogen gas was present; some water vapor may also have been present because of evaporation; and there was *no gaseous oxygen* in the primeval atmosphere.

Syntheses require an input of energy, which was amply available in solar radiation, as well as in electricity from lightning, and other sources. Miller provided an energy source of an electrical discharge and some simple raw materials in a sealed vessel. After several days to a week he opened the vessel and collected the organic precipitate that had formed. When this material was hydrolyzed to its constituent units, various compounds were found including **amino acids** that had polymerized into insoluble polypeptides. Amino acids are an essential component of living systems, so this was a tremendously important line of evidence showing that such syntheses could have occurred on the primeval Earth.

Numerous studies since 1953 have shown that various kinds of biologically important organic molecules can be made abiogenically. Included among these molecules are nucleic acids and ATP (a monomer of nucleic acids as well as an organic energy source). During the first few hundred million years, many such molecules must have accumulated in the

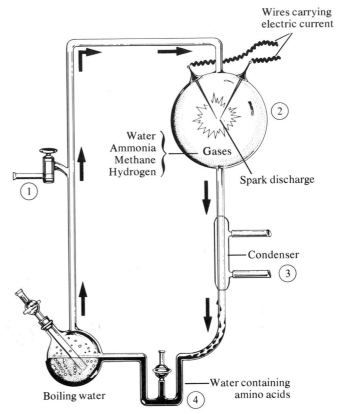

**Figure 12.1**
Diagram of the apparatus used by Stanley Miller in experiments on abiogenic synthesis of organic compounds. Water is added to the flask and the entire apparatus is evacuated of air. Then ammonia, methane, and hydrogen gases are introduced through a valve (1), and these gases ciruclate clockwise in a stream of steam produced when water in the flask is heated to boiling. The steam and gases enter the sparking chamber (2) and are subjected to spark discharges. As the molecules move through the vicinity of a water-cooled condenser (3), condensations occur. Nongaseous substances accumulate in the trap (4), while gaseous substances continue to circulate through the apparatus and past the spark, until the experiment is concluded. Molecules collected from the trap are then analyzed.

primeval seas. There was no life to consume these molecules, and their degradation would have been very slow as we know from the sluggishness of organic reactions in general. The primeval waters would have changed to an "organic soup" as they were filled with interacting molecules essential to life as we know it.

## Coacervates and Proteinoid Microspheres

Life is not an open sea of organic reactions, however. Living systems are separated from one another and their surroundings by selectively permeable *boundaries*. This is an essential prerequisite for life, since it leads to the **open system** which can exchange matter and energy with its surroundings. Cellular open systems exist in a state of flux, or different **steady states,** in which work can be done because equilibrium is not attained (see Chapter 3). Steady states lead to flexibility in chemical reactions, responding to fluctuating environments in ways that allow the system to persist. Furthermore, boundaries allow a system to carry out coordinated reactions amidst external chaos. Some of these coordinated reactions could lead to more retained free energy and to greater probabilities for biosyntheses.

There are two major ideas that attempt to explain the development of bounded systems containing sequestered molecules: (1) **coacervate** formation, first proposed by Alexander Oparin in 1921, and (2) **proteinoid microsphere** formation, originally proposed by Sidney Fox in the 1950s. There is no question that both kinds of systems can be produced under abiogenic conditions; what we don't know is which of these, if either, was the preliminary step leading to cellular life.

Coacervates form spontaneously when polar water molecules become oriented around some molecule or particle with an electrical charge (Fig. 12.2). The film of bound water acts as a boundary which separates the interior of the coacervate from the disordered surroundings. In experiments, it has also been shown that molecules *become concentrated within* the coacervate droplets. For example, in a dilute solution of gelatin or some other protein, more than 95 percent of the gelatin molecules are later found inside the spontaneously formed coacervates. This is an important consideration for originating life in the chaos of molecules and reaction systems of the "organic soup".

When particular molecules are put together in a test system, it is possible for exchanges between coacervate and surroundings to take place, exchanges of matter *and* energy (Fig. 12.3). The test systems can become more complex as new reactions are incorporated, particularly if they are related and coordinated with ones already present. These coacervates have a greater chance for surviving a longer time, and if they survive for a longer time there is a greater chance that biologically significant reactions and molecules can be included in at least some of them. Coacervates could incorporate reaction systems that produced nucleic acids, that stored energy in ATP, that made polypeptides, and so forth. It has been shown that all of these products can be made abiogenically as well as in modern living systems.

Proteinoids are molecules that assemble from amino acids under abiogenic, water-free conditions at high temperatures. If there is some surface, such as clay, on which the polymers can become adsorbed,

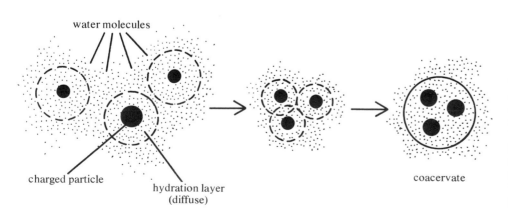

water molecules

charged particle

hydration layer
(diffuse)

coacervate

**Figure 12.2**
Diagram showing the process of coacervation. The exclusion of some of the water molecules from the hydration layer, and the delimitation of the rest of the layer, leads to a coacervate droplet, which contains one or more kinds of electrically charged particles. The dashed or solid lines bounding the droplets represent the boundary layer separating the bound water within the droplet from the free water outside.

then proteinoids assemble and eventually drop off into cooler waters below, where microspheres may form. These microspheres persist for relatively long times.

When seen in the microscope, proteinoid microspheres have definite boundaries (Fig. 12.4). These boundaries provide osmotic properties, allowing substances through at different rates. Osmotic regulation is a known property of membrane-bounded cells and cell compartments.

Each of these prototypes of the cell, or **protobionts** (pre-life forms), has its drawbacks as well as its suitable features. Coacervates are usually short-lived. They form and disperse very quickly. Proteinoid microspheres are longer-lived, but rather homogeneous in their chemical composition. Life must have originated from a persisting, chemically variable protobiont, but neither model fits all the requirements we expect in nonliving ancestors of the first life forms. Each model remains a viable alternative, but we cannot choose between them at present.

### The First Life Forms

Somewhere, at some time in the ancient past, life forms **(eubionts)** arose from protobionts. It is uncertain whether a protein system later incorporated nucleic acids, or whether a nucleic acid system later came to make proteins. In either case, eubionts must have had a primitive genetic machinery which allowed them to transmit their characteristics to descendants, who also were alive and capable of passing on inherited traits from generation to generation. However life may have originated, eubionts must have been able to *evolve* because they contained genetic information that could be passed on to their descendants. When such information is translated into proteins and metabolism takes place, living forms grow, reproduce, and maintain their structures and functions through repair and replenishment. If mistakes appear in genes, by **mutations,** these altered genes must also be transmitted faithfully to descendant generations. Evolution can be defined as *descent with modification*. Descent requires a means of transferring information, while modification leading to new information occurs by mutations of existing genes and by new genes arising.

Primeval life forms did not require elaborate metabolism, because the surrounding waters were filled with organic molecules that had accumulated for hundreds of millions of years. A few genes directing synthesis of a few enzymes would probably have been adequate at the beginning. Most of the nutritional needs for life would have been available among the molecules in the environment, and living systems would have been engaged in exchanging matter and energy with their surroundings. But certain essential molecules would have become depleted as living systems removed them from the waters. These molecules would not be replaced quickly by the ponderously slow organic reactions out in the "soup". How could life continue to exist when raw materials for growth and reproduction became diminished in amount? A reasonable explanation was proposed in 1945 by Norman Horowitz. He postulated the idea of **"evolution backwards"**, which can be summarized as

$$G \xleftarrow{f} F \xleftarrow{e} E \xleftarrow{d} D \xleftarrow{c} C \xleftarrow{b} B \xleftarrow{a} A$$

Eubionts with simple needs and few genes could

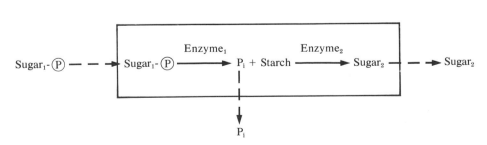

$$Sugar_1\text{-}\textcircled{P} \text{ -- -- } \rightarrow Sugar_1\text{-}\textcircled{P} \xrightarrow{Enzyme_1} P_i + Starch \xrightarrow{Enzyme_2} Sugar_2 \text{ -- -- } \rightarrow Sugar_2$$

$$\downarrow$$

$$P_i$$

**Figure 12.3**
A flow of metabolites can lead to exchange of energy and matter between a coacervate droplet and its surroundings. There is energy exchange during the reactions within the droplet, leading to free-energy differences between reactants and products and to the possibility for retention of some of the free-energy differences within the droplet.

use compound G for metabolism, whatever compound G might be. As life multiplied and G became scarce, many organisms could not survive. But by chance, some organism may have undergone a mutation that allowed it to make enzyme *f*, which catalyzed synthesis of G from compound F. Such a mutant and its descendants could survive and transmit their gene mutation to later generations. As new populations multiplied, compound F would become depleted. Still other mutants existing by chance in these populations would now have an advantage if they could synthesize enzyme *e* which catalyzed formation of F within the cell, using compound E from the environment as a precursor. This and other sequences would continue to occur in the ancient seas.

At each step in the evolutionary sequence, *chance* mutations appear and may confer an advantage depending on the prevailing conditions for existence. Mutations are random changes in genes that occur regardless of advantage or disadvantage; mutations do not arise because the organism needs some ability in order to survive. This is the distinction between evolution as Jean Lamarck saw it in 1809 (organisms change according to their needs), and as Charles Darwin first proposed in 1859 in his theory of evolution by **natural selection.** According to modern understanding of genes and natural selection, some components of variable populations are inherently better adapted to certain living conditions and these are the individuals most likely to survive to reproductive age and therefore to leave descendants like themselves (Fig. 12.5).

New metabolic pathways probably developed as we see them in modern organisms, that is, through coupled reaction systems and coordinated sets of molecules engaged in energy-using and energy-releasing activities. In some metabolic pathways, one end product of a reaction can serve as the precursor for a different sequence, usually coupled by an ADP-ATP energetic link or similar energy-transferring arrangement (Fig. 12.6). Stepwise release of energy and the conservation of energy for cellular work would have been modifications with a high priority for successful life, even as we see it all around us today. Precursors might therefore be made available *internally* as well as

from the external environment, and together they would subsidize metabolism in successful life forms.

### Modifications in Nutrition

Once we accept the idea of "evolution backwards" we also accept the idea that the simpler eubionts must have had relatively few genes and been able to manufacture relatively few of their molecular needs. Such organisms would derive energy and carbon atoms for their growth and reproduction through breakdown of complex organic fuels. Organisms of this type are

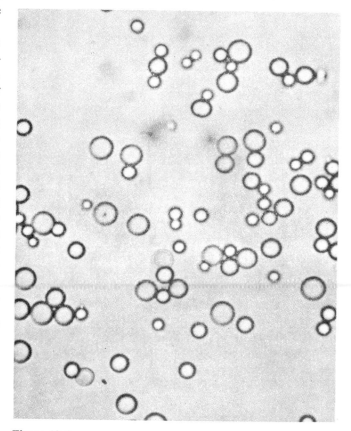

**Figure 12.4**
Microspheres made by cooling a solution of proteinoid from 25°C to 0°C. × 900. (Reproduced with permission from Young, R. S., p. 347, in Fox, S. W. (ed.), *The Origins of Prebiological Systems*. Academic Press, New York, 1965.)

## LAMARCK'S GIRAFFE

Short-necked ancestor

Stretches neck
to reach food
higher up on tree

Keeps on
stretching

and stretching
until neck becomes
progressively
longer in descendants

## DARWIN'S GIRAFFE

Original group
exhibits variation
in neck length

Natural selection
favors longer
necks: better
chance to get
higher food.
Favored character
passed on to
next generation

After many, many
generations the group is
still variable, but shows
a general increase in
neck length

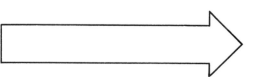

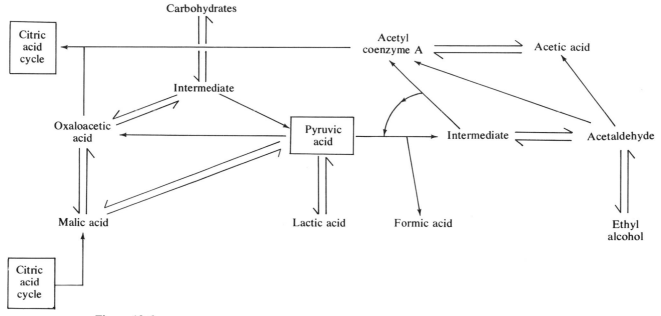

**Figure 12.6**
Pyruvic acid can be diverted to various metabolic pathways in the cell. The multiple uses of
pyruvic acid are economical, since a few kinds of chemical reactions and key organic
compounds can produce a variety of metabolic intermediates and end products.

**heterotrophic** in their nutrition. As evolution proceeded, heterotrophic metabolism would have become more complicated, but energy and carbon would
still be derived by breakdown of organic compounds.
Modern animals, fungi, many protists, and most bacteria are heterotrophic organisms of varying degrees of
complexity. There are two other main types of nutri-

**Figure 12.5**
Comparison of Lamarckian and Darwinian explanations of
the evolutionary events leading to the modern-day giraffe
from shorter-necked ancestors. Lamarck proposed that
inherited variations appeared in response to the needs of the
organism. Darwin proposed that there was differential
reproduction within a genetically diverse population, and the
inherently better fit individuals transmitted their inherited
properties from generation to generation until a new diverse
population evolved. The ideas are stated in Lamarck's
Theory of Inheritance of Acquired Characteristics
(discredited) and Darwin's Theory of Descent by Natural
Selection (accepted).

tion, in which energy and carbon sources are recog-
nizably different from each other and from heterotrophic pathways: **chemotrophic** and **autotrophic**
nutrition.

Chemotrophs obtain energy by oxidizing inorganic
molecules, such as reduced sulfur and iron compounds, and obtain carbon from simple organic acids
and alcohols for the most part. Autotrophs, on the
other hand, derive both energy and carbon from
nonorganic sources. The most familiar autotrophs are
green plants which obtain energy from sunlight and
carbon from atmospheric carbon dioxide. Chemotrophs are found mostly among some bacterial
groups, as are a few autotrophic species. Most bacteria, however, are heterotrophic. There are further
fine distinctions among nutritional types, reflecting the
specific energy or carbon sources they use. For
example, green plants are photoautotrophs, since light
is their energy source.

Although chemotrophs and autotrophs *seem* simple

because their needs are simple, they are enzymatically (and therefore genetically) complex because they make all or most of their cellular materials through biosynthetic pathways. Since the first eubionts must have been genetically and enzymatically very simple, although their raw materials were complex, we must conclude that heterotrophs were the first kind of primeval life. As new genes and new metabolic potential became incorporated in certain populations, chemotrophs and finally autotrophs must have evolved from particular heterotrophic ancestors. The majority of heterotrophic lineages, however, continued in the heterotrophic life style. All three major nutritional modes have persisted to the present day, and have become considerably diversified during evolution. We see many variations on these three basic themes in modern life forms.

As life evolved, therefore, we believe that metabolism became increasingly complex. By "evolution backwards", simpler genetic and enzymatic capacities were expanded to modern levels of genetic machinery and protein products. Primeval life was cellular, genetically and enzymatically simple, but capable of continued evolutionary modification through processes of genetic change and the overseeing forces of natural selection. Those better adapted were more likely to succeed and multiply, while competing and poorly adapted components in primeval populations became reduced in numbers or even extinct. These same events are happening all around us today, too.

## EUKARYOTES FROM PROKARYOTES

We cannot begin to document the numerous evolutionary studies aimed at analyzing species relationships, but we can examine viewpoints concerning the origin of eukaryotic cellular organization. We have a good idea of the similarities and differences between prokaryotic and eukaryotic cell organization and chemistry, and we can make at least three basic points at the start: (1) Prokaryotes existed for about 2 billion years before eukaryotes appeared, according to the history of life preserved in the fossil record. (2) There are basic similarities that clearly show **common**

**descent** for prokaryotes and eukaryotes; for example, they possess the same genetic codon dictionary; their genetic material is DNA; there is a similar ribosomal machinery for protein synthesis; there are common metabolic pathways and enzyme catalysis, and various other features. (3) The air contained only traces of molecular oxygen by about 1.2 billion years ago, and became essentially oxidizing to its present level of about 21 percent $O_2$ perhaps 700 million to 1 billion years ago. Estimates concerning oxygen are derived from studies of fossils and of rock formations of known age.

From these observations we may conclude that prokaryotes were ancestral to eukaryotes, and that the great evolutionary divergence that led to eukaryotes took place while the Earth was relatively but not entirely anaerobic (Fig. 12.7). Since prokaryotes are far more ancient than eukaryotes, and since there are fundamental similarities that are unlikely to have arisen independently by chance, we can safely say that eukaryotes must have evolved from prokaryotic ancestors.

### Origin of the Eukaryotic Cell

Eukaryotic cells have membrane-bounded compartments while prokaryotic cells have only a plasmalemma and in some cases one or more types of infolding of this membrane. These infoldings are usually enzymatically distinct from the remainder of the plasmalemma. Because there is a basic organizational difference between compartmented and noncompartmented cell types, we must look for explanations of permanent membrane systems that are physically separate from the plasmalemma in eukaryotes. Most of the formal hypotheses include the infolding and eventual separation of infoldings from the common plasmalemma source. As these internalized membranes became functionally distinct and as they came to enclose particular sets of molecules and reaction systems, they eventually assumed the special qualities and appearances we see today in the nucleus, ER, lysosomes, microbodies, and other compartments.

There are different views, however, concerning the

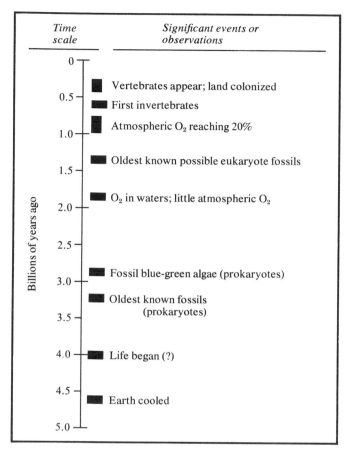

| Time scale | Significant events or observations |
|---|---|
| 0 | |
| 0.5 | Vertebrates appear; land colonized |
| | First invertebrates |
| 1.0 | Atmospheric $O_2$ reaching 20% |
| 1.5 | Oldest known possible eukaryote fossils |
| 2.0 | $O_2$ in waters; little atmospheric $O_2$ |
| 2.5 | |
| 3.0 | Fossil blue-green algae (prokaryotes) |
| 3.5 | Oldest known fossils (prokaryotes) |
| 4.0 | Life began (?) |
| 4.5 | Earth cooled |
| 5.0 | |

(Billions of years ago)

**Figure 12.7**
An evolutionary time scale showing significant changes in organisms and the Earth during the past 5 billion years.

origins of mitochondria and chloroplasts in particular. Each of these organelle types is unique in having its own portion of DNA, RNA, and ribosomal machinery. According to one point of view these organelles evolved through membrane infolding, separation, and differentiation just like other eukaryotic compartments, but with the difference of having captured some piece of the cellular genetic apparatus which then became part of the organelle construction and function. According to the opposing view, mitochondria and chloroplasts originally were prokaryotic organisms in their own right, but later became functioning parts of their host organism. Their genetic ma-

chinery, therefore, is the remnant of the system originally present in their free-living ancestors.

There are common features, however, held by both schools of opinion about mitochondria and chloroplast origins. In each theory it has been postulated that the primary modification in the prokaryotic ancestor of eukaryotes was *loss of the rigid, confining cell wall and acquisition (by mutation) of a mobile cell surface.* This "ameboid" prokaryote could move around by creeping locomotion, but most importantly it could evolve into an *ingesting* organism. Various materials, including foods, would enter the cell by endocytosis. This would be an adaptive change, since it would allow additional sources of food and therefore provide the anaerobic organism with a larger supply of fuel for growth and reproduction. Beyond this point there are major differences in the two evolutionary sequences. The two principal hypotheses explaining mitochondria and chloroplast origins are (1) **endosymbiosis,** and (2) **internalized membrane differentiation.**

### Endosymbiosis

The Earth's atmosphere contained only 1 percent $O_2$ 1.3 to 1.4 billion years ago, the time when the first eukaryotic fossils appear in the fossil record. There must have been some earlier prokaryotic organisms, therefore, which used oxygen in aerobic respiration. If such respiring bacteria were ingested by an "ameboid" prokaryote, and if these bacteria persisted unharmed as symbiotic partners with their host cell, the result would be a respiring "ameboid" prokaryote host (Fig. 12.8). This would be beneficial to the host, allowing more efficient energy extraction during food breakdown, and it would be beneficial to the bacteria because they would have available, ready-made foods provided in the host cell. Such a mutually beneficial association is called **symbiosis.** The bacteria are *endo*symbionts, housed *inside* their permanent host cell.

The theory was first proposed in the late 1800s, but has been revived most recently by Lynn Margulis. Since 1967 she has presented an extensive exposition of the endosymbiosis theory. In her view, the respirer bacteria eventually evolved into mitochondria of the

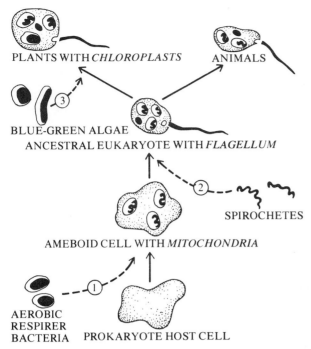

PLANTS WITH *CHLOROPLASTS*

ANIMALS

BLUE-GREEN ALGAE

ANCESTRAL EUKARYOTE WITH *FLAGELLUM*

SPIROCHETES

AMEBOID CELL WITH *MITOCHONDRIA*

AEROBIC
RESPIRER
BACTERIA

PROKARYOTE HOST CELL

**Figure 12.8**
A summary of the Endosymbiosis Theory of the origin of
eukaryotic cells, as proposed by L. Margulis. (Adapted with
permission from *Origin of Eukaryotic Cells*, by L. Margulis,
Yale University Press, p. 58, 1970. Copyright © 1970 by
Yale University.)

eukaryotic cells. The DNA and other genetic apparatus in modern mitochondria are therefore
considered to be remnants of the formerly freeliving
endosymbiotic bacteria. Respiratory enzymes are
essentially similar in mitochondrial membranes and in
modern bacterial plasmalemma. A further similarity
between mitochondria and bacteria is their size and
shape, if we accept this particular model of the
mitochondrion (see Chapter 5).

Margulis proposed that a second endosymbiosis
took place some time afterward, leading to more effective locomotion for the respiring ameboid prokaryote.
In this case it is postulated that spiral-shaped bacteria,
called **spirochetes,** became incorporated as endosymbionts. These were ultimately modified to become cilia and flagella in eukaryotic descendants. Spi-

rochetes have an unusual ultrastructure, and there are
certain protozoa that move about by the locomotor action of their own symbiotic spirochetes. The modern
observation led to the suggestion for the past evolutionary endosymbiosis.

A third endosymbiosis has been suggested by Margulis to have occurred after eukaryotes had evolved;
one or more of these ancient, respiring, flagellated
eukaryotes ingested blue-green algae. These algae
eventually evolved into modern chloroplasts. The
genetic machinery in modern chloroplasts is therefore
explained as being the remains of the genetic system
originally present in the endosymbiotic blue-green
algae. Eukaryotes that did not happen to establish a
symbiotic relationship with blue-green algae remained
nonphotosynthetic, ultimately giving rise to various
types of protists, some of which in turn diverged to
produce fungi and animals. Eukaryotes with chloroplasts gave rise to photosynthetic protists, algae, and
green plants (see Fig. 1.6).

There are numerous modern examples of symbiotic
associations between different organisms. In many
cases it is possible to show that each symbiotic partner
can live independently of the other, but that mutual
benefit keeps the partners together under natural conditions. The scruffy lichens that grow on rocks and
trees and in relatively barren living zones are actually
composed of an alga and a fungus. *Paramecium bur-
saria* is a ciliated protozoan that harbors a population
of unicellular green algae in symbiotic association.
Assorted cellulose-digesting protozoa live in the gut of
termites, providing the essential digestive capacity in
these woodeating animals. There are examples of blue-
green algae that live symbiotically in protozoan and
other host cells, and such algae usually lose their rigid
cell wall when they take up residence in their hosts. In
an experiment in which chloroplasts from flowering
plants were injected into mouse cells grown in culture,
the chloroplasts could be recovered undamaged some
days later. All together, symbiosis is a well known
natural phenomenon, often involving an alga as one of
the partners. There is no case, however, of a prokaryote which houses a symbiont organism.

The endosymbiosis theory is attractive in certain

ways in explaining why mitochondria and chloroplasts have a genetic apparatus, alone among all the types of eukaryotic organelles. In addition, the theory also explains particular similarities between prokaryotes and these two kinds of organelles by the accepted evolutionary doctrine of common descent. For example, they all have naked DNA duplex molecules that are circular in form and that contain all the genes of the system in a single molecule. Their ribosomes respond similarly to various antibiotic inhibitors of protein synthesis while eukaryotic cytoplasmic ribosomes respond differently to the same drugs. Taking these and other factors into consideration, many people have accepted endosymbiosis as the evolutionary explanation for mitochondria and chloroplasts. The rationale for flagellar origin, however, is less well accepted.

### Internalized Membrane Differentiation

There have been various suggestions concerning the origin of one or another of the eukaryotic membranous compartments, particularly in relation to mitochondria. In the mid-1970s, T. Cavalier-Smith presented a relatively comprehensive scheme for eukaryotic cell development during evolution. He suggested that the prokaryotic ancestor was a blue-green alga that had developed a mobile cell surface capable of endocytosis and other surface activities (Fig. 12.9). One of the first membranous compartments to appear would be lysosomes, since powerful digestive enzymes would be advantageous in handling ingested solid foods. These enzymes would digest the cell itself if not sequestered behind a membrane, so any mutations leading to lysosomal compartments would be highly advantageous to a cell with a mobile surface.

In prokaryotes the DNA molecule is attached to the inside surface of the plasmalemma. This would not be a safe place in a cell with an active cell surface, since the DNA could be expelled by exocytosis or be taken into the cell interior by endocytosis and perhaps be digested there by lysosome interactions. Mutations leading to detachment of DNA from the cell membrane would be adaptive, but segregation of replicated DNA during cell divisions requires some rigid attachment

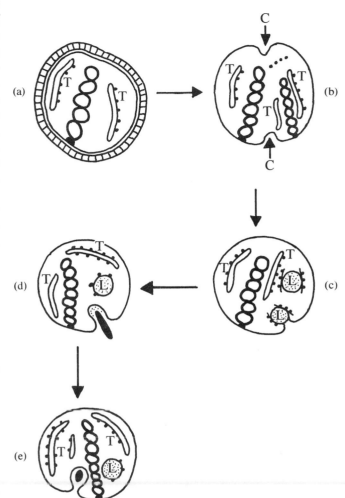

**Figure 12.9**
Origin of cytosis and of internalized membrane compartments in ancestral, prokaryotic blue-green algae. (a) Blue-green alga with cell wall, thylakoids (T), and a twisted molecule of DNA; (b) loss of the cell wall leads to a mobile cell surface with the capacity to invaginate in cytosis (C) events leading to cell division after DNA replicaiton; (c) lysosomes (L) differentiate and act directly at the cell surface at first; (d) lysosomes gradually become internalized structures; and (e) lysosomes eventually are completely internalized and interact with substances brought into the cell in vesicles formed during endocytosis. (Adapted with permission from Cavalier-Smith, T., 1975, *Nature* **256**:463, Fig. 1.)

system. In place of the cell surface, Cavalier-Smith proposes the evolution of a microtubule system which serves to hold the DNA molecules taut during their separation into new cells. This system of DNA and microtubules eventually became enclosed in a nuclear envelope, and differentiated as an internalized system that is physically separate from the plasmalemma (Fig. 12.10). Other membrane differentiations led to cytoplasmic compartments including mitochondria, thylakoid-containing chloroplasts, ER, and others.

According to this theory there is no reason to postulate a different origin for any of the organelles in eukaryotes, or to require a sequence of endosymbioses occurring in a particular order and at particular times in eukaryote evolution. Compartmentation is an effective means for sequestering coordinated reactions and components, and once there was some internalized membrane there would eventually be others during eukaryote evolution. Cavalier-Smith postulates that chloroplasts were lost in some eukaryotic lineages, which we see today as nonphotosynthetic groups of organisms. Modern photosynthetic groups presumably arose from eukaryotes that retained their chloroplast compartments.

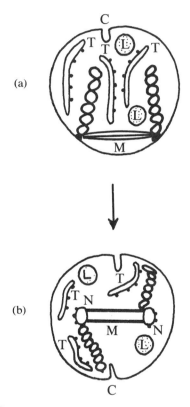

**Figure 12.10**
The evolution of spindle microtubules (M) in the ancestral prokaryote, acting as a device that pushes daughter DNA molecules apart to opposite poles of the cell and leads to one molecule in each cell produced after completion of the cleavage furrow (C). (a) Initially, DNA is attached to the cell membrane; but later (b) the attachment sites are taken into the cell by endocytosis events and eventually become the beginnings of a nuclear envelope system. (Adapted with permission from Cavalier-Smith, T., 1975, *Nature* **256**:463, Fig. 2.)

### Different Origins for Chloroplasts and Mitochondria?

There are so many differences between mitochondrial and bacterial systems, as well as similarities, that it is difficult to decide which is the more important aspect in evolutionary origins. For example, according to three-dimensional reconstructions of mitochondria, they may not be the same size and shape as bacteria (see Fig. 5.21). Mitochondrial ribosomes vary from 55S to 80S in size whereas bacterial ribosomes are 70S. When ribosome monomers are constructed from one kind of mitochondrial subunit and one kind of bacterial subunit, the hybrid ribosomes cannot sponsor protein synthesis. This shows a considerable difference between ribosomes from the two sources even though they respond similarly to various drugs that inhibit protein synthesis. Mitochondrial DNA varies from 5 $\mu$m in animal cells to 30 $\mu$m in plant cells, with a range of sizes in between these extremes among protists and fungi.

Chloroplasts, on the other hand, resemble prokaryotic systems in many more ways. Chloroplast ribosomes are 70S, like bacterial ribosomes, and hybrid ribosome monomers made from subunits of each source are functional in protein synthesis. These ribosomes are therefore very similar in more than size.

Chloroplasts resemble blue-green algae internally as well as externally, the resemblance sometimes being uncanny. Chloroplast DNA molecules have been shown to be 40–45 $\mu$m in contour length, regardless of their source. While not in the same league with a 1300 $\mu$m-long DNA molecule as in *E. coli,* chloroplast DNA is large enough to code for perhaps 100 different proteins. This is a substantial amount of potential information, considering that most of chloroplast structure and function is coded by nuclear genes anyway.

At present it is not easy to decide whether one or both organelle types evolved from endosymbionts or whether both developed from differentiated, internalized membranes. It is a very interesting evolutionary topic and should be kept open until we have resolved the mode(s) of origin.

Regardless of their origin, there is an advantage to the cell if chloroplasts and mitochondria code for some of their polypeptides. These genes may have been part of the cellular genetic machinery originally, or remains of endosymbiont genes. In either case, organelle genes code for polypeptides that cannot pass across the membrane barrier very easily. These organelle-coded polypeptides have unusually high amounts of hydrophobic amino acids. If the polypeptides are made within the organelle on organelle ribosomes, the finished chains can easily be positioned directly within the organelle thylakoids or inner membrane after they are made in the matrix (stroma). In those cases that have been studied in some detail, organelle-specified polypeptides are parts of proteins whose remaining polypeptide chains are made in the cytoplasm from nuclear coded information. The whole proteins assemble once the two sets of polypeptides come together inside the mitochondrion or chloroplast (Fig. 12.11). Three known organelle enzymes coded partly by the organelle DNA and partly by nuclear DNA are known to be made in this way. Two of these enzymes are parts of mitochondrial inner membrane construction and functions: **cytochrome oxidase** and the **ATPase** of oxidative phosphorylation. Similarly, **RuDP carboxylase** of the photosynthetic dark reactions is made from combined genetic instructions of chloroplast and nucleus.

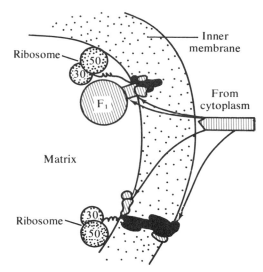

**Figure 12.11**
Hypothetical scheme illustrating cooperation between mitochondrial and cytoplasmic protein synthesis in the assembly of cytochrome oxidase and ATPase of the inner mitochondrial membrane in yeast. Mitochondrially-synthesized enzyme subunits (solid black) are formed on mitochondrial ribosomes (dotted) close to their site of deposition within the membrane. Cytoplasmically-synthesized enzyme subunits (cross-hatched) are imported from the cytoplasm outside the mitochondrion and are situated close to the membrane surface. The two curved lines signify the thickness of the mitochondrial inner membrane. (From Ebner, E., T. L. Mason, and G. Schatz, 1973, *J. Biol. Chem.* **248**:5369-5378, Fig. 12.)

## SUMMARY

1. Life arose on Earth about 3.5 to 4 billion years ago, from suitable protobiont systems which contained abiogenically-synthesized organic compounds in a structure bounded by a selectively permeable boundary. Whether the structure was a coacervate or a proteinoid microsphere system, pre-life forms could exchange matter and energy with the surrounding organic soup. Eubiont (true life) forms eventually arose when some protobiont system contained a genetic system coordinated with a simple metabolism, allowing growth and reproduction of like forms in successive generations.

2. Life forms became increasingly complex as they incorporated additional metabolic activities, directed by new or mutated genes. Through natural selection acting on

genetically variable populations, "evolution backwards" led to more complex heterotrophs and finally to chemotroph and autotroph descendant species. In addition to newly evolved modes of nutrition and improvements in heterotroph activities, life became predominantly aerobic about 1 billion years ago, primarily as the result of cellular adaptations to molecular oxygen produced by photosynthetic species.

3. Prokaryotes are ancestral to all modern life forms, including eukaryotes which first appeared about 1.4 billion years ago, according to the fossil record. According to both prevailing theories on eukaryote origin, a significant early or first modification in the prokaryotic ancestry of eukaryotes was loss of the rigid cell wall and development of a mobile cell surface. The Endosymbiosis Theory states that prokaryotic ancestors could engulf respiring and photosynthetic prokaryotes, which later evolved into mitochondria and chloroplasts, respectively. The opposing theory states that all eukaryotic cell compartments, including mitochondria and chloroplasts, could have arisen through processes of internalized membrane differentiation. In this theory it is proposed that a second major innovation which led to eukaryotes was the evolution of a microtubular apparatus to which cellular DNA became attached and which later was surrounded by membranes, leading to the nucleus that characterizes eukaryotic cells.

## STUDY QUESTIONS

1. How old is the Earth? When do we think life originated on Earth? What is abiogenic synthesis? How did Stanley Miller show that organic molecules could have been synthesized on the primeval Earth before life originated? How did organized, boundaried systems arise in the primeval oceans (organic soup)? What possible models have been examined as protobiont systems? What is a coacervate? How do coacervates and proteinoid models differ? In what ways are they similar? How does either model of a protobiont fit our definition of an open system in a thermodynamic sense?

2. What are the essential components of eubionts that are missing in protobionts? Why do eubionts reproduce and evolve? How did life forms develop increasingly complex metabolisms during primeval evolutionary times? Could the same mechanisms be operating today? What do we mean by "evolution backwards"? How can we explain "evolution backwards" according to Lamarck's theory? according to natural selection as proposed by Darwin?

3. What is a heterotroph, and what is an autotroph? Why is it more likely that primeval life was heterotrophic and only later gave rise to autotrophic organisms?

4. What essential change is presumed to have occurred in the ancestral prokaryotic lineage that eventually gave rise to eukaryotes? What is the advantage of a mobile cell surface in relation to nutrition and reproduction? Why are we sure that prokaryotes are more ancient than eukaryotes as evolutionary groups? What were the conditions on the Earth 4 billion years ago? 2 billion years ago? 1 billion years ago? 500 million years ago?

5. How does the Endosymbiosis Theory explain the origin of eukaryotic cells? What are the proposed endosymbiont ancestors for mitochondria? flagella? chloroplasts? What kind of prokaryotic ancestor is postulated in this theory? How does it differ from the postulated ancestor in Cavalier-Smith's theory of internalized membrane differentiation? How does each theory explain the origin of plants and of animals? What is unique about chloroplasts and mitochondria among all other organelles in eukaryote cytoplasm?

## SUGGESTED READINGS

Avers, C. J. *Evolution*. New York: Harper & Row (1974).

Barghoorn, E. S. The oldest fossils. *Scientific American* **224:**30 (May 1971).

Bonen, L., and Doolittle, W. F. Partial sequences of 16S rRNA and the phylogeny of blue-green algae and chloroplasts. *Nature* **261:**669 (1976).

Calvin, M. Chemical evolution. *American Scientist* **63:**169 (1975).

Cavalier-Smith, T. The origin of nuclei and of eukaryotic cells. *Nature* **256:**463 (1975).

Clarke, B. The causes of biological diversity. *Scientific American* **233:**50 (Aug. 1975).

De Witt, W. *Biology of the Cell. An Evolutionary Approach*. Philadelphia: Saunders (1977).

Kubai, D. F. The evolution of the mitotic spindle. *International Review of Cytology* **43:**167 (1975).

Mahler, H. A., and Raff, R. A. The evolutionary origin of the mitochondrion: A nonsymbiotic model. *International Review of Cytology* **43:**1 (1975).

Margulis, L. Symbiosis and evolution. *Scientific American* **225:**48 (Aug. 1971).

Margulis, L. *Origin of Eukaryotic Cells*. New Haven: Yale Univ. Press (1970).

Miller, S. L., and Orgel, L. E. *The Origins of Life on the Earth*. Englewood Cliffs, N. J.: Prentice-Hall (1975).

Schopf, J. W., and Oehler, D. Z. How old are the eukaryotes? *Science* **193:**47 (1976).

Uzzell, T., and Spolsky, C. Mitochondria and plastids as endosymbionts: A revival of special creation? *American Scientist* **62:**334 (1974).

Whittaker, R. H. New concepts of kingdoms of organisms. *Science* **163:**150 (1969).

# Glossary

**acetyl coenzyme A.** a high-energy intermediate in energy-transferring reactions in metabolism; activated acetate linked to coenzyme A.

**actin.** a major protein of muscle; the principal protein of thin filaments of striated muscle and many nonmuscle cells; involved in cell movements.

**active site.** the region of an enzyme at which a substrate binds and reacts with the enzyme to form a temporary enzyme-substrate complex.

**active transport.** the energy-requiring assisted passage of molecules across a membrane, in the direction of higher concentration of the molecules.

**adenosine triphosphatase.** ATPase; the enzyme that hydrolyzes ATP to form ADP and inorganic phosphate and catalyzes the reverse reaction of ATP synthesis.

**adenosine triphosphate.** ATP; a nucleoside triphosphate which is a high-energy intermediate in energy-transferring metabolism and one of the precursors for ribonucleic acid synthesis.

**aerobic.** an organism, cell, or process that utilizes molecular oxygen.

**allele.** one of the alternative forms of a gene.

**amino acid.** the basic building block of proteins; a carboxylic acid with one or more $NH_2$ groups.

**aminoacyl-tRNA synthetase.** one of a group of enzymes which catalyzes the activation of an amino acid to the aminoacyl form and the joining of the aminoacyl group to specific transfer RNA carriers.

**amniocentesis.** a medical procedure during which a sample of amniotic fluid is removed from around the fetus using a hypodermic syringe inserted into the mother's abdomen.

**amphipathic.** molecules having spatially separated hydrophilic and hydrophobic regions.

**anaerobic.** an organism, cell, or process that does not use or require molecular oxygen.

**anaphase.** stage of mitosis or meiosis in which the chromosomes move to opposite poles of the cell.

**antibody.** a protein which is secreted by plasma cells (activated *B* cells) and which interacts with a specific invading antigen; an immunoglobulin.

**antigen.** a foreign protein that stimulates the body to secrete specific antibodies or to proliferate activated *T* cells sensitized to the antigen.

**aster.** the region at the poles of dividing cells, including microtubules surrounding a clear zone within which a pair of centrioles is located.

**ATP.** *see* **adenosine triphosphate.**

**ATPase.** *see* **adenosine triphosphatase.**

**autoradiography.** a method for localizing radioactive atoms in microscope slide preparations by exposing a photographic emulsion to radioactive atoms incorporated into the specimen material.

**autosome.** any chromosome of the complement that is not a sex chromosome.

**autotroph.** an organism that obtains its energy and carbon for growth from inorganic sources.

**bacteriophage.** any virus which requires a bacterial host for its replication.

**Barr body.** any condensed or inactivated X chromosome in the interphase nucleus; also called sex chromatin.

**basal body.** *see* **centriole.**

**B cell.** a differentiated lymphocyte derivative that secretes antibodies; plasma cells of the blood system.

**bioenergetics.** thermodynamics applied to living systems.

**bivalent.** a synapsed pair of homologous chromosomes seen in Meiosis I.

**breakage and reunion.** a mechanism of crossing over between linked genes giving rise to gene recombinants; a mechanism of exchange between sister chromatids of a single chromosome.

**C-banding.** a method for staining chromosomes differentially showing locations of heterochromatin in the stained banded regions of the chromosome.

**C₃ cycle.** part of the dark reactions of photosynthesis; $CO_2$ is reduced to carbohydrate via a 3-carbon intermediate, 3-phosphoglyceric acid; also known as the Calvin cycle.

**C₄ cycle.** an accessory $CO_2$-reducing pathway in photosynthesis; occurs in plants lacking photorespiration; also known as the Hatch-Slack pathway.

**Calvin cycle.** *see* **C₃ cycle.**

**carotenoid.** an accessory photosynthetic pigment.

**carrier.** a transport protein within the membrane that binds temporarily with a molecule to be assisted across a membrane.

**catalyst.** any agent that modulates the rate of a chemical reaction without altering the equilibrium point of that reaction (*see also* enzymes).

**cell cycle.** the sequence of events in dividing cells in which an interphase consisting of $G_1$, *S, and* $G_2$ phases separates one mitosis from the mitosis of a previous or successive cell cycle.

**cell division.** formation of two daughter cells from a parent cell by enclosure of the two nuclei in separate cell compartments (*see also* **cell plate formation, furrowing**).

**cell-mediated immune response.** interaction between sensitized *T* cells and invading antigens.

**cell plate formation.** a cell division process typical of higher plants; new wall materials are laid down first in the cell center and then continue to be laid down in an outward direction to the cell periphery.

**cell surface.** the outer covering of the cell; the cell wall in plants, fungi, and bacteria, and the plasma membrane in animal cells and various protists.

**cell theory.** a generalization which states that the cell is the ultimate structural unit of the organism; first stated by Schleiden and Schwann in 1838–1839.

**cell wall.** rigid or semirigid structure encasing the living protoplast of plant, algal, fungal, and prokaryotic cells.

**centriole.** a microtubule-containing, cylindrically-shaped organelle located at the spindle poles in dividing cells or embedded at the cell periphery and forming the basal portion of a cilium or flagellum.

**centromere.** the region of the chromosome to which the spindle fibers attach and which is required for chromosome movement to the poles at anaphase.

**chiasma.** site of a previous exchange between two chromatids of a bivalent, visible in diplotene, diakinesis, and metaphase I of meiosis as a crossover figure.

**chlorophyll.** the principal photosynthetic light-capturing pigment, located in chloroplast thylakoids or in prokaryotic photosynthetic folded membranes.

chloroplast. the chlorophyll-containing photosynthetic organelle in eukaryotes.

cholesterol. a major lipid constituent of animal cell plasma membrane.

chromatid. one half of a replicated chromosome, joined to the other chromatid at the centromere region.

chromatin. the deoxyribonucleoprotein material of the chromosomes.

chromatin fiber. the continuous deoxyribonucleoprotein strand of a chromosome.

chromosome. the gene-containing structure in the nucleus.

cilium (pl. cilia). a whiplike locomotor organelle produced by a centriole.

cisterna (pl. cisternae). a flattened membranous sac filled with fluid contents.

cleavage. see furrowing.

coacervate. a droplet formed when a film of bound water molecules encloses one or more electrically-charged particles in suspension.

codon. a triplet of nucleotides that specifies an amino acid in a protein.

coenzyme. a small organic molecule associated with the protein portion of an enzyme, and which is weakly bound but required for enzyme activity (see also prosthetic group).

coenzyme A. a small organic molecule that participates in energy-transfer reactions, usually as a carrier of activated metabolites (e.g., acetate).

colchicine. a drug derived from a flowering plant, used to disrupt microtubules or to prevent their assembly from tubulin monomers.

co-linearity. the spatial correlation between codons in DNA and amino acids in the polypeptide translated from the DNA blueprint.

complementary base pairing. specific hydrogen bond interactions between a particular purine and a particular pyrimidine component in nucleic acids; for example, guanine and cytosine or adenine with thymine or uracil.

constitutive. constant or unchanging; for example, a constitutive enzyme which is synthesized at a constant rate and not subject to regulation; or constitutive heterochromatin which is permanently condensed nuclear chromatin.

contact inhibition. cessation of cell movement upon cell contact with other cells.

cooperativity. interaction between subunits of an enzyme and the substrate, leading to increase or decrease in enzyme activity in relation to fluctuating concentrations of the substrate molecules.

co-repressor. a metabolite that combines with repressor protein and blocks transcription of messenger RNA specifying a repressible enzyme synthesis.

coupling factor. $F_1$ factor; the headpiece of the mitochondrial inner membrane subunit which has ATPase activity and is a catalyst in mitochondrial oxidative phosphorylation during electron transport.

covalent bond. interaction between atoms which involves sharing of their electrons.

cristae. infoldings of the mitochondrial inner membrane and the site of enzymes of coupled oxidative phosphorylation and electron transport during respiration.

crossing over. exchange of homologous chromosome segments; may lead to recombinations between linked genes.

cycloheximide. an organic molecule that inhibits protein synthesis on cytoplasmic ribosomes of eukaryotic cells.

cytochalasin B. a microfilament-disrupting drug extracted from certain fungi.

cytochrome oxidase. cytochrome $a$-$a_3$; the terminal enzyme of aerobic respiration; transfers electrons that reduce molecular oxygen.

cytochromes. electron-transport enzymes containing heme or related prosthetic group components which undergo oxidation-reductions through valency change of their iron atom.

cytogenetics. the study of biological systems using the combined methods of cytology and genetics.

cytokinesis. see cell division.

cytology. the study of cells and their parts using microscopy.

cytoplasm. the protoplasmic contents of the cell, exclusive of the nucleus.

cytosol. the granular portion of the cytoplasm in which the organelles are bathed; the cytoplasmic matrix.

dalton. unit of molecular weight approximately equal to the weight of a hydrogen atom.

denaturation. the process of weakening or disrupting secondary or tertiary structure, or both, of proteins and nucleic acids, leading to loss of function (see also melting).

density-dependent inhibition. cessation of cell division in a layer of cells whose edges touch one another in a culture dish; postconfluence inhibition.

deoxyribonucleic acid. DNA; the genetic material.

desmosome. a 7-layered region of differentiated plasmalemma; serves as a site of adhesion between contiguous cells.

diakinesis. last of the stages of prophase in Meiosis I.

**dictyosome.** a stack of cisternae that forms part of the Golgi apparatus.

**diploid.** a cell or individual or species having two sets of homologous chromosomes in the nucleus.

**diplotene.** a stage of prophase in Meiosis I when synapsed chromosomes "open out" and remain associated only at the chiasmata.

**Down syndrome.** a clinical condition in humans, usually associated with an extra chromosome-21 in the nucleus; trisomy-21, formerly called "mongolism".

**dyad.** a replicated chromosome consisting of two chromatids.

**dynein.** a protein component, with ATPase activity, of the arms of subfiber-A in microtubule doublets of a cilium or flagellum.

**effector.** a modulator or regulatory metabolite that activates or inhibits an enzyme by binding at an allosteric site on the enzyme rather than at the active site where the substrate binds.

**electron transport chain.** a group of electron carriers, such as cytochromes, which transfer electrons during oxidation-reductions, with an accompanying release of energy at each transfer step along the chain.

**endocytosis.** intake of solutes or particles by formation of a vesicle made from a part of the plasma membrane, bringing these materials into the cell.

**endomembrane system.** concept that states there is a physical continuity among the membranes of the eukaryotic cell, each membrane type being a differentiated region of a single cellular membrane system.

**endoplasmic reticulum.** ER; sheet(s) of folded membrane distributed within the cytoplasm of the eukaryotic cell; functions as sites of protein synthesis and transport (*see also* **rough ER, smooth ER**).

**endosymbiont.** a free-living organism that establishes residence in a host cell and maintains a mutually beneficial association with its host.

**end product repression.** a control mechanism in which the synthesis of an enzyme is inhibited when the final product of the metabolic pathway is present, thereby stopping further pathway reactions.

**enzymes.** the unique protein catalysts of living systems.

**equilibrium density gradient centrifugation.** a method used to separate macromolecules and cell components on the basis of differences which cause them to come to rest at equilibrium in a region of the gradient that has a density of solute corresponding to their own buoyant density in the solute; the gradation of densities is produced by centrifugation at very high speeds.

**ER.** *see* **endoplasmic reticulum.**

**eubiont.** a true life form, containing coordinated genetic and metabolic systems.

**euchromatin.** noncondensed, active chromosomes or chromosome regions of the interphase nucleus.

**eukaryotes.** organisms with a well-defined nucleus enclosed in a nuclear envelope and usually having one or more other membranous subcellular compartments; any cellular organism that is not prokaryotic.

**excitation.** raising of the energy level of an atom or molecule when one of its electrons moves to an orbital of higher accessible energy level; excitation induced by absorption of light is called photoexcitation.

**exocytosis.** a mode of transport of substances out of a cell by enclosure in a portion of the plasmalemma and subsequent expulsion of the contents of the vesicle to the outside.

**facilitated diffusion.** assisted passage (transport) of molecules across the membrane toward their lower concentration along a gradient.

**FAD, FADH$_2$.** *see* **flavin adenine dinucleotide.**

**fatty acids.** long hydrocarbon chain components of many lipids; may be saturated (lacking double-bonded carbon atoms) or unsaturated (having one or more double bonds between adjacent carbon atoms in the chain).

**feedback inhibition.** a control mechanism that regulates enzyme activity through inhibition of an enzyme sequence by the end product in the sequence (usually by inhibiting the first enzyme in a sequence), thereby stopping the sequence and any further production of the end product.

**fermentation.** oxidation of carbohydrate in oxygen-independent pathways (*see also* **glycolysis**).

**fibroblast.** a cell type found in connective tissue in animals.

**First Law of thermodynamics.** energy can be neither created nor destroyed; statement of the principle of the conservation of energy in the universe.

**flagellum** (pl. flagella). whiplike locomotor organelle produced by a centriole; ultrastructurally identical to but usually much longer than a cilium.

**flavin adenine dinucleotide.** FAD; an electron carrier molecule that acts in energy-transfer reactions as a coenzyme portion of an enzyme; the reduced form of the redox couple is FADH$_2$.

**fluid mosaic membrane.** model of cellular membranes; postulates the distribution of a mosaic of proteins in and on a phospholipid bilayer with the consistency of a light oil which permits movements of particles laterally within the plane of the membrane.

**formylmethionyl-tRNA.** fmet-tRNA; the first amino acid brought to the ribosome at the initiation of polypeptide chain synthesis in prokaryotes; met-tRNA is the initiating aminoacyl-tRNA in eukaryotes.

**free diffusion.** the unassisted passage of molecules from a region of their higher concentration towards a region of their lower concentration along a concentration gradient.

**free energy.** the usable energy in biological systems; energy that is released in chemical reactions and becomes available to do work in the cell.

**freeze-fracture method.** procedure for preparing materials for electron microscopy by rapid freezing and sectioning to induce fracture formation; the exposed fracture faces are treated by physical methods before being observed and photographed in the electron microscope.

**furrowing.** a cell division mechanism typical of animal groups and which involves a pinching in, or cleavage, to form two daughter cells from the parent cell.

**G-banding.** a method for staining chromosomes with Giemsa stain; reveals patterns of deeply stained bands that are separated by lightly stained regions.

**gamete.** a reproductive cell that can develop only after uniting with another reproductive cell to produce the new individual of the next sexual generation; eggs and sperm are types of gametes.

**gap junction.** a region of differentiation involving portions of plasma membranes of adjacent cells, and which contains a space between the two cells that permits cell-cell communication.

**gene amplification.** differential replication of some genes, producing many copies of these genes, at the same time that other genes in the chromosome set do not replicate.

**gluconeogenesis.** synthesis of carbohydrates from noncarbohydrate precursors, such as fats, proteins, and other substances.

**glyceride.** *see* **neutral fats**.

**glycolysis.** the oxidation of sugar to lactic acid in fermentation reactions which are independent of oxygen, and typical of animal cell metabolism.

**glycoprotein.** a conjugated protein containing one or more sugar residues; a component of the plasma membrane.

**glyoxylate cycle.** a metabolic pathway in which intermediary metabolites can be replenished, and which involves five enzymes including three also active in the Krebs cycle.

**glyoxysome.** *see* **microbody**.

**Golgi apparatus.** a region of smooth endoplasmic reticulum that functions in processing and packaging cell secretions and other kinds of proteins.

**haploid.** cell or individual having one copy of each chromosome in the set.

**Hatch-Slack pathway.** *see* $C_4$ **cycle**.

**heme.** an iron-containing porphyrin derivative which serves as a prosthetic group in hemoglobins and in enzymes such as catalase and cytochromes.

**heterochromatin.** chromatin or chromosomes found in the interphase nucleus and which is either condensed at all times (constitutive) or, in some cells, at some times (facultative).

**heterotroph.** an organism that obtains its energy and carbon for growth through oxidation of organic compounds.

**histone.** a major protein component of the chromosome, having a high content of the basic amino acids arginine and lysine.

**homologous.** having the same or similar gene content.

**humoral immune response system.** plasma cells (activated *B* cells) which synthesize and secrete antibodies in response to invading antigens.

**hydrogen bond.** a weak chemical interaction between a covalently-bonded hydrogen atom and two oxygens, two nitrogens, or one of each, also covalently bonded in a molecule.

**hydrolysis.** the process by which one molecule is converted to two smaller molecules by the addition of water, usually leading to the release of energy.

**hydrophilic.** molecules or parts of molecules that readily interact with dipolar water molecules; usually containing polar groups that form ionic or hydrogen bonds with water.

**hydrophobic.** molecules or parts of molecules that do not readily associate with water; usually nonpolar, poorly soluble in water, or insoluble.

**immune response.** any one of a number of host cell activities that interact with invading antigens and defend the host against infection or cancer.

**immunoglobulin.** an antibody; one of a class of proteins that interact specifically with antigens, usually at the cell surface.

**independent assortment.** a Mendelian principle of inheritance for genes on different chromosomes; leads to random combinations of parental alleles in their progeny.

**inducible enzyme.** a type of enzyme which is synthesized

only in the presence of its inducing substrate, for example, $\beta$-galactosidase.

**ingestion.** process of engulfing solids, including solid foods.

**inner membrane subunits.** IMS; particulate components lining the inner surface of the mitochondrial inner membrane, and consisting of a headpiece with ATPase activity and a stalk.

**interferon.** a protein synthesized by a host cell in response to invading virus, leading to host cell production of antiviral protein which then inhibits viral multiplication in the host cell.

**interphase.** the state of the eukaryotic nucleus when it is not engaged in mitosis or meiosis; consists of $G_1$, $S$, and $G_2$ periods in cycling cells.

**isomers.** alternative molecular forms of a chemical compound.

**isotopes.** alternative nuclear forms of an atom, differing in neutron number but all having the same number of protons; radioactive isotopes are unstable and emit radiation, whereas heavy isotopes are stable forms having one or more extra neutrons in the atomic nucleus; $^3H$ and $^{14}C$ are radioactive isotopes of $^1H$ and $^{12}C$, respectively, while $^{15}N$ is a heavy isotope of ordinary $^{14}N$.

**junction.** differentiation of plasma membrane of adjacent cells involved in cell-cell communication phenomena; major types include gap junction, tight junction, and septate junction.

**karyotype.** an arrangement of photographed chromosomes from a cell or individual and showing chromosomes in pairs and in order of decreasing size.

**Kleinfelter syndrome.** a clinical condition in a human male who has one or more extra X chromosomes.

**Krebs cycle.** most common pathway for oxidative metabolism of pyruvic acid which is an end product of glucose fermentation; part of the pathway of aerobic respiration.

**lampbrush chromosomes.** giant bivalents with extensively looped-out regions of chromatin fiber, especially prominent in amphibian oocyte nuclei during diplotene of Meiosis I prophase.

**leptotene.** the first of the prophase substages in the first of the two divisions of meiosis, before chromosome pairing begins.

**ligase.** enzyme that joins together the parts of single strands of DNA between the 5′ end of one strand and the 3′ end of another.

**linked genes.** genes on the same chromosome.

**lipids.** a heterogeneous class of organic compounds that are poorly soluble in water but soluble in nonpolar, nonaqueous solvents such as ether.

**lymphocyte.** a type of white blood cell that functions in immune response systems; gives rise to $T$ and $B$ cells.

**lysosome.** a membrane-bounded cytoplasmic organelle in eukaryotic cells that contains a variety of acid hydrolytic enzymes capable of digesting almost all biologically important organic compounds.

**matrix.** the essentially unstructured substance of a cell or organelle, consisting of a suspension of particles and molecules in a watery medium.

**meiocyte.** any cell capable of undergoing meiosis; oocytes, spermatocytes, and sporocytes are different kinds of meiocytes.

**meiosis.** the reduction division of the nucleus in sexual organisms; produces daughter nuclei having half the number of chromosomes as the original meiocyte nucleus; consists of two successive divisions.

**melting.** the separation of two strands of duplex DNA to form single strands upon disruption of hydrogen bonds between the paired strands (*see also* **denaturation**).

**meromyosin, heavy.** the portion of the myosin molecule with ATPase activity and $Ca^{2+}$-binding properties, produced by trypsin digestion of myosin and used in assays to identify actin filaments with which it binds in "arrowhead" displays.

**messenger RNA.** mRNA; the complementary copy of DNA that is made during transcription and which codes for protein synthesized during translation.

**metabolite.** a molecule which undergoes change in a chemical reaction.

**metaphase.** the stage of mitosis or meiosis when chromosomes are aligned along the equatorial plane of the spindle, and the time when the chromosomes are most contracted.

**microbody.** a membrane-bounded organelle with varied enzyme contents and functions; usually contains catalase and may also contain the unique enzymes of the glyoxylate cycle; glyoxysomes and peroxisomes are major types of microbodies.

**microsome.** a membrane fragment of the endoplasmic reticulum produced during centrifugation and usually having ribosomes attached to the outer surface of the microsome vesicle.

**microtubule.** an unbranched, hollow cylindrical assembly of tubulin monomers; is involved in cell movement phenomena; spindle fibers and subfibers of cilia and centrioles are all microtubules.

**microvilli.** fingerlike projections of plasma membranes of animal epithelial cells, particularly of the gut.

**mitochondrion.** the double-membrane cytoplasmic organelle found in eukaryotic cells and characterized by inner membrane invaginations called cristae; center of aerobic respiration metabolism in which partly oxidized sugars are processed to $CO_2$ and $H_2O$, with free energy conserved in ATP.

**mitosis.** the division of the nucleus that produces two daughter nuclei exactly like the original parent nucleus; somatic nuclear division.

**mole.** grams molecular weight of a substance.

**molecular hybridization.** formation of a double-stranded structure, DNA-DNA, DNA-RNA, or RNA-RNA, by hydrogen bonding of complementary single-stranded molecules or parts of molecules; serves as a test for complementarity.

**monomer.** the basic unit of a larger functional molecule or particle or cell structure.

**mRNA.** *see* **messenger RNA.**

**myofibril.** a unit of the multinucleated muscle fiber of striated muscle; contains bundles of myofilaments.

**myofilament.** individual thick (myosin) or thin (actin) filament of the myofibril.

**myosin.** the muscle protein making up the thick filaments of striated muscle and of a few other cell systems; has ATPase activity (*see also* **meromyosin**).

**NAD⁺, NADP⁺.** *see* **nicotinamide adenine dinucleotide.**

**neutral fats.** glycerides; fatty acid esters of the alcohol glycerol; a major storage form of fats.

**nicotinamide adenine dinucleotide.** $NAD^+$; an electron carrier molecule that acts in energy-transferring reactions as a coenzyme portion of an enzyme; the reduced form of the redox couple is NADH; nicotinamide adenine dinucleotide phosphate, or $NADP^+$, acts in a similar manner in oxidation-reductions during biosynthesis; its reduced form is NADPH.

**nondisjunction.** faulty separation of homologous chromosomes or of sister chromatids during nuclear division, producing cells or individuals with more or less than the expected number of chromosomes.

**NOR.** *see* **nucleolar-organizing region.**

**nuclear envelope.** the double membrane surrounding the eukaryotic nucleus.

**nucleic acid.** polymer of nucleotides in an unbranched chain; DNA and RNA.

**nucleoid.** in prokaryotic cells, the region of DNA which is not separated from the surrounding cytoplasm by a membrane.

**nucleolar-organizing chromosome.** a chromosome in the set which contains various genes, including the genes for ribosomal RNA; has the capacity to produce a nucleolus at the site of the ribosomal RNA genes.

**nucleolar-organizing region.** NOR; the specific part of the nucleolar-organizing chromosome where ribosomal RNA genes are situated and where the nucleolus is produced.

**nucleolus.** a discrete structure in the nucleus that is associated with synthesis of ribosomal RNA and of ribosomal subunit precursor particles.

**nucleoplasm.** the matrix portion of the nucleus in which the chromosomes and nucleoli are bathed.

**nucleoside.** molecule containing a nitrogenous base linked to a pentose sugar.

**nucleotide.** a nucleoside phosphate; any of the monomeric units that make up DNA and RNA.

**nucleus.** the major membrane-bounded compartment of the eukaryotic cell, housing the chromosomes and nucleoli.

**operator.** a specific nucleotide sequence in the operon which binds repressor and thereby exerts control over transcription of its adjacent structural gene(s).

**operon.** a cluster of associated genes and recognition sites that participate in regulating and specifying transcription and translation of coded DNA.

**organelle.** a structural differentiation of the cell; contains particular enzymes and performs particular functions for the whole cell or individual.

**oxidant.** an oxidizing agent, which accepts electrons or hydrogens from a reducing agent, or reductant.

**oxidation.** reaction involving loss of electrons or hydrogens.

**oxidative phosphorylation.** synthesis of ATP during coupled electron transport in aerobic respiration.

**pachytene.** the stage of prophase I of meiosis when synapsis of homologous chromosomes is completed.

**peptide bond.** the universal link between amino acids in proteins; formed when the amino group of one monomer joins with the carboxyl group of the adjacent amino acid in a dehydration reaction.

**permease.** a type of carrier protein situated within the membrane and involved in transport of specific molecules across the membrane.

**peroxisome.** *see* **microbody.**

**pH.** measure of hydrogen ion concentration in aqueous solutions.

**phage.** *see* **bacteriophage.**

**phosphate group.** $-O-P{=}O$ ; $-PO_4{}^{2-}$.

**phosphoryl group.** $-P{=}O$ ; involved in energy transfer reactions.

**photophosphorylation.** process of ATP synthesis in the light, coupled to electron transport along a chain of carriers between PS II and PS I.

**photorespiration.** uptake of oxygen and release of carbon dioxide by photosynthetic cells or individuals in the light.

**photosynthesis.** manufacture of sugar from $CO_2$ and $H_2O$ in the light and in the presence of chlorophyll in eukaryotic chloroplasts or within prokaryote photosynthetic membranes.

**photosystem I.** PS I; a photochemical reaction system in photosynthesis; produces NADPH but does not evolve $O_2$.

**photosystem II.** PS II; a photochemical reaction system in photosynthesis; is linked in series to PS I by an electron transport chain; $O_2$ is evolved when water donates electrons to reaction center chlorophyll *a*.

**phycobilin.** accessory photosynthetic pigment present in red and blue-green algae.

**plasmalemma.** plasma membrane of the cell; the membrane that surrounds the living protoplast.

**plasmodesmata.** cytoplasmic channels between adjacent plant cells.

**plastid.** eukaryotic organelle of one or more types but most often referring to the chloroplast.

**polymer.** an association of monomer units in a large molecule.

**polymerase.** enzyme catalyzing the synthesis of DNA or RNA from nucleoside triphosphate precursors.

**polypeptide.** a polymer of amino acids in a long unbranched chain.

**polyploid.** cell or individual having one or more whole sets of chromosomes in excess of the usual number for the species or species group.

**polysome.** polyribosome; an aggregation of ribosomes which is actively engaged in protein synthesis when connected by a strand of messenger RNA.

**polytene chromosome.** a multistranded, repeatedly replicated chromosome.

**pore.** an opening in a membrane or other structure; usually referring to the opening of the nuclear pore complex in the nuclear envelope.

**primary structure.** the sequence of amino acids in a polypeptide chain.

**procentriole.** an immature centriole.

**prokaryotes.** organisms of the bacteria and blue-green algae groups lacking a nucleus separated from the cytoplasm by a membrane; any cellular organism that is not a eukaryote.

**prometaphase.** stage of nuclear division when chromosomes move toward the equatorial plane of the spindle, but before they align there.

**promoter.** a specific nucleotide sequence in the operon to which RNA polymerase binds, and which is therefore a regulatory component of the operon.

**prophase.** the first stage of mitosis or meiosis, after DNA replication and before chromosomes align on the equatorial plane of the spindle.

**proplastid.** an immature plastid.

**prosthetic group.** a relatively small molecule that remains very tightly bound to the active site of an enzyme and which is required for the enzyme to interact with its substrate (*see also* **coenzyme**).

**proteinoid.** presumptive protobiont structural form, made from amino acids and having a selectively permeable boundary.

**protist.** any eukaryotic organism not classified as a fungus, plant, or animal; usually a unicellular organism, such as a protozoan or euglenoid.

**protobiont.** pre-life form that could have evolved into a true life form, or eubiont.

**protoplasm.** the living material of the cell.

**protoplast.** the living structure of the cell, made of protoplasm, and contained within but including the plasma membrane.

**PS I, PS II.** *see* **photosystem I, photosystem II.**

**puff.** a region of expanded chromosome undergoing active transcription, usually observed in giant polytene chromosomes.

**pumps, ion.** systems that underwrite active transport of molecules across a membrane by expelling one substance out of the cell and thereby helping to drive many kinds of molecules into the cell along an energy gradient generated by the ions.

**purine.** parent compound of the nitrogen-containing bases adenine and guanine.

**pyrimidine.** parent compound of the nitrogen-containing bases cytosine, thymine, and uracil.

**quantum.** the energy of a photon, its amount being inversely proportional to the wavelength of emitted radiation.

**quaternary structure.** specific assemblages of different polypeptide chains which, when combined into the protein, do not have the same structural or chemical properties as in the individual chain.

**rDNA.** *see* **ribosomal DNA.**

**reannealing.** renaturation; specifically, the restoration of duplex DNA regions through complementary base pairing of single-stranded DNA molecules.

**redox couple.** compounds that occur in both the oxidized and reduced states and which are participants in oxidation-reductions, such as $NAD^+/NADH$.

**redox potential.** *see* **standard electrode potential.**

**reductant.** a reducing agent, which loses electrons or hydrogens in oxidation-reductions.

**reduction.** reaction involving gain of electrons or hydrogens.

**regulation.** modulation of metabolism or gene action through control mechanisms.

**repetitious DNA.** repeated sequences of nucleotides that may occur in hundreds, thousands, or millions of reiterated units in a chromosome or set of chromosomes.

**replication fork.** a site within a replicating duplex DNA molecule at which synthesis of complementary strands is proceeding at that moment.

**repressible enzyme.** type that is synthesized in the absence of its substrate which represses synthesis of the enzyme.

**repressor.** a protein product of the regulator gene of the operon; binds to the operator site and thereby regulates transcription of structural genes.

**respiration.** the principal energy-yielding reactions of aerobic cells, involving transfer of electrons from organic fuel molecules to molecular oxygen.

**ribonucleic acid.** RNA; nucleic acid polymers that function in translation of coded DNA; transcripts of complementary DNA.

**ribosomal DNA.** rDNA; the genes at the nucleolar-organizing region that code for ribosomal RNA.

**ribosomal RNA.** rRNA; ribonucleic acids that are part of the ribosome structure and which function during protein synthesis.

**ribosome.** a complex structure composed of RNA and proteins, and which is the site of protein synthesis in the cytoplasm, chloroplasts, and mitochondria.

**RNA.** *see* **ribonucleic acid.**

**rough ER.** that portion of the endoplasmic reticulum with attached ribosomes.

**rRNA.** *see* **ribosomal RNA.**

**S period.** the interval during the cell cycle when DNA replicates.

**S value.** *see* **sedimentation coefficient.**

**sarcolemma.** the plasma membrane of the muscle fiber.

**sarcoplasmic reticulum.** the endoplasmic recticulum of the muscle fiber.

**secondary constriction.** any pinched-in site along a chromosome other than the primary constriction at the centromere region.

**secondary structure.** the local structure of the polymer molecule, resulting from interactions between closely neighboring residues.

**Second Law of thermodynamics.** statement that all systems tend toward an equilibrium, all systems tend to minimize their free energy content.

**secretion.** product or process of cell synthesis in which the molecule acts elsewhere than at its site of origin.

**sedimentation coefficient.** S; a quantitative measure of the rate of sedimentation of a given substance in a centrifugal field, expressed in Svedberg units.

**self assembly.** the organization of structure in the absence of a template or parent structure.

**semiconservative replication.** the usual mode of duplex DNA synthesis resulting in daughter duplex molecules which contain one parental strand and one newly formed strand.

**septate junction.** *see* **junction.**

**sex chromatin.** *see* **Barr body.**

**sex chromosome.** any chromosome involved in sex determination; the X and Y chromosomes are sex chromosomes.

**sister chromatid exchange.** result of crossover between sister chromatids of a single replicated chromosome.

**sliding filament mechanism.** a proposal applied to explain cellular movement phenomena, particularly to striated muscle contraction and to ciliary bending (as sliding microtubule mechanism).

**smooth ER.** that portion of the endoplasmic reticulum which lacks attached ribosomes.

**spindle.** an aggregate of microtubules seen during nuclear division; it functions in the alignment and movement of chromosomes during metaphase and anaphase.

**spindle fiber.** a microtubule in dividing cells which extends from one pole to an attachment in the centromere region of a chromosome, or which extends from pole to pole in mitosis and meiosis.

**standard electrode potential.** $E_0$; the oxidation-reduction potential of a substance relative to a hydrogen electrode; expressed in volts.

**standard free-energy change.** $\Delta G^0$; a thermodynamic constant representing the difference between the standard free energy of the reactants and the standard free energy of the products of a reaction; energy-releasing reactions have a negative $\Delta G$ while energy-requiring reactions have a positive $\Delta G$.

**stem cell.** precursor cell type that gives rise to one or more types of differentiated cells.

**stroma.** the granular matrix of the chloroplast.

**substrate.** molecule that undergoes chemical change in a chemical reaction.

**synapsis.** the specific pairing of homologous chromosomes, typically during zygotene of prophase I in meiosis.

**synaptinemal complex.** a three-layered structural component formed between a pair of synapsing chromosomes during early prophase of Meiosis I; stabilizes the paired chromosomes and holds them in register.

*T* **cell.** a white blood cell type that functions in the cell-mediated immune response, and which is sensitized to a specific invading antigen; derived from *T* lymphocytes.

**T system.** invaginations of the sarcolemma in muscle fibers of striated muscle, producing a system of transverse tubular infoldings that maximize signal reception in the entire muscle fiber.

**telophase.** the stage of nuclear division when nuclear reorganization occurs.

**tertiary structure.** the three-dimensional folding of a polymer into a particular shape according to interactions between residues at some distance from one another in the primary structure of the molecule.

**thermodynamics.** the branch of physical science that deals with energy exchange in collections of matter (*see also* **bioenergetics**).

**thylakoid.** a closed membrane sac that may be disc-shaped in grana or may be greatly elongated in a chloroplast or a prokaryotic cell; contains the systems active in the light-dependent reactions of photosynthesis.

**tight junction.** *see* **junction.**

**transcription.** a process by which the base sequence of DNA is copied into a complementary single-stranded RNA molecule.

**transfer RNA.** tRNA; the RNA molecule that carries an amino acid to a specific codon in messenger RNA during protein synthesis at the ribosomes.

**translation.** the process by which the sequence of amino acids is assembled into a polypeptide at the ribosomes, under the direction of the coded base sequence in the messenger RNA copy of the gene.

**transmembrane control system.** a coordinated cytoskeletal framework of microfilaments and microtubules which regulate redistribution of peripheral and integral proteins in the cell membrane.

**transport.** assisted passage of molecules across a membrane (*see also* **active transport, facilitated diffusion**).

**trisomy.** condition of a cell or individual having three copies of a particular chromosome, as in trisomy-21, or Down syndrome.

**tritium.** $^3H$; a radioactive isotope of hydrogen.

**tropomyosin.** a regulatory protein bound to the actin filaments in striated muscle; it interacts with troponin.

**troponin.** a regulatory protein bound to the actin filaments in striated muscle; it acts as a $Ca^{2+}$-dependent switch in muscle contraction.

**tubulin.** the protein monomer of microtubule construction.

**Turner syndrome.** clinical condition in a human female having only one X chromosome (*45,X*).

**unit membrane.** any membrane showing a dark-light-dark pattern of electron density in the electron microscope; a model of membrane structure proposing that a phospholipid bilayer is coated on both surfaces by proteins in extended conformation.

**unwinding protein.** structural protein that binds to single-stranded regions of duplex DNA during replication and recombination.

**vacuole.** a region in the cytoplasm surrounded by a membrane and filled with substances in a watery medium; particularly characteristic of mature plant cells.

**vesicle.** a small, spherical membranous element filled with protein in solution.

**zygote.** product of the fusion of two gametes; the cell from which the new individual develops in each sexual generation.

**zygotene.** stage during prophase of Meiosis I when homologous chromosomes undergo synapsis.

**zymogen.** a digestive enzyme precursor lacking catalytic activity in this form; a cell secretion of a certain type.

# Index

Numbers in boldface indicate pictorial information.